Principles of
Architecture

G. Muthu Shoba Mohan

Lecturer in Architecture,
Sun College of Engineering & Technology,
Kanyakumari District

OXFORD

UNIVERSITY PRESS

To my appa
my friend, philosopher, guide, and teacher
and
my beloved amma
without whom nothing would have been possible

Foreword

In the field of architectural practice, group culture is necessary to establish aesthetically appealing and structurally sound buildings. An architect is a co-coordinator in the planning and execution of any project. He should be conversant with other related disciplines such as structural design, air conditioning, heating and ventilation, mechanical and electrical engineering, acoustics, lift technology, boiler technology, and landscaping to be able to interact with experts in these disciplines. In the same way, the engineers connected with the construction must learn, understand, and appreciate the basics of architecture to interact with and appreciate the perspective of architects and other experts.

In this direction, this book is a long felt need and I am happy to be a part of this book through this Foreword. To the best of my knowledge, this is the first attempt at simplifying architecture for students and experts of other disciplines, especially students of engineering and technology in India. This book teaches them the basics and takes them to greater heights of understanding. The students of engineering are often found to be non-appreciative of the usefulness of aesthetic aspects and creativity; this book attempts to bridge that gap. Its simple language and explanations through illustrative sketches will facilitate the comprehension of these concepts.

I am confident that most students will be able to appreciate the necessity of aesthetics in architecture after going through this book, the entire credit for which would be due to the author. I congratulate the author for coming up with this book and extend my best wishes to her.

A. MOHAMMED HARIS
(FORMER DEAN AND DIRECTOR,
SCHOOL OF ARCHITECTURE AND PLANNING, ANNA UNIVERSITY, CHENNAI)

Preface

While teaching an introductory course in architecture to civil engineering students at Anna University, I felt the need for a basic introductory single text. Students and teachers need to refer to many textbooks to find material pertaining to the different units given in the curriculum. This book aims to overcome this difficulty and render the course contents and difficult terminology simpler for the students, thus making the teaching–learning process simple and efficient.

About the Book

The aim of this text is to help students appreciate architectural concepts and assimilate the basic principles involved. It encompasses the basic theory of architecture, design aspects, building regulations, and services required among other topics. It covers all the minute details under each topic to facilitate easy understanding. It includes sufficient number of illustrations, which make the explanations simpler. As drawing is said to be the language of engineering, by referring to these illustrations, students can easily grasp concepts that will otherwise need pages and pages of written explanation. It is important, therefore, for the reader to carefully study the text as well as the illustrations.

As with any other discipline, the study of architecture involves the study of its history. By studying the architectural history of buildings, one is able to know past experiences and reflect on the various concepts explained through them. Accordingly, the book quotes many historical examples.

Contents and Structure

The book is aimed at civil engineering students who wish to master the basic architectural concepts as quickly as possible. It must be noted that the illustrations are not supplements to the text but in a very real sense the text itself. This illustrative approach will facilitate the understanding of important design concepts for those students who do not have enough time to digest lengthy descriptions.

Chapter 1—Fundamentals of Architecture—covers the basic definitions, elements, and principles of design.

Chapter 2—Aesthetic Components of Design—discusses the aesthetic qualities of buildings such as unity, proportion, scale, balance, symmetry, and rhythm with the aid of historic examples.

Chapter 3—Aesthetic Relationships, Character, and Style in Buildings—explains the meaning of character and style in traditional and modern architecture as well as the effect of the aesthetic impact of harmony, contrast, dominance, punctuation, and climax on the overall composition of a building.

Chapter 4—Factors Influencing Architectural Design—addresses the factors to be considered in architectural design.

Chapter 5—Contemporary Architecture—presents the biographies, themes, philosophies, and works of six famous architects.

Acknowledgements

This being a textbook for the benefit of students' learning, I had to draw heavily on the already available knowledge base. I acknowledge my indebtedness to all those whose works I have referred to in this book. I am grateful to the authors of all the books mentioned in the bibliography.

Several people have contributed to the development of this book. I would like to thank all those who have directly or indirectly contributed to it. I am grateful to all my teachers for the knowledge they gave me. I am honoured by the Foreword to this book by Professor A. Mohammed Haris, former Dean and Director, School of Architecture and Planning, Anna University, Chennai. I am extremely thankful to him. I express my sincere thanks to my friend Shanmugha Priya for her valuable suggestions and timely help. I am grateful to Professor K.V. Marthandan, former Principal of Sun College of Engineering & Technology, Kanyakumari, who initiated me into teaching. Thanks are also due to Dr N. Subramanian for encouraging me to write this book.

This book would never have been possible without the encouragement and support of my dear husband and my sons who had tremendous patience to put up with my work. I am grateful to my mother-in-law for her patience and assistance during the processing of this book. I thank all the members of my family for the encouragement they provided to me. Special thanks are due to Oxford University Press for their support.

I would appreciate any feedback from the readers of this book towards the further improvement of its content and presentation

G. MUTHU SHOBA MOHAN

Contents

Foreword   *v*

Preface   *vii*

1. Fundamentals of Architecture   **1**
    Nature of Architecture   *1*
    Definitions of Architectural Terms   *5*
    Understanding the Basic Elements of Design   *8*
    Understanding the Principles of Design   *32*
    Summary 51
    Review Questions 52

2. Aesthetic Components of Design   **54**
    Aesthetic Qualities   *54*
    Unity and Elements of Unity   *54*
    Proportion   *60*
    Scale   *80*
    Balance   *87*
    Symmetry   *91*
    Rhythm or Repetition   *92*
    Summary 92
    Review Questions 93

3. Aesthetic Relationships, Character, and Style in Buildings   **95**
    Character and Style in Buildings   *95*
    Aesthetic Impact   *126*
    Summary 140
    Review Questions 141

4. Factors Influencing Architectural Design   **142**
    Meaning of Architectural Design   *142*
    Factors to be Considered in Architectural Design   *145*
    Environmental Factors   *203*
    Litigation Factors Considered in Architectural Design   *245*

Summary 252
Review Questions 253

5. **Contemporary Architecture**   **255**
   Theme, Philosophy, and Works of Famous Architects *255*
   Frank Lloyd Wright (1867–1959), USA *256*
   Le Corbusier (1887–1965), France *267*
   Ludwig Mies van der Rohe (1886–1969), Germany *275*
   Louis Isadore Kahn (1901–74), USA *285*
   Charles Correa (1930), Hyderabad, India *289*
   Balkrishna V. Doshi (1927), Ahmedabad, India *298*
   Summary 306
   Review Questions 307

Glossary   **309**

Bibliography   **310**

Model Question Papers   **311**

Fundamentals of Architecture

1.1 Nature of Architecture

Human beings have certain physical needs without which they cannot survive. Some of these, such as air and water, are naturally available. The others are food, shelter, and clothing. With minor exceptions, these are not readily available in nature and have to be procured from the environment. Food material has to be cultivated, processed, and cooked; shelter has to be constructed; clothing has to be manufactured. These basic requirements have to not only be produced but also provide a degree of satisfaction.

As the human society evolved, great changes took place. The activities required for mere survival became less important and civilizing activities began to develop. Man refined his food, shelter, and clothing to provide not only physical satisfaction but also *aesthetic pleasure*. The following changes took place. Different types and preparations of food were available. Clothes were no longer just fabrics wrapped around the body, but tailored and embroidered according to personal tastes. Shelters, in the form of buildings, were given conscious forms and decorated. Aesthetics—beauty and taste—became important. Objects which had only utility values also came to have emotional values. These are nothing but the psychological needs of human beings.

What is Aesthetics?

The word aesthetics is derived from a Greek word meaning 'a perceiver' or 'sensitive'. It is a branch of philosophy dealing with the nature of beauty. The word aesthetics was first used by German philosopher Alexander Gottlieb Baumgarten. It can be used as a noun meaning 'that which appeals to the senses'.

One way of understanding the meaning of 'aesthetic' is by comparing it to its antonym, 'anaesthetic'. The word 'anaesthetic' implies something that tends to dull the senses or causes sleepiness. In contrast, aesthetic may be thought of as anything that tends to enliven or invigorate or wake one up.

With these changes, every object—clothing, dwelling, vessel, etc.—became both useful and beautiful. Beauty did not exist separately but was merged or fused with the primary needs.

1.1.1 Technology, art and craft

For a better understanding of the concept of aesthetics, we must understand the difference between technology, art, and craft/design. Let us consider three kinds of people:

- Technologist
- Artist
- Craftsman

Technologist

A technologist is a person who is concerned with the scientific aspect of a project/activity. For example, when engineers are given the task of designing a bridge, the various factors they consider are

- purpose of construction
- strength
- cost
- functional requirements such as site conditions and materials

Based on these factors, they make a series of calculations and estimate the cost of the proposed structure. They perform analysis and use logic, based upon scientific and mathematical laws. At no stage are the engineers concerned with the aesthetic appearance of the bridge. The concepts of beauty and ugliness do not enter into any of their calculations. The engineer sees the bridge as an object for satisfying primary human needs. This kind of basic activity is done by the technologist. Similarly, a civil engineer, mechanical engineer, physicist, physician, chemical engineer, etc. thus does not take into account the need for aesthetics.

Artist

When an artist paints, the theme he selects may or may not be useful to his client or the general public. He paints to convey his inner feelings about a subject and not to please anyone. His concern is not whether the painting will be sold, because his only aim is to express his emotions and thoughts through the medium of painting. Thus, he will ignore all subjective conditions such as the viewer's satisfaction, the cost involved, and the utility of his work. The effect of the painting will be judged by the emotional response it stimulates in the observer. Other fields concerned with emotional response are music, poetry, and sculpture.

Craftsman

When a craftsman is given the problem of designing a piece of furniture, say, a kind of chair, he considers its utility or functional aspects such as
- type of wood
- size of the chair
- overall strength
- joinery details
- cost

After considering the functional requirements he thinks about giving a *form* to the chair. Deciding on the form is an aesthetic problem and the carpenter becomes an artist, as form creates emotional feelings.

The chair, however beautiful, could be functionally defective if the carpenter chooses to make it with excessively slender legs. It will fail when put to use— in other words, it will break.

In this example, two criteria are involved

Utility	$\Rightarrow$	Utility	$\Rightarrow$	Functional	
+		+		+	
Emotional		Beauty		Aesthetic	

That is, in designing a chair, both the functional and the aesthetic requirements should be satisfied. The chair must not only look beautiful, it must also perform well. A chair which is aesthetically pleasing may be uncomfortable to sit on. Thus, the choice of material, size, shape, form should satisfy both criteria. For example, a coarse-grained wood though technically satisfying may not be aesthetically satisfying.

As they move up the social scale, human beings want to see utility and beauty brought together in items of personal use, e.g., furniture, clothing, and housing. Modern life is so complex that the physical needs are satisfied by the technical specialist and the emotional needs are satisfied by the aesthetic specialist.

Craftsmanship includes, among others, the fields of pottery, carpentry, blacksmithy, interior design, architecture, and dress design. All these fields deal with the concept of *design*.

1.1.2 Definition of Architecture

The meaning of architecture has undergone many changes during the evolution of human civilization. Earlier, architecture was identified with large elaborate buildings such as palaces built by rulers to house their families and armies, and to impress their subjects. The aim was to display status and prestige. Shelters for the common people were not considered fit for the architect to design. As religion became organized, architecture became important for spiritual needs. To satisfy spiritual needs, temples, churches, and mosques were constructed.

In modern society, architecture is concerned with every building task. From public toilets to individual homes, all constructions have become objects of architectural design. The scope of architecture has broadened to include much more than individual buildings. It includes

- town planning
- regional planning
- urban design
- urban planning
- landscape architecture

Architecture is a form of craftsmanship intended to serve a basic human need. These are

- need for shelter
- need for an environment

Both should be designed to be technically and aesthetically satisfactory.

The various building types which involve architectural design are

- dwellings
- schools
- hospitals
- places of worship
- railway stations
- cinemas
- stadiums
- playgrounds

The common factor in this list of diverse buildings and areas is the concept of *space*. Space is a specific volume intended for a specific form of human activity. Nature has provided natural spaces such as valleys, caves, and groups of trees, which have served in the past for human activities. These are not architectural. Architectural space is man made, it must result from the deliberate use of materials; it must be technically efficient and aesthetically satisfying. The aesthetic aspect should not be coincidental as in an electric motor (which could be made beautiful, but costly and less efficient), but should reflect the art of creativity.

Architecture can be defined as the conscious creation of spaces for utility, constructed from materials in such a way that the whole is both technically and aesthetically satisfying. It is a fusion of art and technology.

Some examples of aesthetic statements commonly used are a red wall or a rough-textured surface, a brilliant red, a delightful contrast, and a beautiful proportion. To define something as aesthetically satisfying is similar to describing physical qualities in emotional terms. Architecture involves the selection of forms and spaces.

An architect deals with the design of the built environment, i.e., design of the actual building, its enclosed space, and the surrounding space. On the other hand, a civil engineer is mainly concerned with the structural aspect of the building. The main objective of this course is to sensitize students of civil engineering to the relation between architecture and engineering.

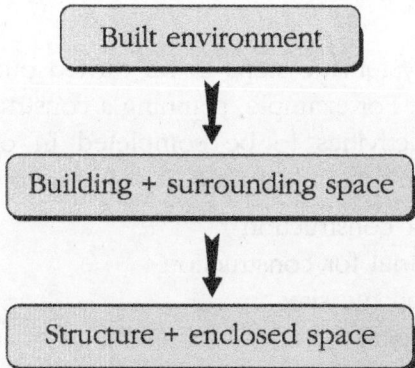

Architecture is a design process which results in functionally efficient, economically viable, and aesthetically pleasing buildings.

1.2 Definitions of Architectural Terms

The theory of architecture involves many technical terms. The definitions of some of the important terms are given in this section.

1.2.1 Aesthetics

Aesthetics concerns beauty or the appreciation of beauty. In other words, it refers to the philosophy behind a pleasing appearance. The set of principles involved in the work of people in fields concerned with design, to make the end product beautiful, is known as aesthetics. It is directly influenced by the artistic taste of the person involved.

Aesthetics is concerned with bringing art into the daily aspects of life, such as the taste and presentation of food, and the colour and design of clothes. Similarly, humans also desire beautiful buildings to live and work in. The form of a building is decided mainly by aesthetics. When a building is designed, the aesthetic aspects can be satisfied by using elements (Fig. 1.1) such as sloped roofs, decorative columns, roofs for window elements, and semicircular or segmental arches.

The elements of aesthetics are:
- mass and space
- proportion
- symmetry
- balance
- contrast
- pattern
- decoration
- massing

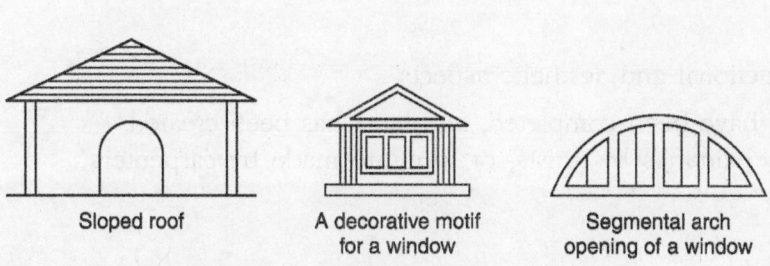

Fig. 1.1 An example of aesthetic elements

1.2.2 Planning

Planning involves the systematic steps to be carried out in order to achieve a given objective or target. For example, planning a construction means identifying the various steps or activities to be completed in order to carry out the construction in the given amount of time. It could involve the following steps:

- clearing the site for construction
- procuring the material for construction
- marking the plan on the site
- the construction process
- finishing and handing over the project

In the smaller, individual sense, planning is the process of preparing for personal requirements. In the larger sense, local governments control building and development through town planning.

1.2.3 Designing

Designing [Figs 1.2(a) and (b)] is the process of procuring a preliminary sketch of an object that is to be physically constructed later. The original idea behind the sketch is called the concept of design. It can exist in two different forms:

- mental idea
- representational idea

The *mental idea* helps to

- identify the purpose of the object (building)
- analyse the aesthetic and functional aspects
- think of a good solution (design)

The *representational idea* simply reproduces the mental map of the building in the form of sketches or diagrams or as a miniature model.

1.2.4 Creating

Creating [Fig. 1.2(c)] is the actual process of execution, where the design is converted into a physical reality. The process of creating an object involves the processes of

- designing
- planning
- incorporating the functional and aesthetic aspects

When all these processes have been completed, an object has been created.

Buildings by architects, paintings by artists, or furniture made by carpenters are examples of creations.

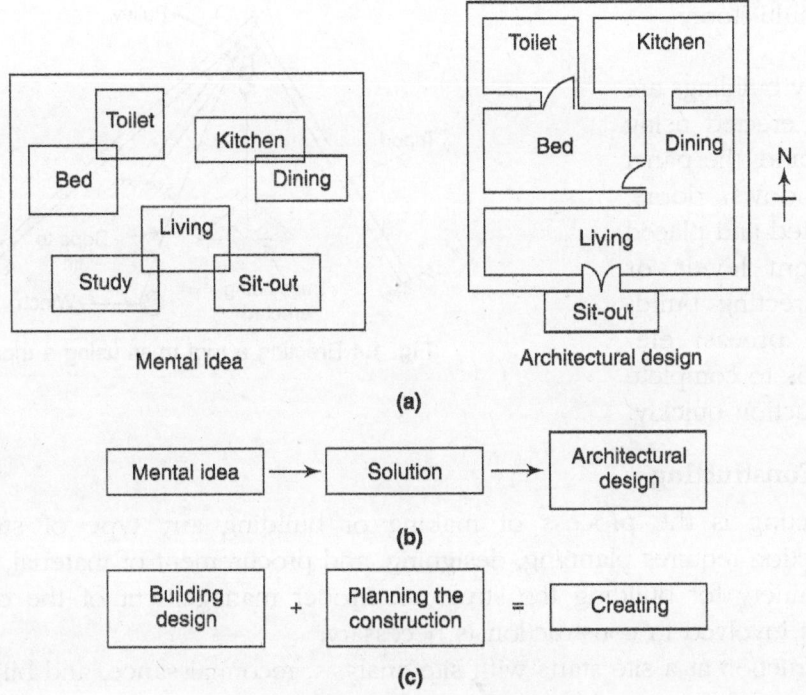

Fig. 1.2 Designing and creating

1.2.5 Erecting

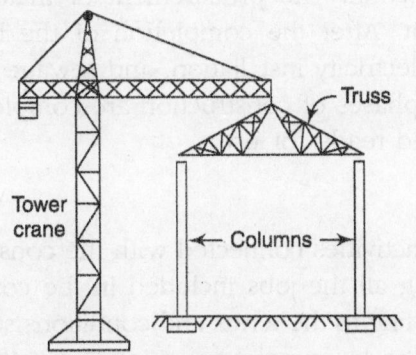

Fig. 1.3 Erecting a roof truss using a tower crane

Erecting is the process of assembling or putting into place the completed or fabricated components of building. Consider the following examples.

(a) Erecting roof trusses in industrial buildings and workshops: The roof trusses are lifted by cranes and placed in position exactly on the

walls (Fig. 1.3). The trusses are then connected to the walls or columns by bolting the base plate.

(b) Erecting precast elements of a building:
The partition walls, heavy beams, columns, etc. are manufactured in a factory or at the construction site by prefabrication. Then these units are brought to the site and erected. This is done by lifting the precast elements using pulley blocks, a tripod, and winches or cranes (Fig. 1.4).

(c) Erecting multi-storey buildings:

Multi-storey buildings are nowadays erected using a tower crane; the panel walls, windows, doors, etc. are lifted and placed on different levels or storeys. Erecting buildings with precast elements helps to complete the construction quickly.

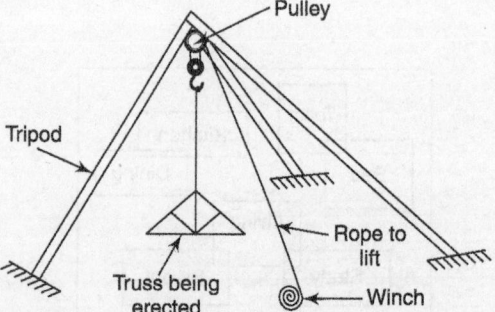

Fig. 1.4 Erecting a roof truss using a tripod

1.2.6 Constructing

Constructing is the process of making or building any type of structure. Construction requires planning, designing, and procurement of material, labour, or machinery for building the structure. Proper management of the different activities involved in construction is necessary.

Construction at a site starts with site analysis, reconnaissance, and building a temporary structure at the site for storage of material, security, road laying, water and electricity connections, etc.

The type of construction (whether brick masonry or framed structure) is identified depending on the function of the building and the total design is completed before starting the construction. The procurement of material and the construction work starts after that. After the completion of the building, other services such as water supply, electricity installation, and sewage disposal system are put in place. Once all the phases of construction are completed, the building is complete in all respects and ready for use.

1.2.7 Executing

Executing is the completion of all the activities connected with the construction. Executing a contract means completing all the jobs included in the contract of the project or the building work, according to the terms and conditions stipulated in the contract. Executing simply means to complete any job, be it the entire contract or any part of it.

1.3 Understanding the Basic Elements of Design

This section discusses the basic or primary elements of design, namely, point, line, plane, and volume (Fig. 1.5). A design is nothing but the creation of a form. A pictorial form begins with the 'point' that sets itself in motion. The point moves and the line, which is one-dimensional, comes into being. If the line shifts to form a 'plane', we obtain a two-dimensional element. In the movement from planes to spaces, the clash of planes gives rise to a volume (three-dimensional element). In this section, each element of design or form is explained in the order of growth, as the sum of the kinetic energies which move—a point to produce a line; a line to produce a plane; a plane to produce a volume.

Each element is described first as a conceptual element and then as a visual element in the vocabulary of architectural design.

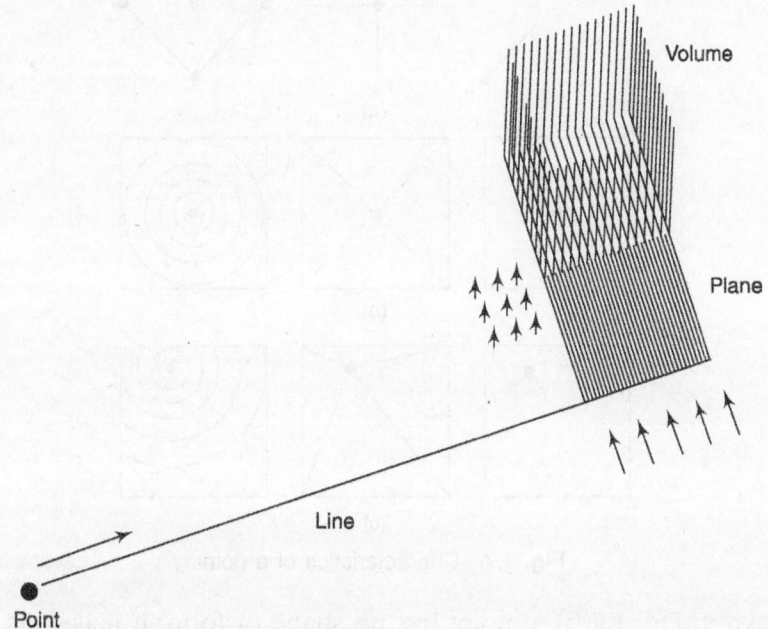

Fig. 1.5 The primary elements of form are point, line, plane, volume

1.3.1 Point

A point is the prime generator of form. It indicates a position in space. It has no length, width, or depth and, therefore, is static, directionless, and centralized (Fig. 1.6).

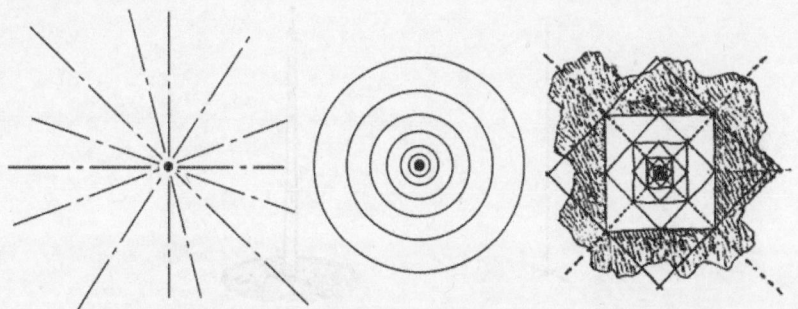

Fig. 1.6 A point is static, directionless, and centralized

As shown in Fig. 1.7(a), a point can serve to mark
- the two ends of a line,
- the intersection of two lines,
- the meeting of lines at the corner of a plane or volume, and
the centre of a field or environment.

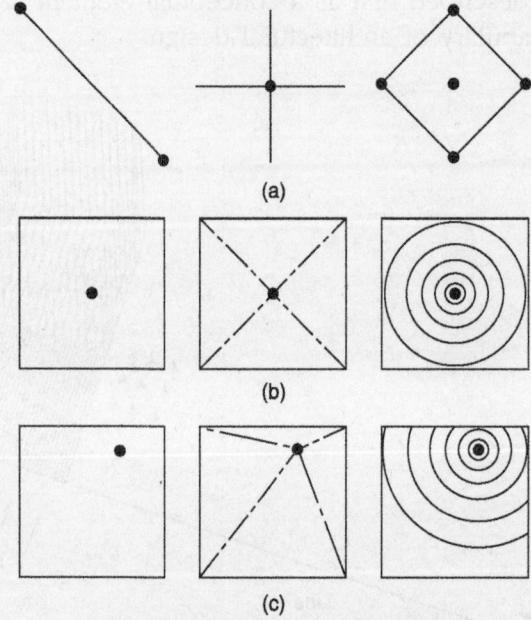

(a)

(b)

(c)

Fig. 1.7   Characteristics of a point

As shown in Fig. 1.7(b), a point has no shape or form. It makes its presence felt when placed within a visual field. When the point is at the centre, it is stable, at rest and dominates its field. As shown in Fig. 1.7(c), when it is moved off-centre, its field becomes more aggressive and a visual tension is created between the point and its field.

A point has no dimension. To visibly mark a position in space or on the ground plane, a point must be projected onto a vertical linear element such as a column or a tower, as shown in Fig. 1.8. The column in a plan marks a point.

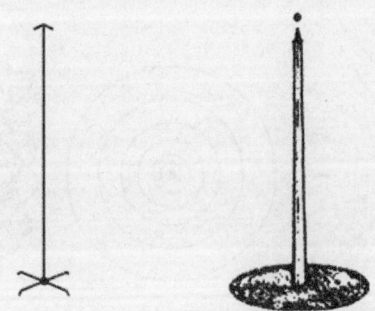

Fig. 1.8 Point projected into a linear element:
column, obelisk, or tower

The circle, cylinder, and sphere are point-generated forms (Fig. 1.9).

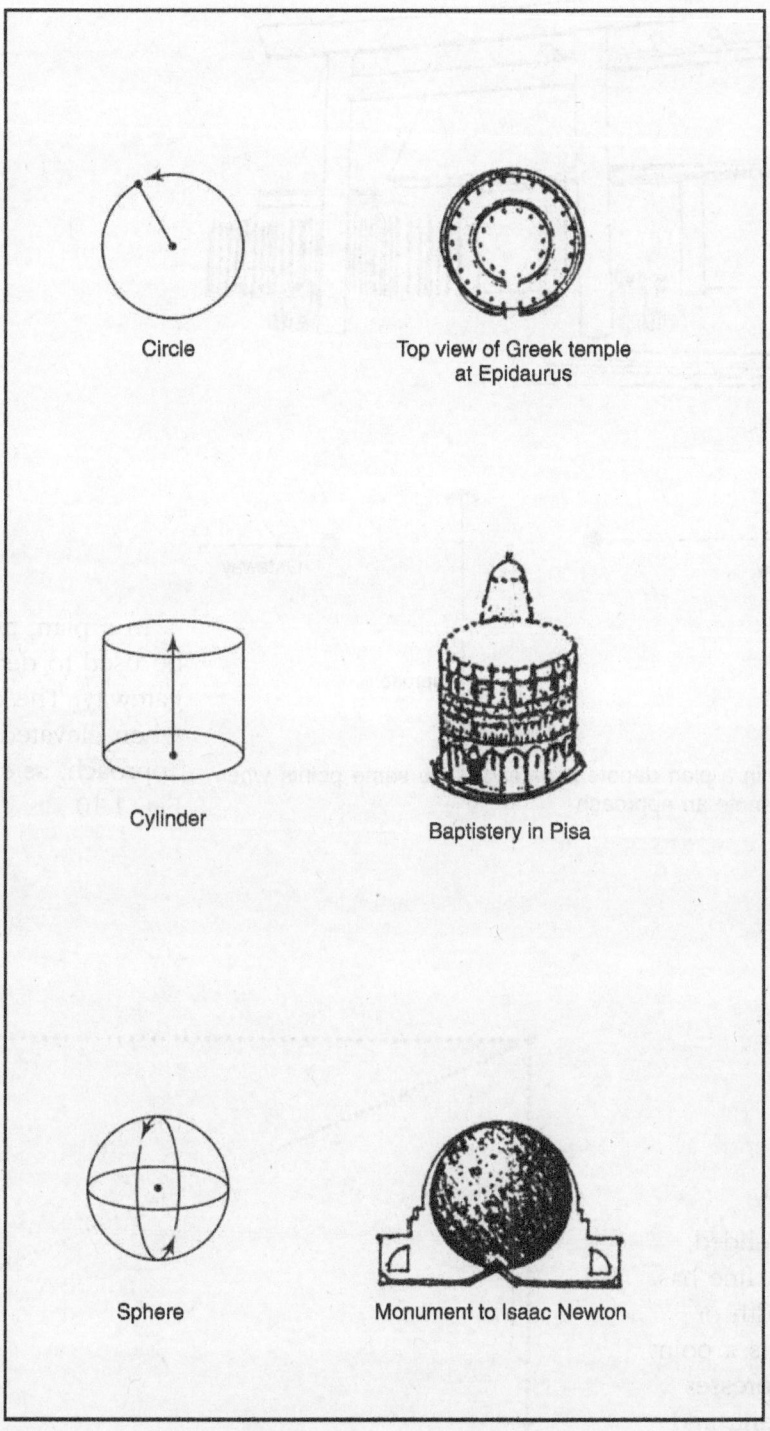

Fig. 1.9 Point-generated forms

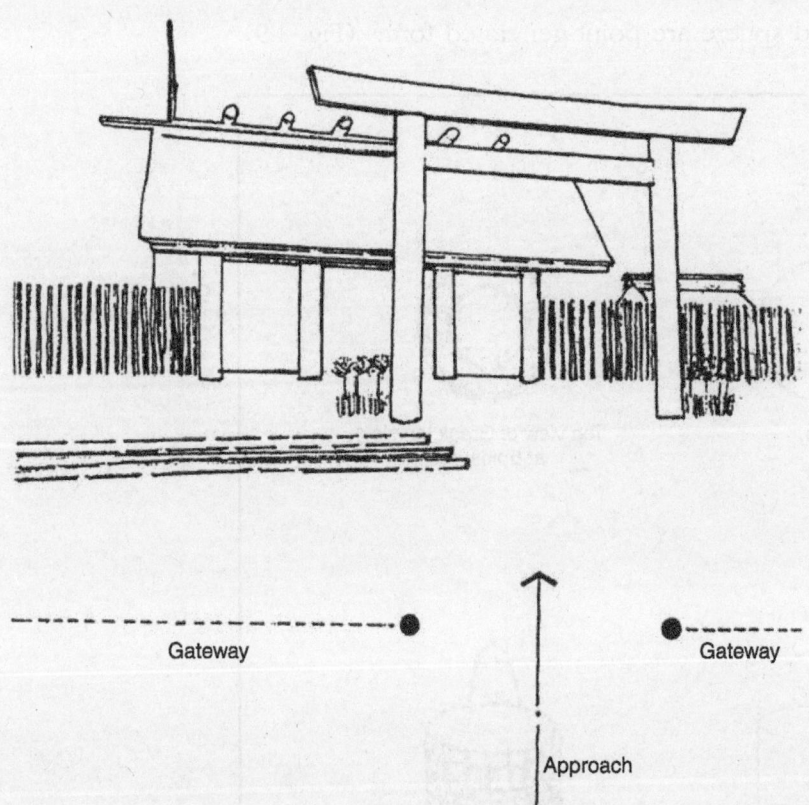

Gateway | Approach | Gateway

Fig. 1.10 Two points in a plan denote a gateway; the same points when elevated denote an approach

In a plan, two points can be used to denote a gateway. These two points when elevated denote an approach, as shown in Fig. 1.10.

1.3.2 Line

A point when extended becomes a line. A line has length, but no width or depth. It represents a point in motion and expresses direction, movement, and growth visually (Fig. 1.11).

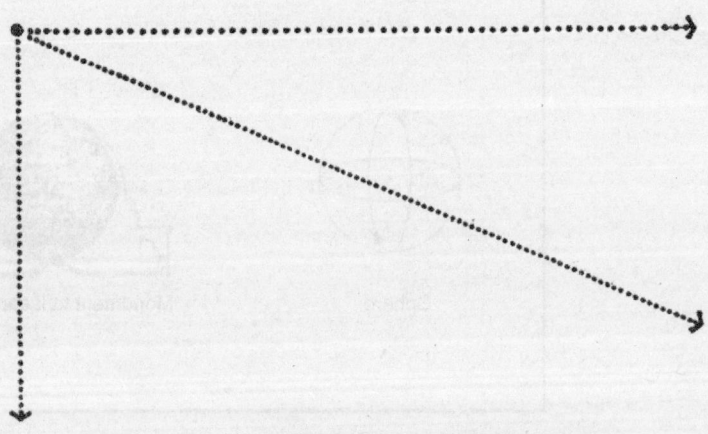

Fig. 1.11 A point when extended becomes a line

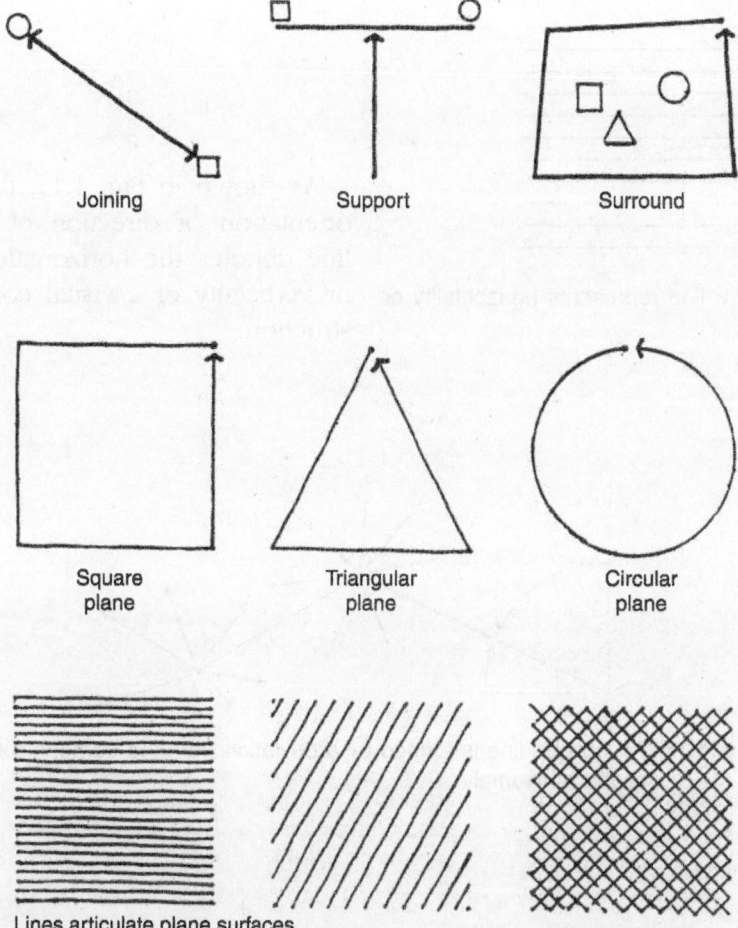

Fig. 1.12 Lines help in the formation of various visual elements

A line can serve to (See Fig. 1.12):

- join, link, support, surround, or intersect other visual elements,
- describe the edges of and give shapes to planes, and
- articulate the surfaces of planes.

Since a line has only one dimension, it must have some thickness or character (see Fig. 1.13) to become visible. A line that is thick enough to be considered a rectangle is still considered a line, as its length dominates its width. The character of line, whether bold or graceful, is determined by our perception of its length-to-width ratio and its contour.

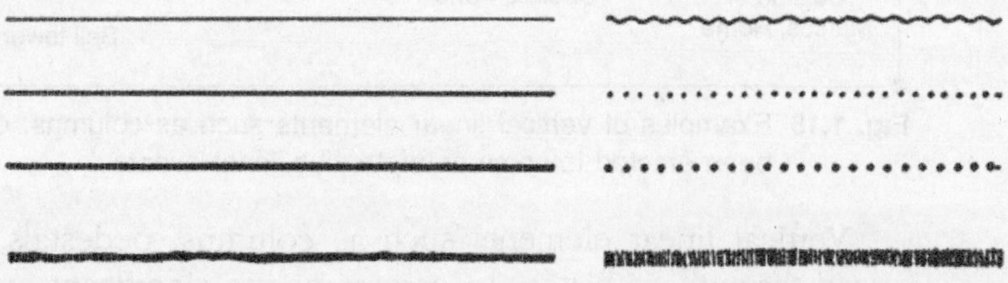

Fig. 1.13 The character of line—bold, graceful, repetitive, etc.

Fig. 1.14 The orientation or direction of a line represents horizontality or verticality

As shown in Fig. 1.14, the orientation or direction of a line denotes the horizontality or verticality of a visual construction.

Another orientation is denoted by the oblique line, which is formed by a deviation from the perpendicular or the horizontal. In other words, it is a falling vertical line or a rising horizontal line. This is illustrated in Fig. 1.15.

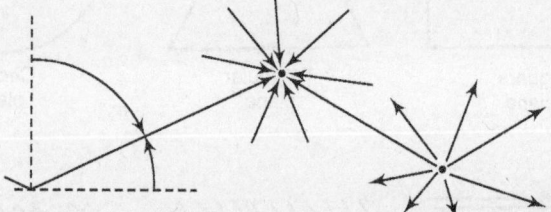

Fig. 1.15 An oblique line is formed by a deviation from the perpendicular or the horizontal

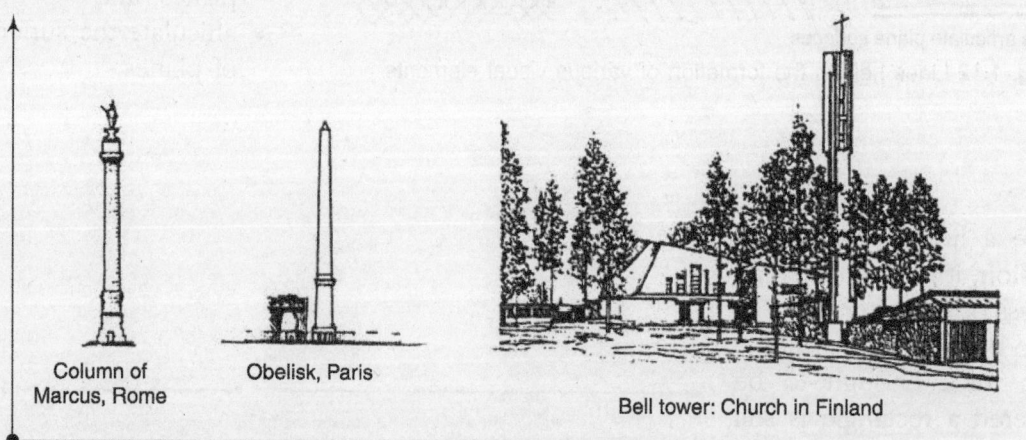

Column of Marcus, Rome

Obelisk, Paris

Bell tower: Church in Finland

Fig. 1.16 Examples of vertical linear elements such as columns, obelisks, and towers that have been erected to commemorate significant events

Vertical linear elements such as columns, pedestals, and towers have been used throughout history to commemorate significant events. Some examples of such structures are shown in Fig. 1.16.

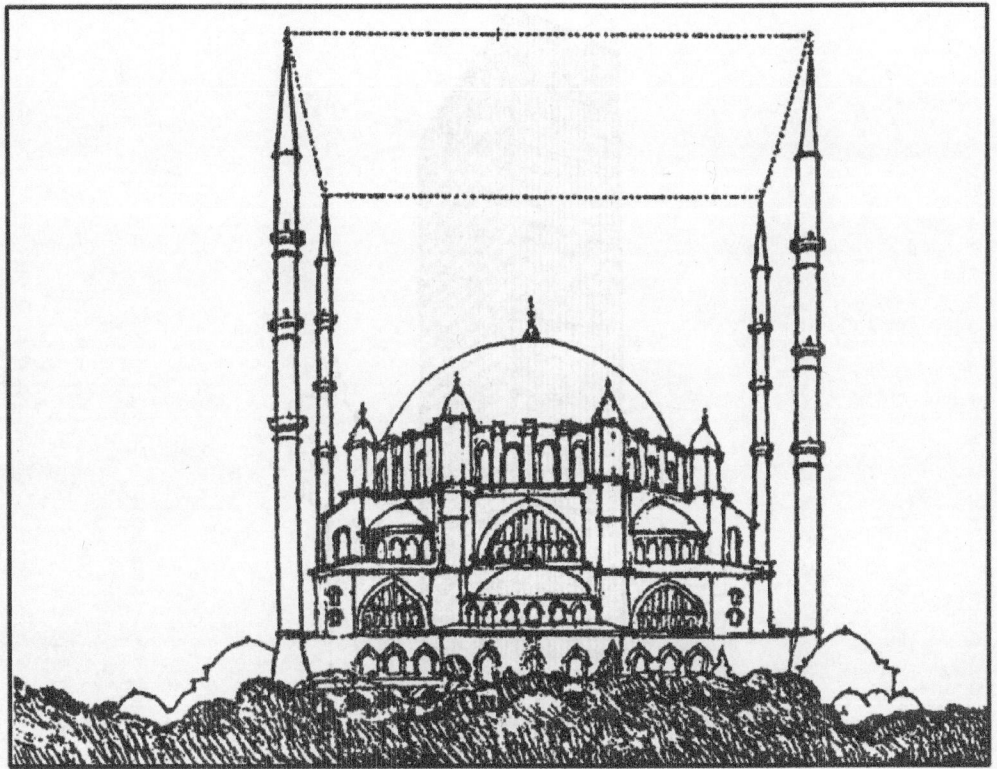

Fig. 1.17 Vertical linear elements define transparent volumes of space

As shown in Fig. 1.17, vertical linear elements are also used to define transparent volumes of space. In this figure, the four minaret towers define a spatial field from which the dome rises.

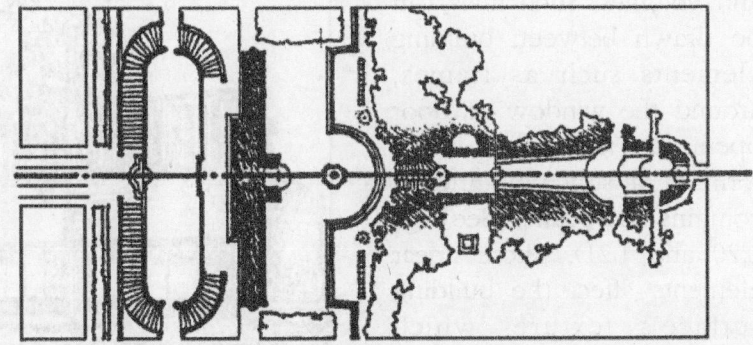

In Fig. 1.18, the linear element, the path, which is a visible element, is simply an axis about which elements are symmetrically arranged.

Fig. 1.18 A villa that has an axis about which elements are symmetrically arranged

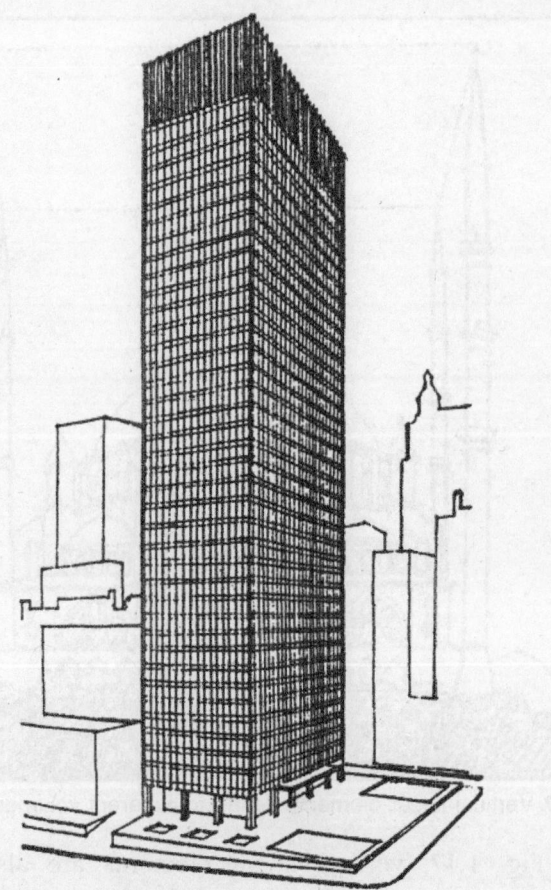

Fig. 1.19 The Seagram Building in New York built by Miles van der Rohe in 1958

Lines also articulate the edges and surfaces of planes and volumes. Such lines can be drawn between building elements such as frames, around the window or door openings (Fig. 1.19), or in the form of a structural grid of columns and beams (See Figs 1.20 and 1.21). These linear elements affect the building surface's texture, which depends on the direction and spacing of these elements.

Fig. 1.20 Town hall, Finland

Fig. 1.21 Crown Hall, Illinois Institute of Technology, Chicago articulates a structural grid of columns and beams

Fig. 1.22 A bridge (in Switzerland)—linear elements suggest movement across space

Linear elements are used to express movement across space as shown in Fig. 1.22, provide support for the overhead plane as shown in Fig. 1.23, and form a three-dimensional structural frame for architectural space as shown in Fig. 1.24.

Fig. 1.23 The Erecthion in Athens—linear elements (columns) provide support for the overhead plane

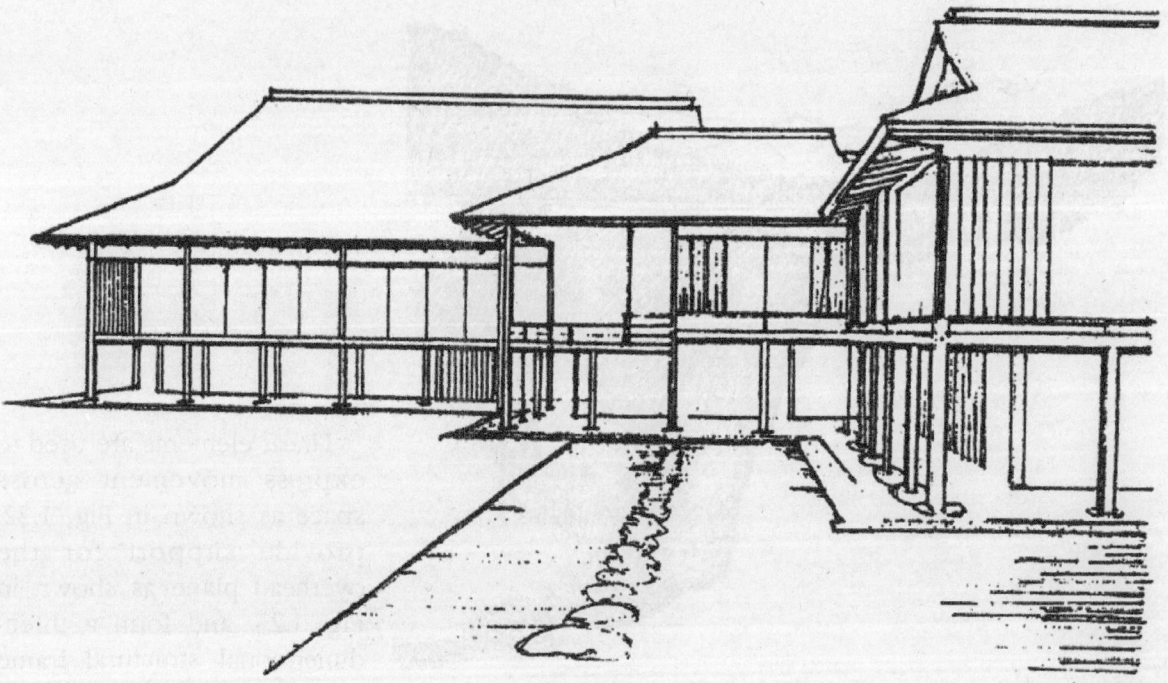

Fig. 1.24 Imperial Villa, Japan—linear elements form a three-dimensional structural frame for architectural space

Linear elements defining planes When a line is shifted or moved, it becomes a plane with the following properties:

- length and width
- shape
- surface
- orientation
- position

Two parallel lines can visually describe a plane. A row of columns represents an open wall with discontinuities at several places.

Figure 1.25 illustrates the transformation of a row of round columns (lines). The row initially supports a portion of a wall (plane); the columns then become square piers (part of the wall plane); finally, remnants of the original column occur as a relief along the surface of the wall.

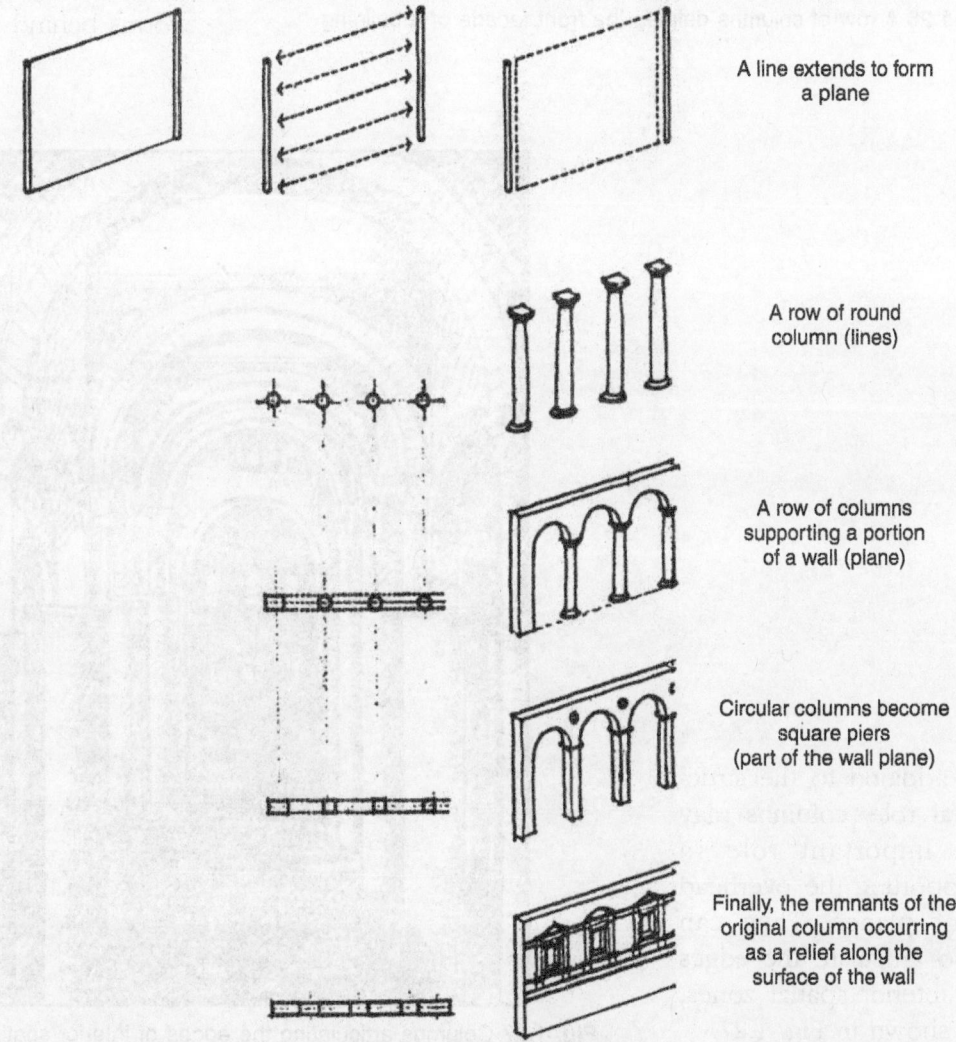

A line extends to form a plane

A row of round column (lines)

A row of columns supporting a portion of a wall (plane)

Circular columns become square piers (part of the wall plane)

Finally, the remnants of the original column occurring as a relief along the surface of the wall

Fig. 1.25 From line to plane

Fig. 1.26 A row of columns defines the front façade of a building

Linear elements define planes:

- A row of columns can be used to define the front façade of buildings—particularly public buildings as shown in Fig. 1.26.
- A colonnaded façade can easily be penetrated for entry and forms a semi-transparent screen—'A public face' that unifies the individual building forms behind it.

- In addition to the structural role, columns play an important role in supporting the overhead roof planes. They can also articulate the edges of interior spatial zones, as shown in Fig. 1.27.

Fig. 1.27 Columns articulating the edges of interior spatial zones

Fig. 1.28 A house in California illustrating horizontal overhead linear members

• Horizontal overhead linear members provide for a moderate degree of enclosure of outdoor space, allowing filtered sunlight and breeze to penetrate. See Fig. 1.28.

Figure 1.29 illustrates two contrasting examples—columns articulating the edges of a building form in space as well as the edges of an exterior space defined within a building form.

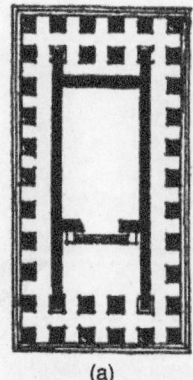

(a)

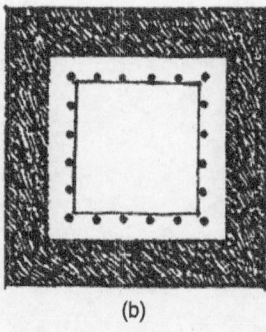

(b)

Fig. 1.29 (a) Columns articulating the edges of a building form in space; (b) columns articulating the edges of an exterior space defined within a building form

Fig. 1.30 Columns articulating the edges of an exterior space defined within a building form

Figure 1.30 clearly illustrates the second case. It shows columns lining an external space that is defined within a building form.

Vertical and horizontal linear elements together can define a volume of space. This is illustrated in Fig. 1.31.

Fig. 1.31 Vertical and horizontal linear elements defining a volume of space

1.3.3 Plane

A line when extended becomes a plane (Fig. 1.32). A plane has a length and a width but no depth. The primary character of a plane is its shape, which is determined by the contour of the line forming the edges of plane. The surface properties of a plane are colour and texture. Planes in architecture define three-dimensional volumes of form and space. Different planes can be manipulated in architectural design.

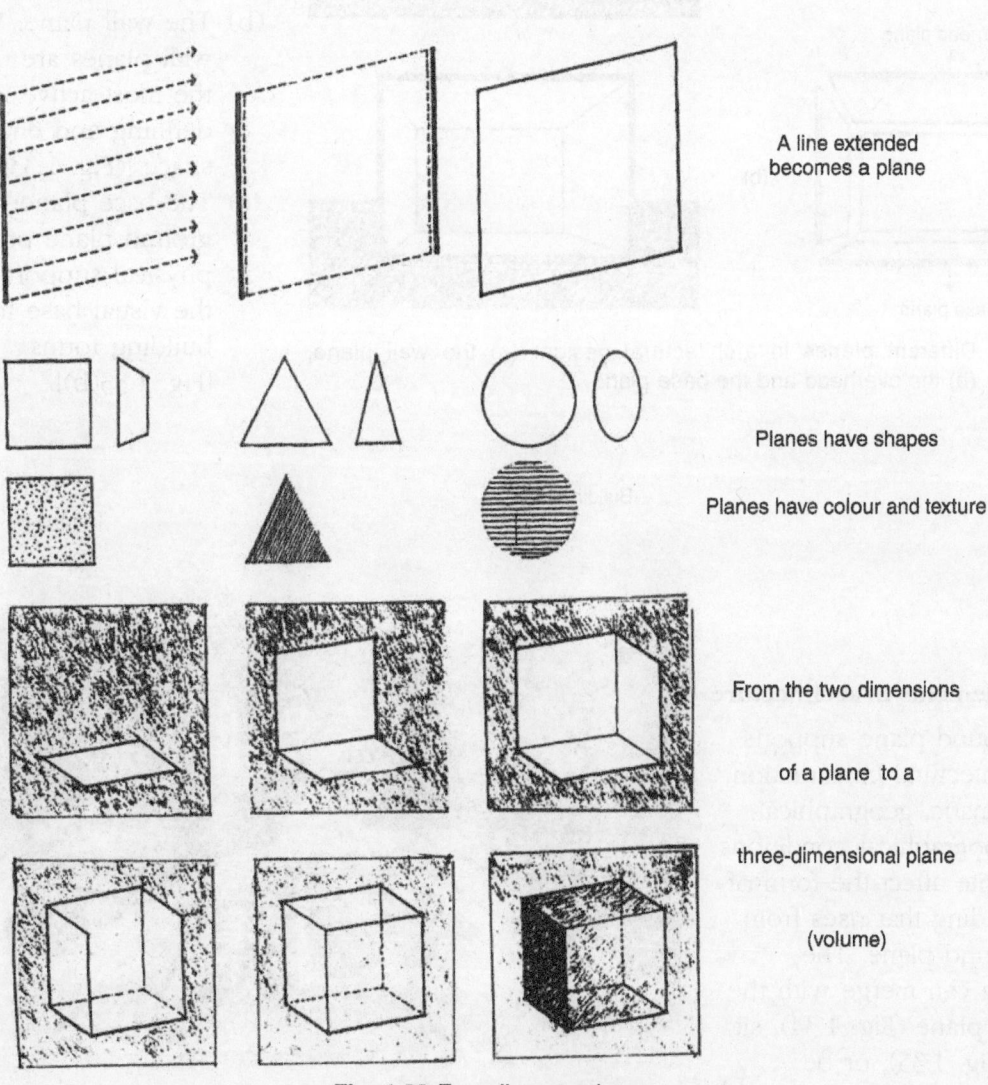

A line extended
becomes a plane

Planes have shapes

Planes have colour and texture

From the two dimensions

of a plane to a

three-dimensional plane

(volume)

Fig. 1.32 From lines to planes

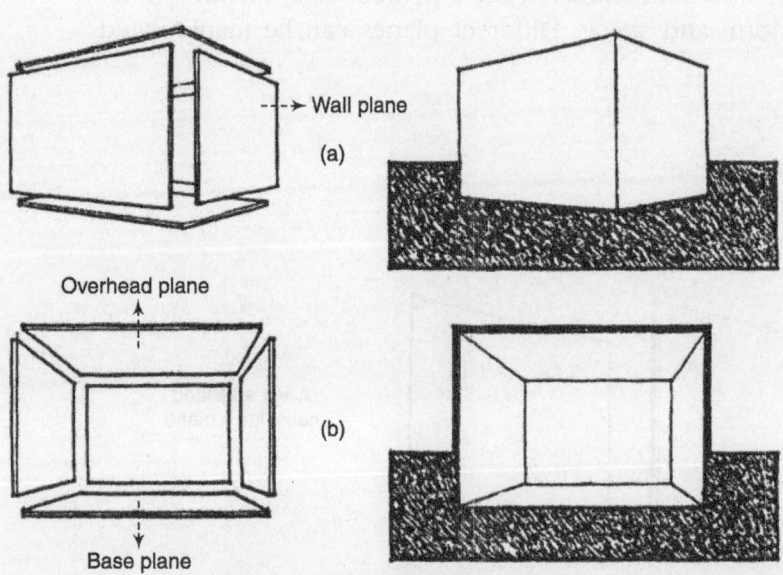

Fig. 1.33 Different planes in architectural design: (a) the wall plane, (b) the overhead and the base plane

There are three different types of planes in architectural design.

(a) The overhead plane: This can either be the roof plane, a building's pri-mary protection against climatic elements, or the ceiling plane, the sheltering element in architectural space [(Fig. 1.33(b)].

(b) The wall plane: Vertical wall planes are usually the most active in defining and enclosing space [(Fig. 1.33(a)].

(c) The base plane: The ground plane provides physical support and the visual base for building forms [Fig. 1.33(b)].

Plane elements in architecture

The ground plane supports all architectural construction. The climatic, geographical, and topographical conditions of the site affect the form of the building that rises from the ground plane. The building can merge with the ground plane (Fig. 1.34), sit on it (Fig. 1.35), or be elevated above it (Fig. 1.36).

Fig. 1.34 The building can merge with the ground plane

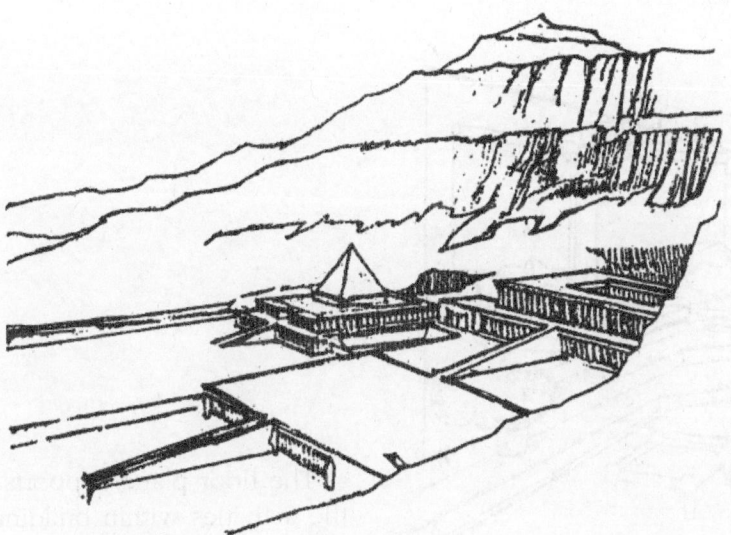

Fig. 1.35 Temple in Thebes—The building sits on the ground plane

The ground plane could be elevated to honour a sacred place. It could be carved or terraced to provide a suitable platform on which the building can be built. It could be stepped to allow changes.

Fig. 1.36 The Spanish Steps (in Rome)—The building is elevated above the ground plane

Fig. 1.37 The floor plane supports the activities inside a space

The floor plane supports the activities within buildings (Fig. 1.37). The interior wall planes define and enclose the building space or rooms.

Like the ground plane, the floor plane can be stepped or terraced, creating platforms for sitting, viewing performances, etc. (Fig. 1.38).

Fig. 1.38 The floor plane can be manipulated—stepped or terraced

A wall plane can merge with the floor or ceiling or an isolated plane. The distribution of the openings in the walls will determine the quality of space. It can be a neutral backdrop for the other elements in the space. The wall and floor are physically close to the people using the building. The ceiling plane is usually more distant and almost a purely visual element. It can correspond to the form or be the under-surface of the roof or floor plane above, and express its structure. It can also be a detached lining within the space.

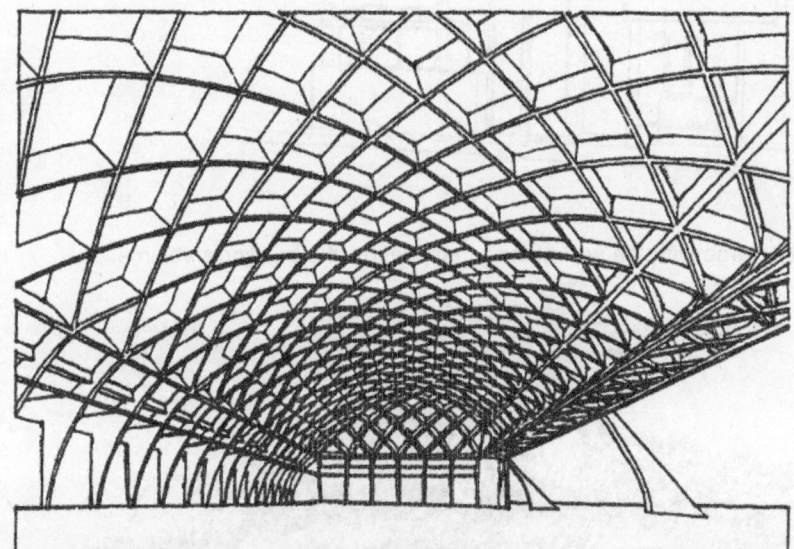

Fig. 1.39 A vaulted sky plane, where the ceiling plane merges with the wall plane

As a detached lining, the ceiling plane can be manipulated to symbolize a vaulted sky plane (Fig. 1.39). It can be raised or lowered to alter the scale of a space or to define the zones of the space within a room.

A roof plane, which is the building's prime sheltering element, protects the interior from climatic elements (Fig. 1.40). The roof can merge with the building's walls (Fig. 1.41).

Fig. 1.40 Robin House, Chicago, built by Frank Lloyd Wright. The roof plane projects out of the building to protect the openings in the walls from sun or rain.

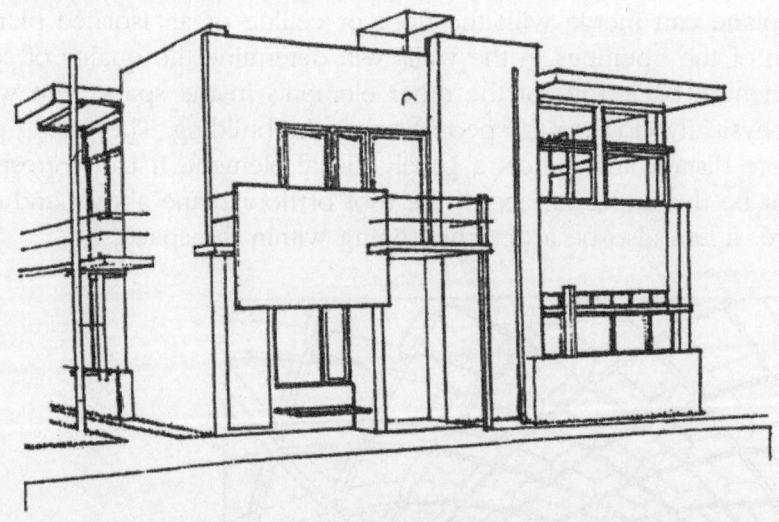

Fig. 1.41 The roof plane can merge with the wall plane to emphasize the building's volume

Ancient megalithic stone structures have been found in burial grounds. They consisted of three vertical stones, which supported the 'roof', which was a fourth horizontal slab (Fig. 1.42).

Fig. 1.42 Dolmen (in Italy)—A burial place

Fig. 1.43 Kaufmann House 'Falling water' Pennsylvania (built by Frank Lloyd Wright)—The building form reflects the planar quality

The roof plane has a significant impact on the building's silhouette.

The building's overall form acquires a planar quality (Fig. 1.43) by differentiating between vertical and horizontal planes, by using different colours, textures, and materials, and by introducing openings between planes and at corners and visually exposing their edges.

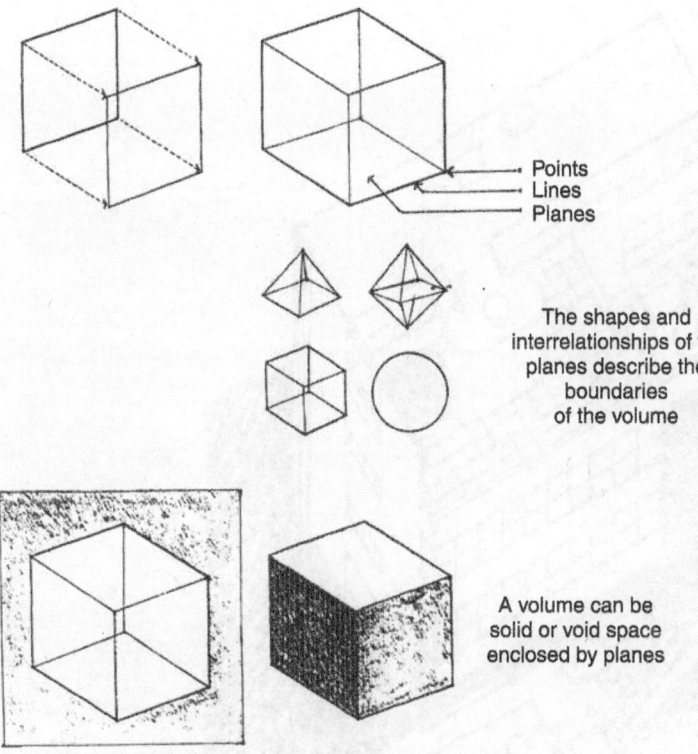

Points
Lines
Planes

The shapes and
interrelationships of the
planes describe the
boundaries
of the volume

A volume can be
solid or void space
enclosed by planes

Fig. 1.44 A plane when extended becomes a volume

1.3.4 Volume

A plane extended perpendicular to itself becomes a volume (Fig. 1.44). A plane has only two dimensions, length and width, whereas a volume has three dimensions—length, width, and depth.

All volumes can be analysed to consist of the following:

- points
- lines
- planes

A point (or vertex) is formed where several planes meet; a line (or edge) is formed where two planes meet. Planes (or surfaces) define the limits or boundaries of volumes. A form is a primary characteristic of a volume. It is determined by the shapes and interrelationships of the planes that form the boundaries of the volume.

As a three-dimensional element in the vocabulary of architectural design, a volume can be either solid space displayed by mass or void space contained or enclosed within planes (Fig. 1.45).

Fig. 1.45 Building forms defining volumes of space (a) Doric temple

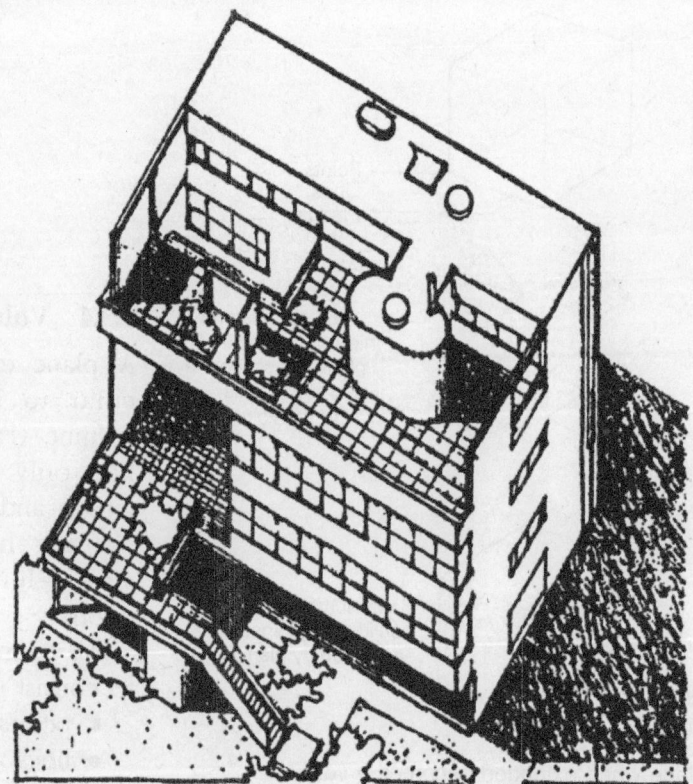

Fig. 1.45 (*contd*) (b) Villa at Gardens (in France) (built by Le Corbusier)

Fig. 1.45 (*contd*) (c) Barn (in Ontario, Canada)

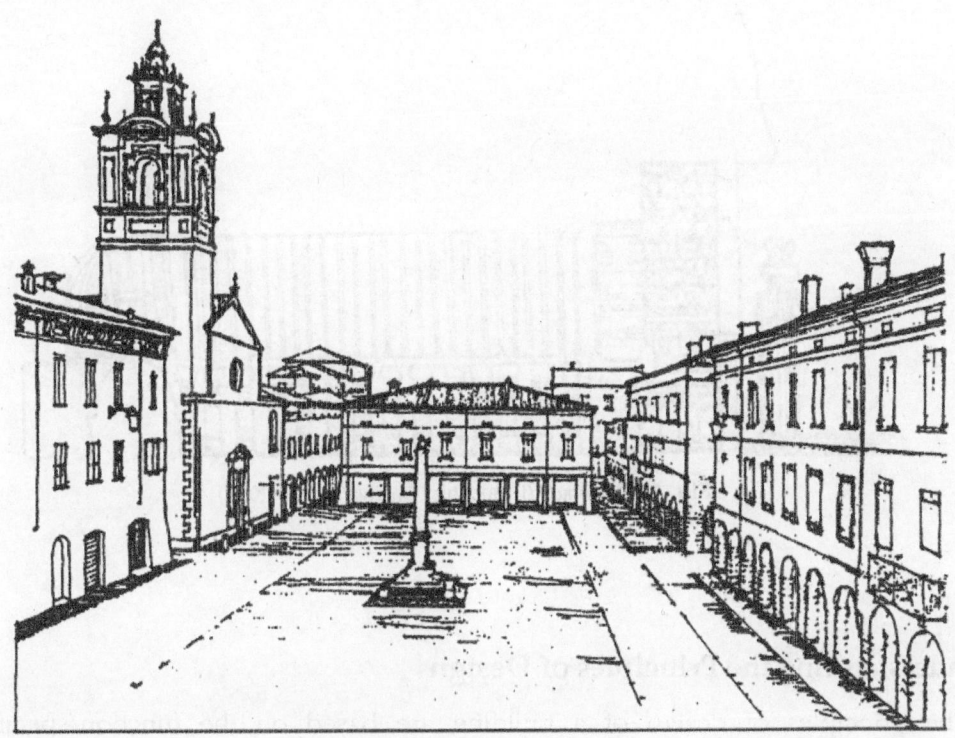

Fig. 1.45 (*contd*) (d) Piazza (in Italy)

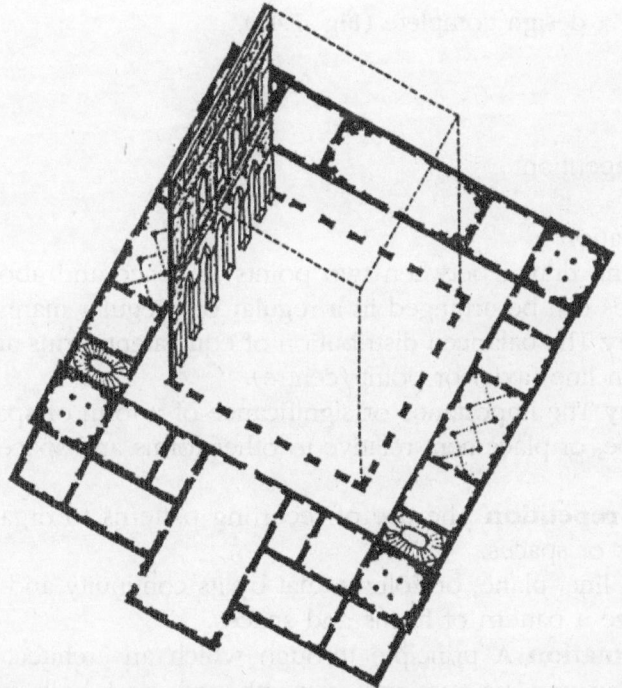

Fig. 1.45 (*contd*) (e) Palazzo (in Italy)

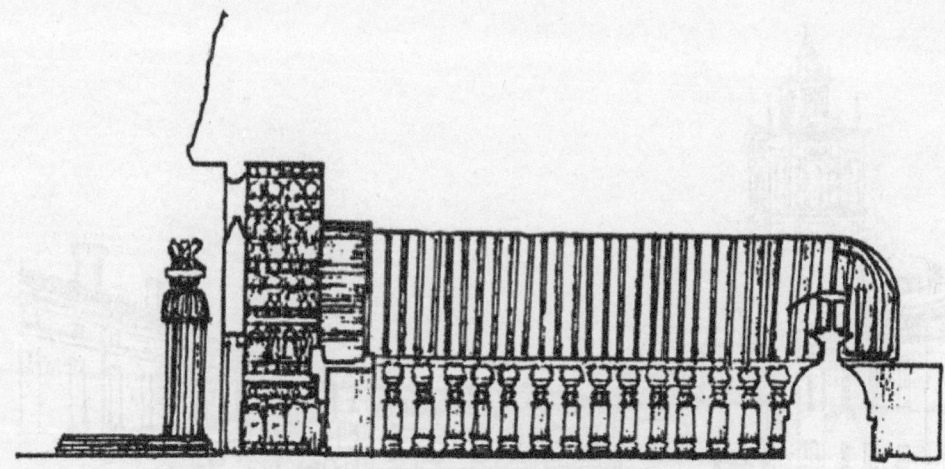

Fig. 1.45 (*contd*) (f) Buddhist Chaitya Hall (in Karli)

1.4 Understanding the Principles of Design

The principles of design of a building are based on the function/specific purpose and the varied needs of the building. The following 'ordering principles' are visual methods that allow the form and space of a building to co-exist in order to make a design complete (Fig. 1.46).

- axis
- symmetry
- hierarchy
- rhythm/repetition
- datum
- transformation

(a) **Axis** A line drawn between two points in space and about which forms and spaces can be arranged in a regular or irregular manner.

(b) **Symmetry** The balanced distribution of equivalent forms and spaces about a common line (axis) or point (centre).

(c) **Hierarchy** The importance or significance of a form or space based on its size, shape, or placement relative to other forms and spaces of the organization.

(d) **Rhythm/repetition** The use of recurring patterns to organize a series of like forms or spaces.

(e) **Datum** A line, plane, or volume that by its continuity and regularity helps to organize a pattern of forms and spaces.

(f) **Transformation** A principle through which an architectural concept or organization can be retained, strengthened, and built upon through a number of manipulations.

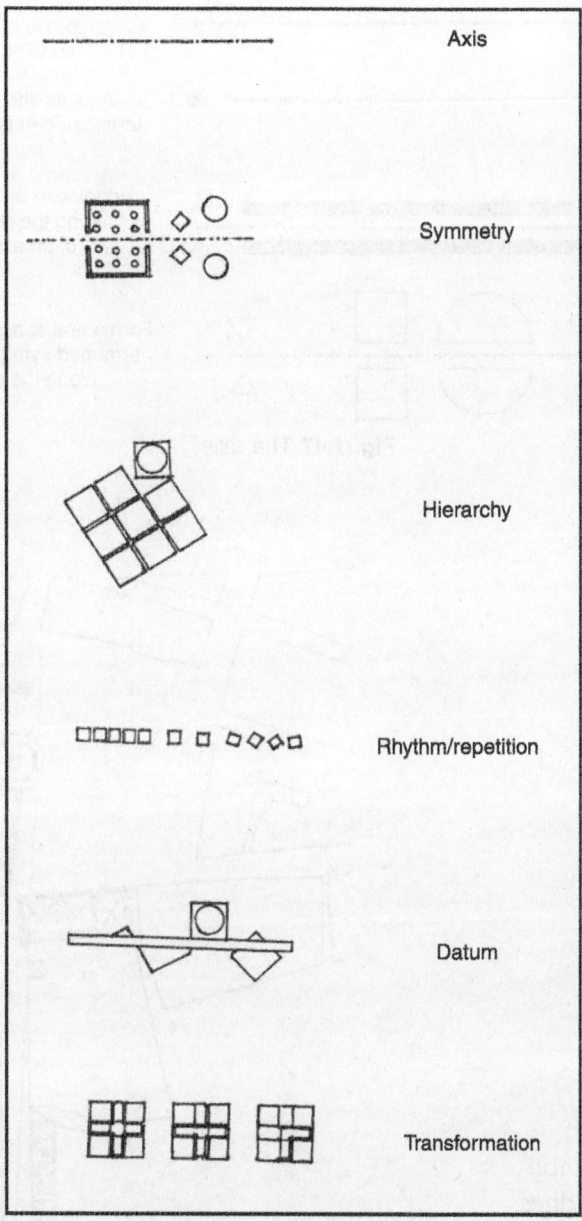

Fig. 1.46 Principles of design

1.4.1 Axis

The axis is the most elementary means of organizing forms and spaces (Fig. 1.47). Although imaginary and not visible, an axis is a powerful dominating and regulating device. It helps to establish symmetry and balance.

(a) An axis is a linear condition (line) which has length and direction; it allows for views and movement along its path.

(b) To define an axis, it should be terminated at both ends.

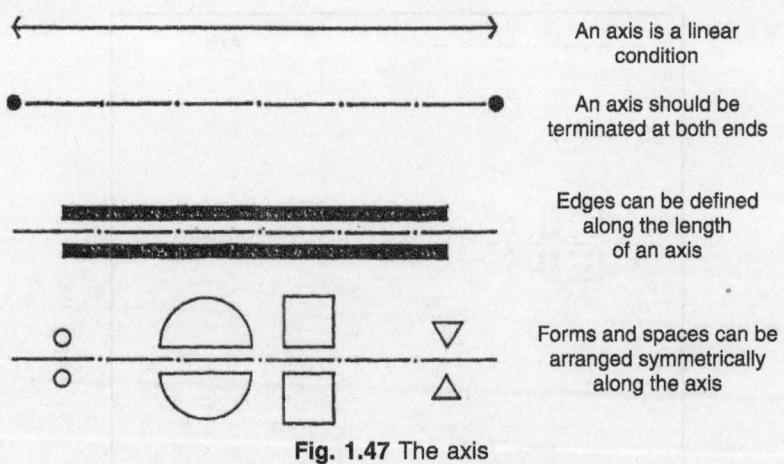

Fig. 1.47 The axis

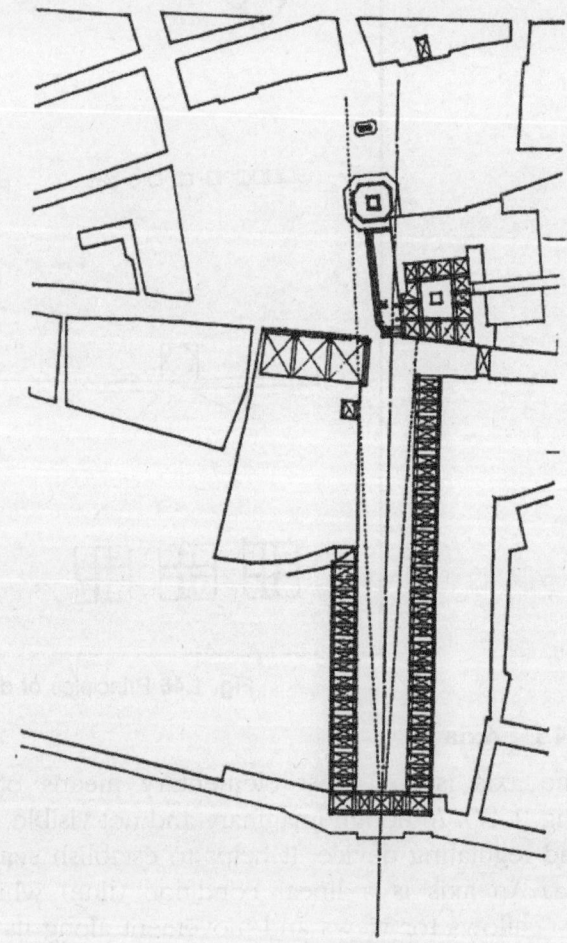

(c) An axis can be established by defining edges
along its length. These
edges can be simply lines
on the ground plan, or
vertical planes that define
a linear space (Fig. 1.48).

(d) An axis can also be
established by a symme-
trical arrangement of
forms and spaces.

Fig. 1.48 Edges (buildings) are defined along an axis

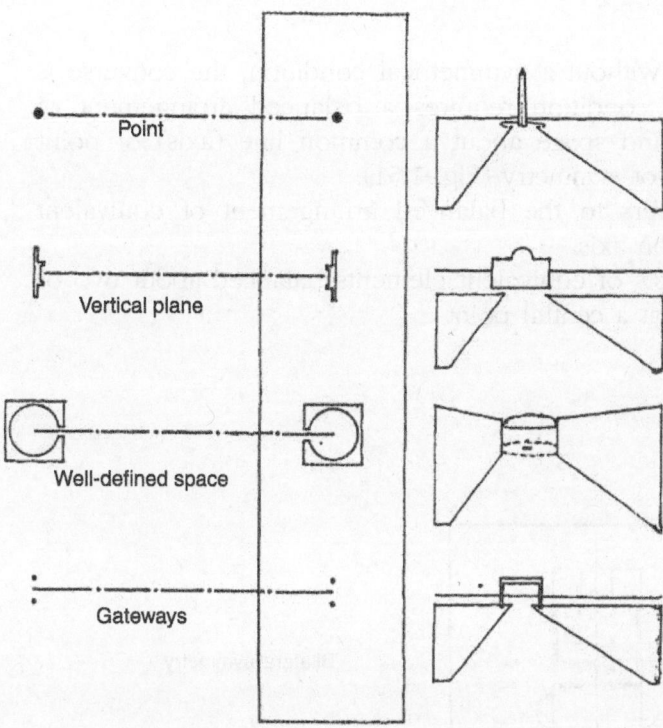

Point

Vertical plane

Well-defined space

Gateways

Fig. 1.49 Different terminating elements of an axis

The terminating elements of an axis serve to both send and receive its visual thrust. These terminating elements can be any of the following (Fig. 1.49).

(a) **Points** in space, which are (represented) established by vertical or linear elements

(b) **Vertical planes** such as a symmetrical building façade

(c) **Well-defined space,** which can be centralized

(d) **Gateway,** which is open towards a view or vista beyond (Fig. 1.50)

Fig. 1.50 A row of columns viewed through the entrance arch defines an axis

1.4.2   Symmetry

An axial condition can exist without a symmetrical condition, the converse is not possible. A symmetrical condition requires a balanced arrangement of equivalent patterns of form and space about a common line (axis) or point (centre). There are two types of symmetry (Fig. 1.51):

(a) **Bilateral symmetry** refers to the balanced arrangement of equivalent elements about a common axis.

(b) **Radial symmetry** consists of equivalent elements balanced about two or more axes that intersect at a central point.

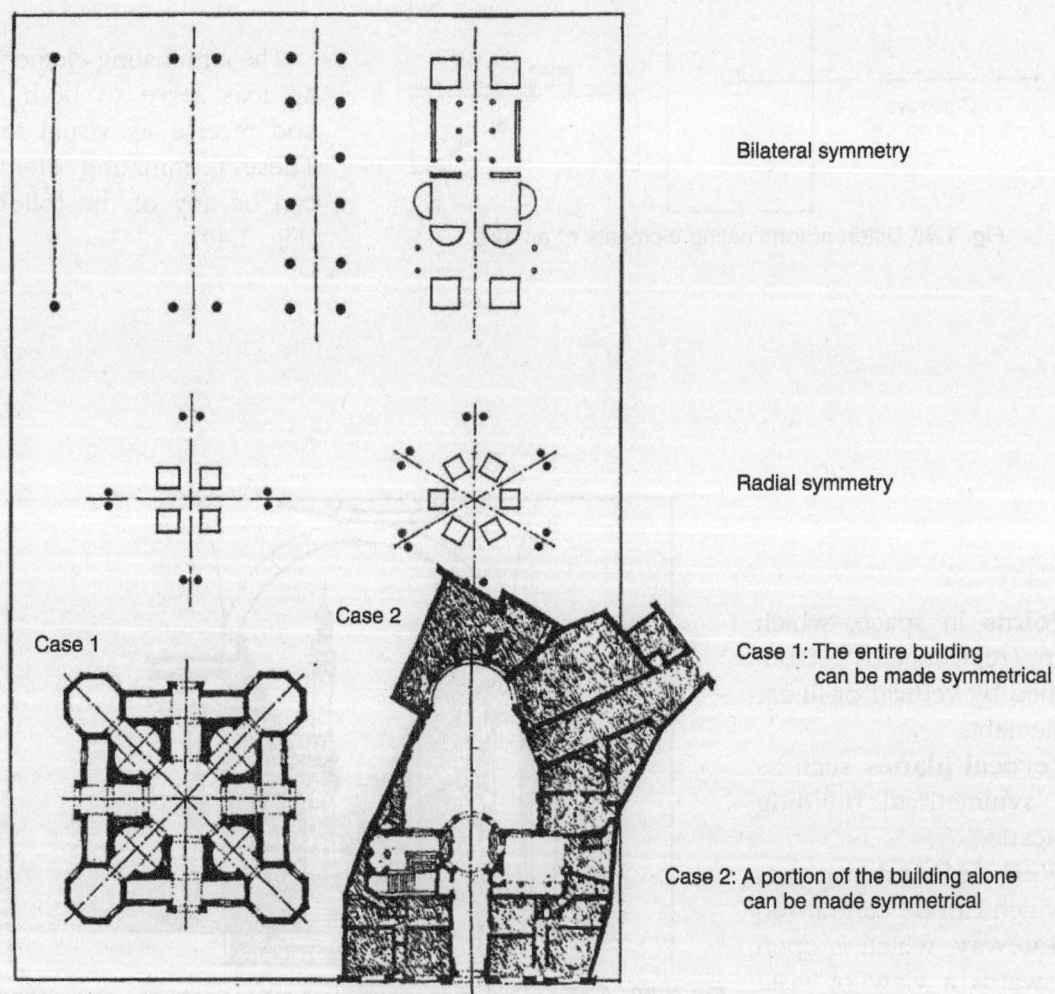

Bilateral symmetry

Radial symmetry

Case 1

Case 2

Case 1: The entire building can be made symmetrical

Case 2: A portion of the building alone can be made symmetrical

Fig. 1.51 Sysmmetrical condition

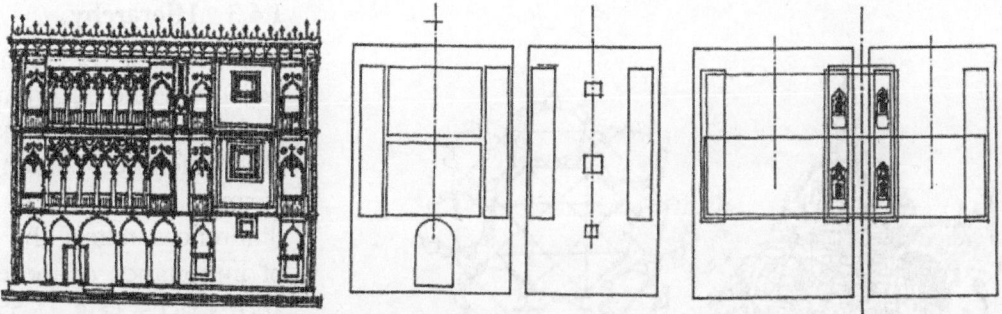

Fig. 1.52 A building in Venice, Austria. Two axes can be established about which the arched openings and the elements in the elevation are symmetrical

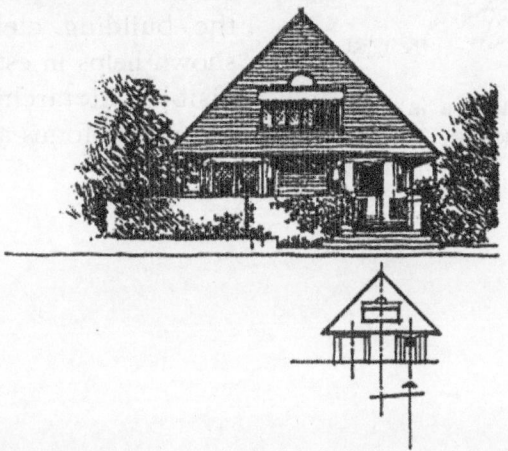

Fig. 1.53 Oak Park, Illinois (built by F.L. Wright). A vertical axis can be established.

An architectural composition can utilize symmetry to organize forms and spaces in two ways.

Case 1 Forms and spaces can be organized symmetrically through an entire building (see Figs 1.52 and 1.53).

Case 2 A symmetrical condition can occur only in a portion of the building (Fig. 1.54), and an irregular pattern of forms and spaces can be organized in the rest of the building. This method helps to satisfy site conditions.

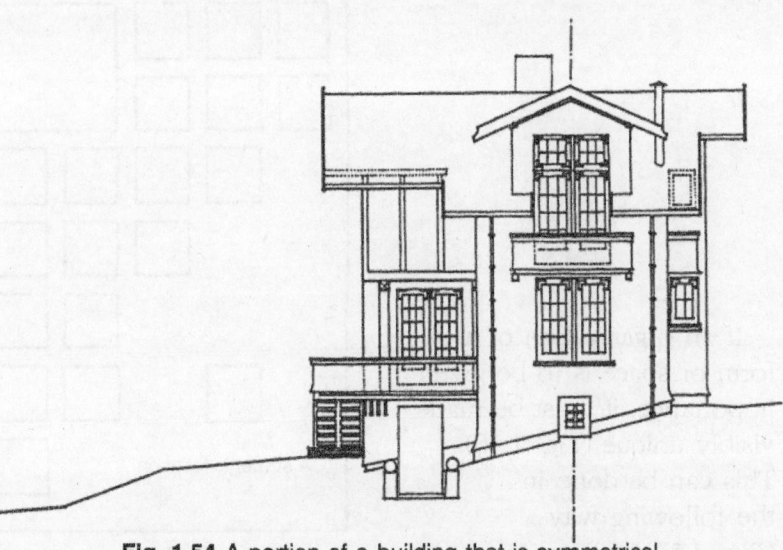

Fig. 1.54 A portion of a building that is symmetrical

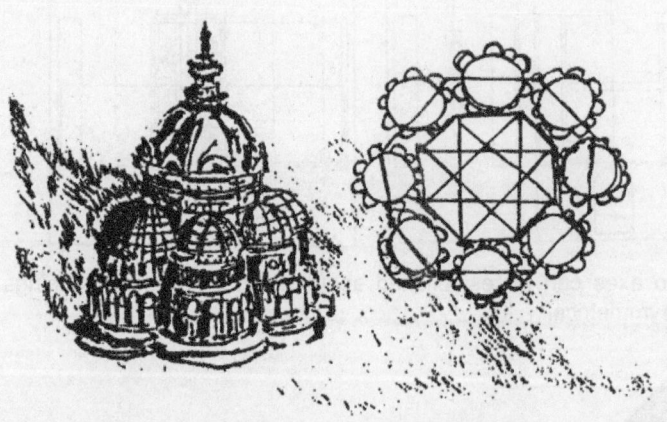

Fig. 1.55 Principle of hierarchy—the central structure is given more importance by varying its height and structure

1.4.3 Hierarchy

The principle of hierarchy is based on the fact that most architectural buildings have real differences among their forms and spaces. These differences reflect the degree of importance of these forms and spaces (Fig. 1.55). The way in which the functional or symbolic differences among the building elements are shown helps in establishing a visible, hierarchical order among the forms and spaces.

If an organization of a form or space is to be given importance, it must be made visibly unique (Fig. 1.56). This can be done in the following ways (Figs 1.57–1.59).

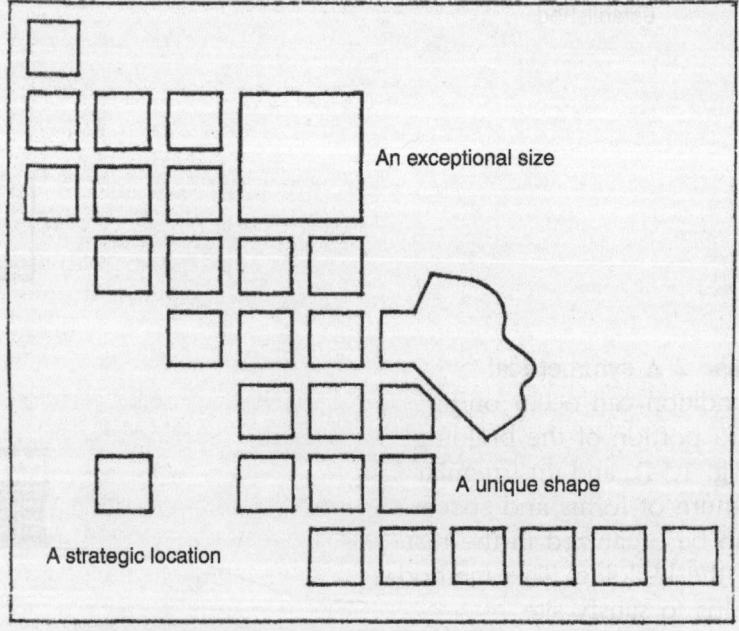

An exceptional size

A unique shape

A strategic location

Fig. 1.56 Form or space articulated to be visually unique

(a) A form or space may be shown to dominate an architectural composition by being made significantly smaller or larger than other elements.

(b) Forms and spaces can be made visually dominant or important by clearly differentiating their shapes from the other elements in the composition. A contrast in shape can also be used.

(c) Forms and spaces can be given importance by allocating to them an important location, by terminating a linear sequence or axial organization, and by focusing on the centralized or radial organization.

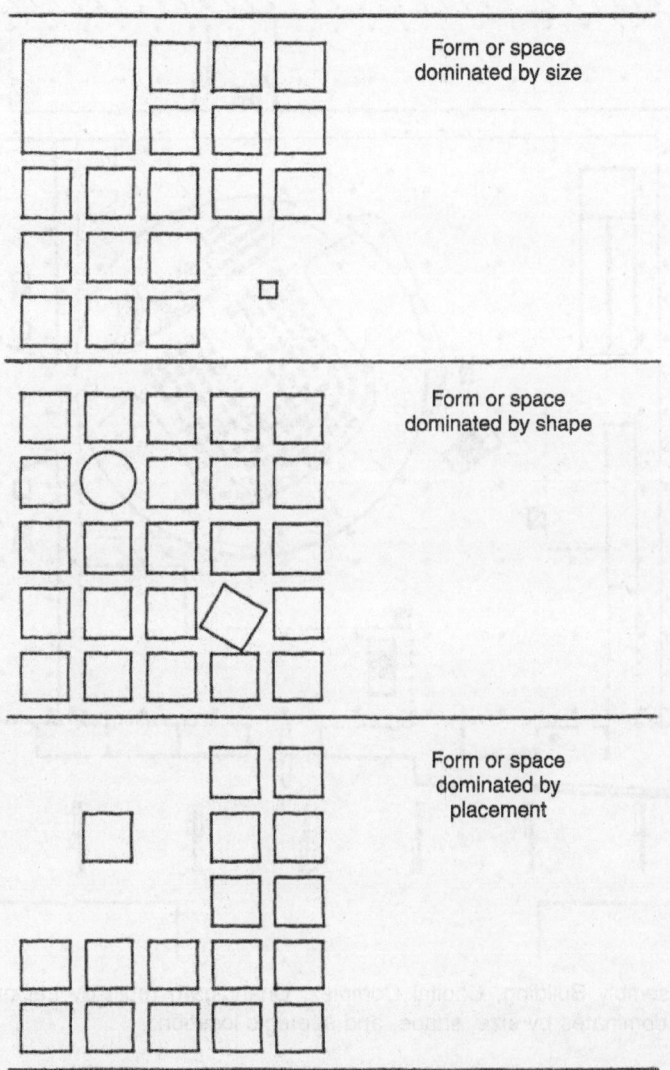

Form or space dominated by size

Form or space dominated by shape

Form or space dominated by placement

Fig. 1.57 Form or space can be made visibly unique

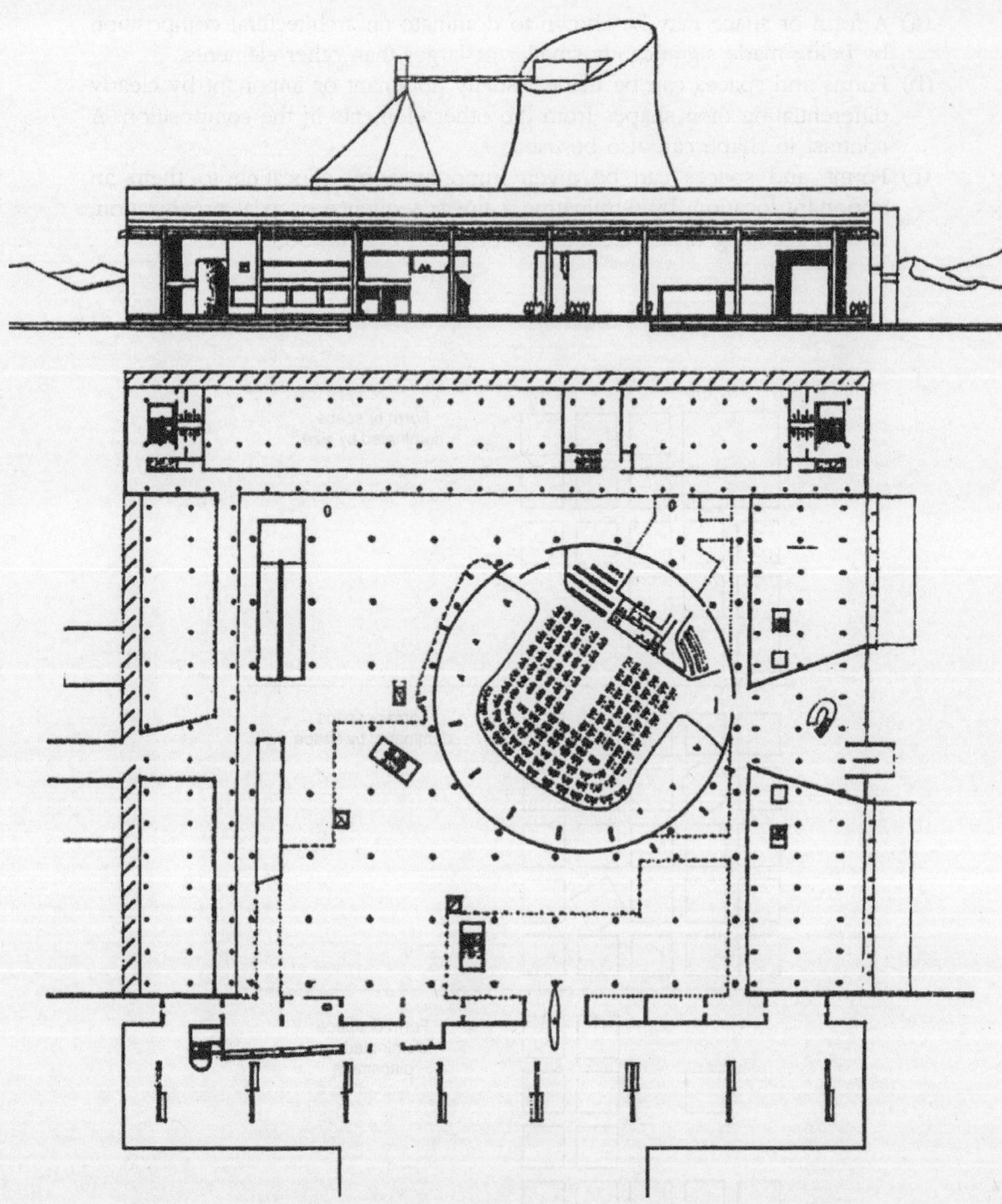

Fig. 1.58 Legislative Assembly Building, Capital Complex, Chandigarh (built by LeCorbusier). The roof of the assembly hall dominates by size, shape, and strategic location.

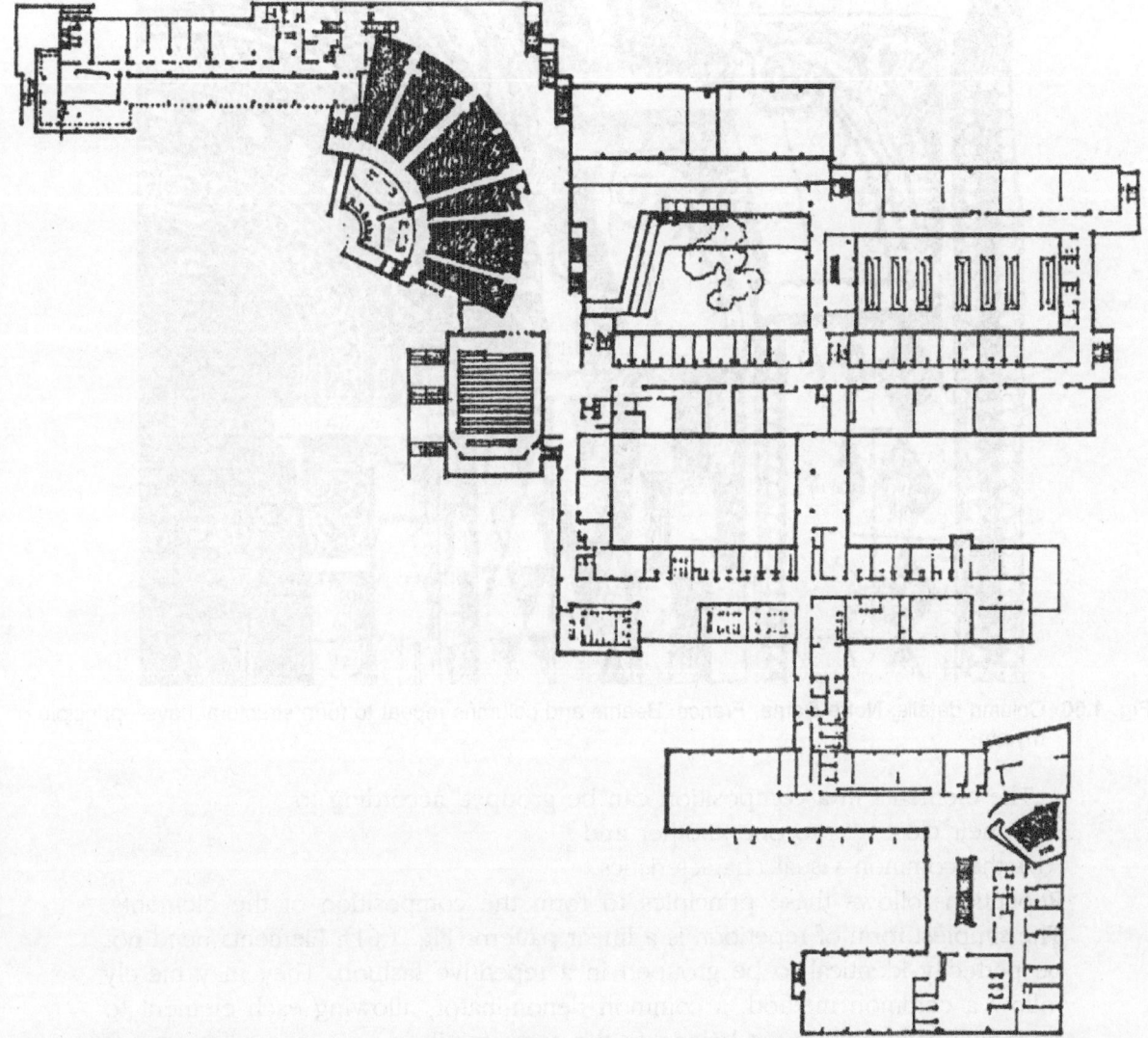

Fig. 1.59 Institute of Technology, Finland. The auditorium block dominates by size, shape, and strategic location.

1.4.4 Rhythm or repetition

Rhythm refers to the regular recurrence of lines, shapes, forms or colours. The concept of rhythm is commonly used to organize forms and spaces in architecture.

All building types incorporate elements that are repetitive:

(a) Beams and columns repeat themselves to form repetitive structural bays (Fig. 1.60) and modules of space.

(b) Windows and doors are built into walls at regular intervals to allow light, air, and people to enter its interiors, and to provide various viewpoints to the surounding landscape.

Fig. 1.60 Column details: Notre Dame, France. Beams and columns repeat to form structural bays—principle of rhythm.

The elements in a composition can be grouped according to
(a) their closeness to one another and
(b) the common visual characteristics.
Repetition follows these principles to form the composition of the elements. The simplest form of repetition is a linear pattern (Fig. 1.61). Elements need not be perfectly identical to be grouped in a repetitive fashion. They may merely follow a common method, a common denominator, allowing each element to be individually unique yet belong to the same family.

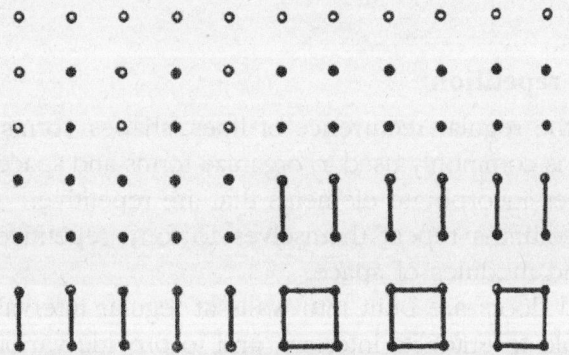

Fig. 1.61 Harmonious recurrence of lines, shapes, or forms

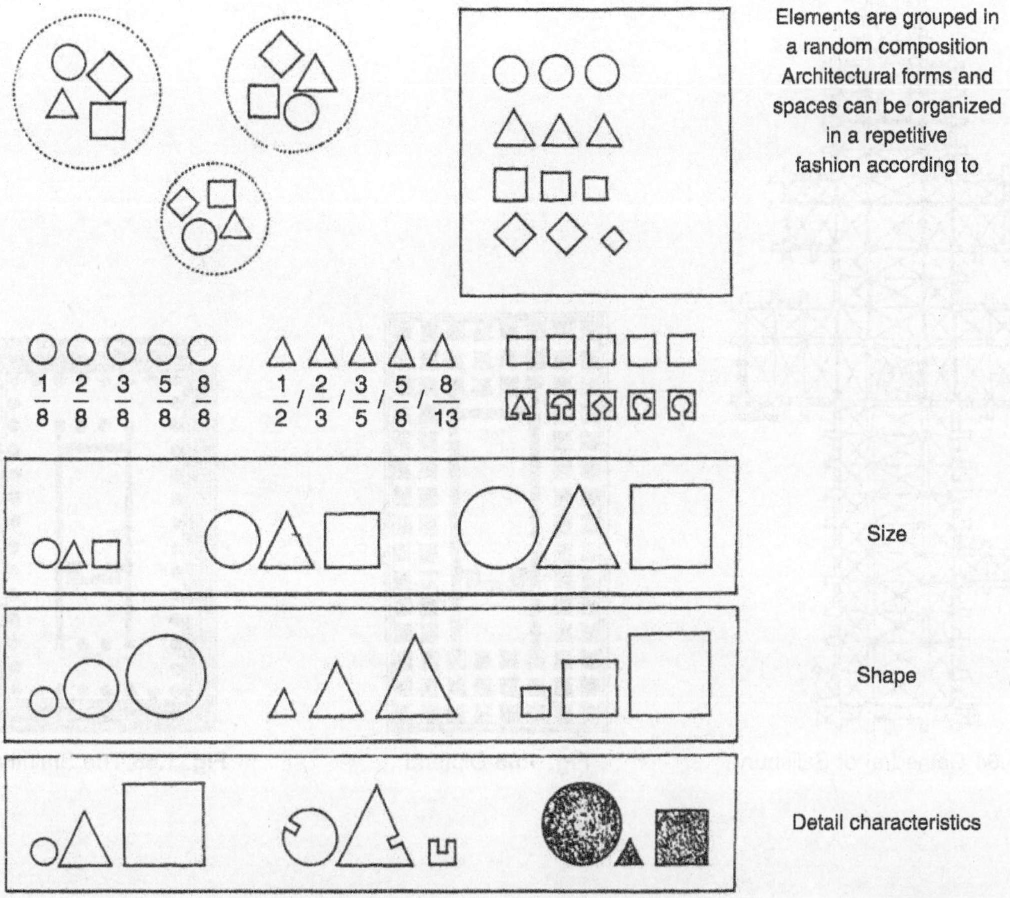

Elements are grouped in
a random composition
Architectural forms and
spaces can be organized
in a repetitive
fashion according to

$$\frac{1}{8} \quad \frac{2}{8} \quad \frac{3}{8} \quad \frac{5}{8} \quad \frac{8}{8}$$

$$\frac{1}{2} / \frac{2}{3} / \frac{3}{5} / \frac{5}{8} / \frac{8}{13}$$

Size

Shape

Detail characteristics

Fig. 1.62 Visual rhythms created by repetitive patterns

Forms and spaces can be arranged in a repetitive order
according to (Fig. 1.62)

- size
- shape
- detail characteristics

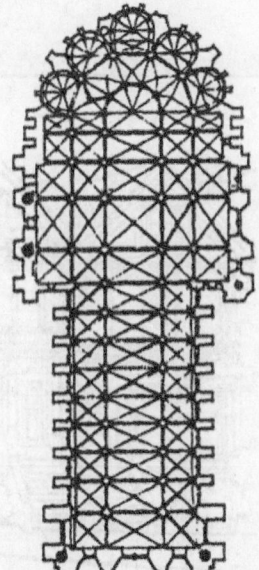

Fig. 1.63 Reims Cathedral, France

Figures 1.63–1.70 illustrate various rhythmic and repetitive
patterns.

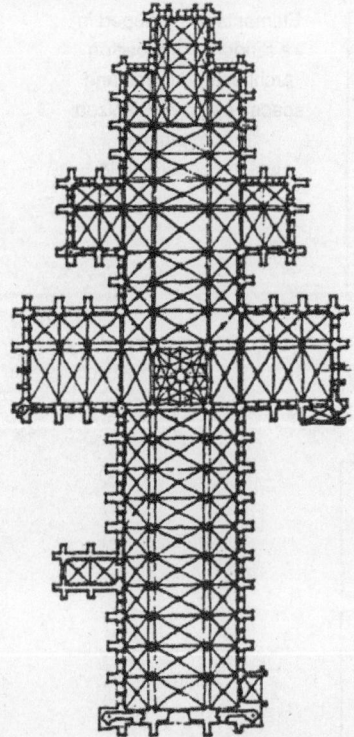

Fig. 1.64 Cathedral of Salisbury

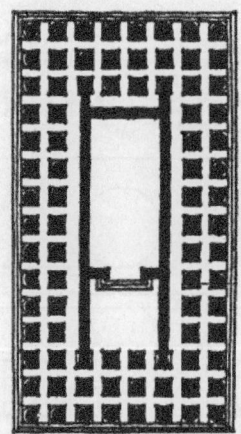

Fig. 1.65 Dipteral

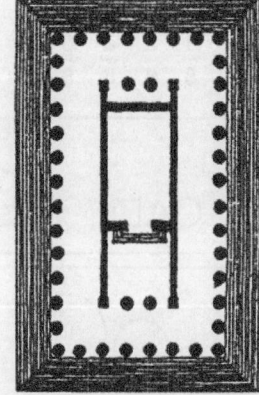

Fig. 1.66 The Smintheum

Fig. 1.67 The Victorian façade facing a San Francisco street. Observe the repetition of the elevation façade.

Fig. 1.68 View of a villa in Spain. The rhythmic pattern is observed in the form of the buildings.

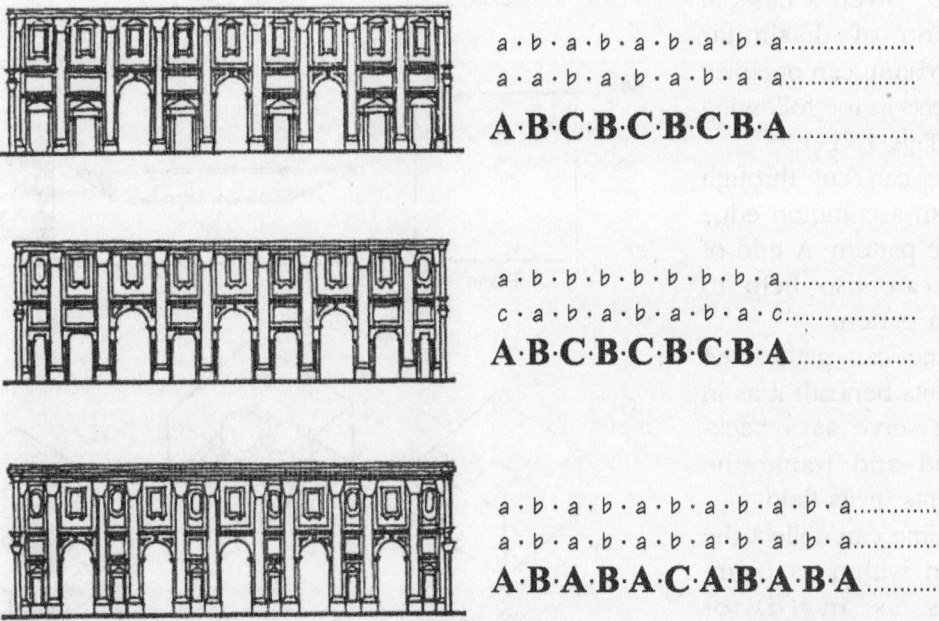

a · b · a · b · a · b · a · b · a....................
a · a · b · a · b · a · b · a · a....................
A·B·C·B·C·B·C·B·A...............

a · b · b · b · b · b · b · b · b · a....................
c · a · b · a · b · a · b · a · c....................
A·B·C·B·C·B·C·B·A...............

a · b · a · b · a · b · a · b · a · b · a.............
a · b · a · b · a · b · a · b · a · b · a.............
A·B·A·B·A·C·A·B·A·B·A...........

Fig. 1.69 Studies on the internal façade of a basilica. The repetition of the columns, archways, and windows in the elevation is marked as a, b, c, ...

Fig. 1.70 Sydney Opera House, Australia (built by John Utson). The roof structure forms a rhythmic pattern.

1.4.5 Datum

A series of elements can be arranged along an axis which serves as a datum. The datum need not be a straight line, it can also be planar or volumetric in form. Given a random organization of dissimilar elements, a datum can organize these elements in the following ways. See Figs 1.71–1.75.

(a) A line can cut through or form a common edge for the pattern. A grid of lines can also help to form a pattern.

(b) A plane can gather the elements beneath it as in (b) or serve as a background and frame the elements in its field.

(c) A volume can collect the pattern within its boundaries as in (d) or organize them along its perimeter as in (e).

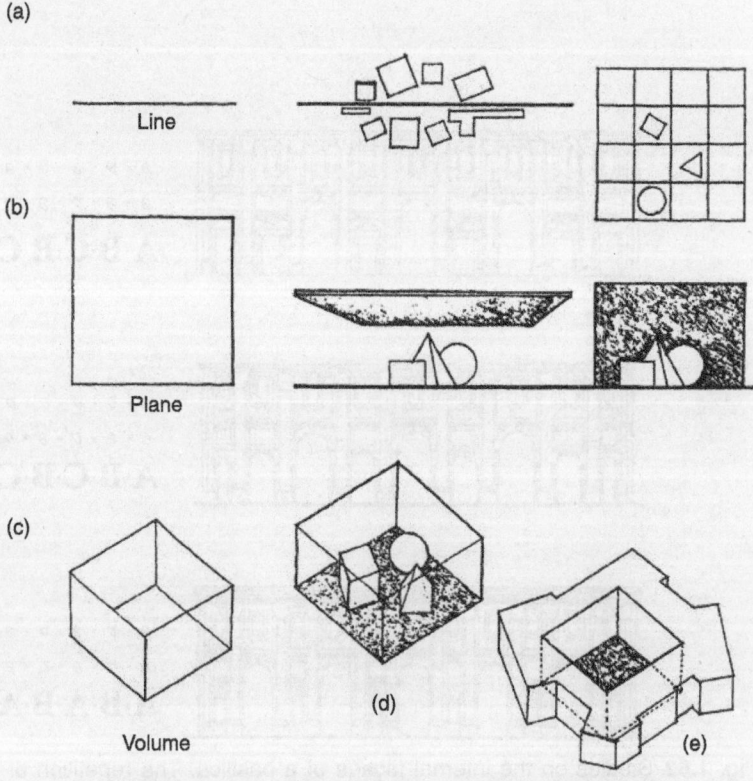

Fig. 1.71 Ways in which random objects can be united

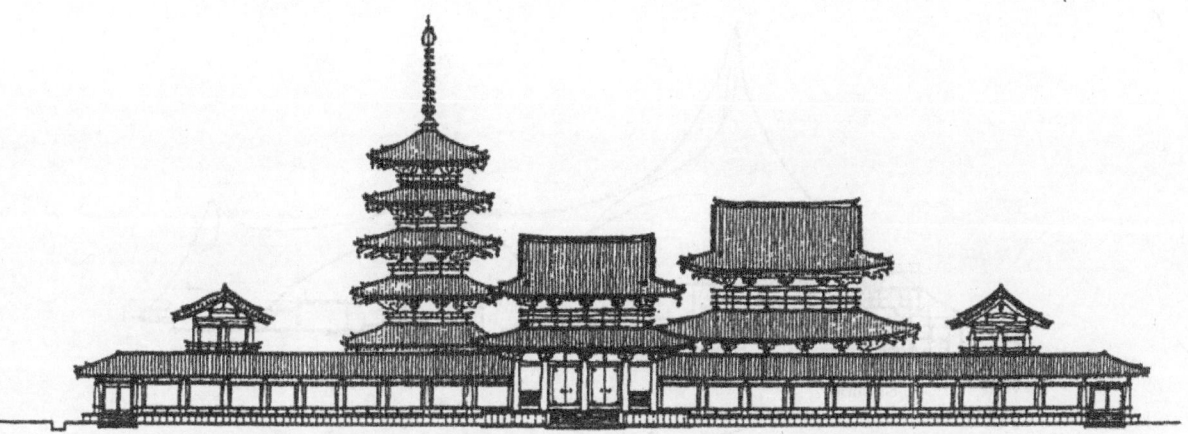

Fig. 1.72 Temple group, Japan—random patterns of elements are organized

Fig. 1.73 Arcades unify the facades of houses that face the Town square of Czechoslovakia

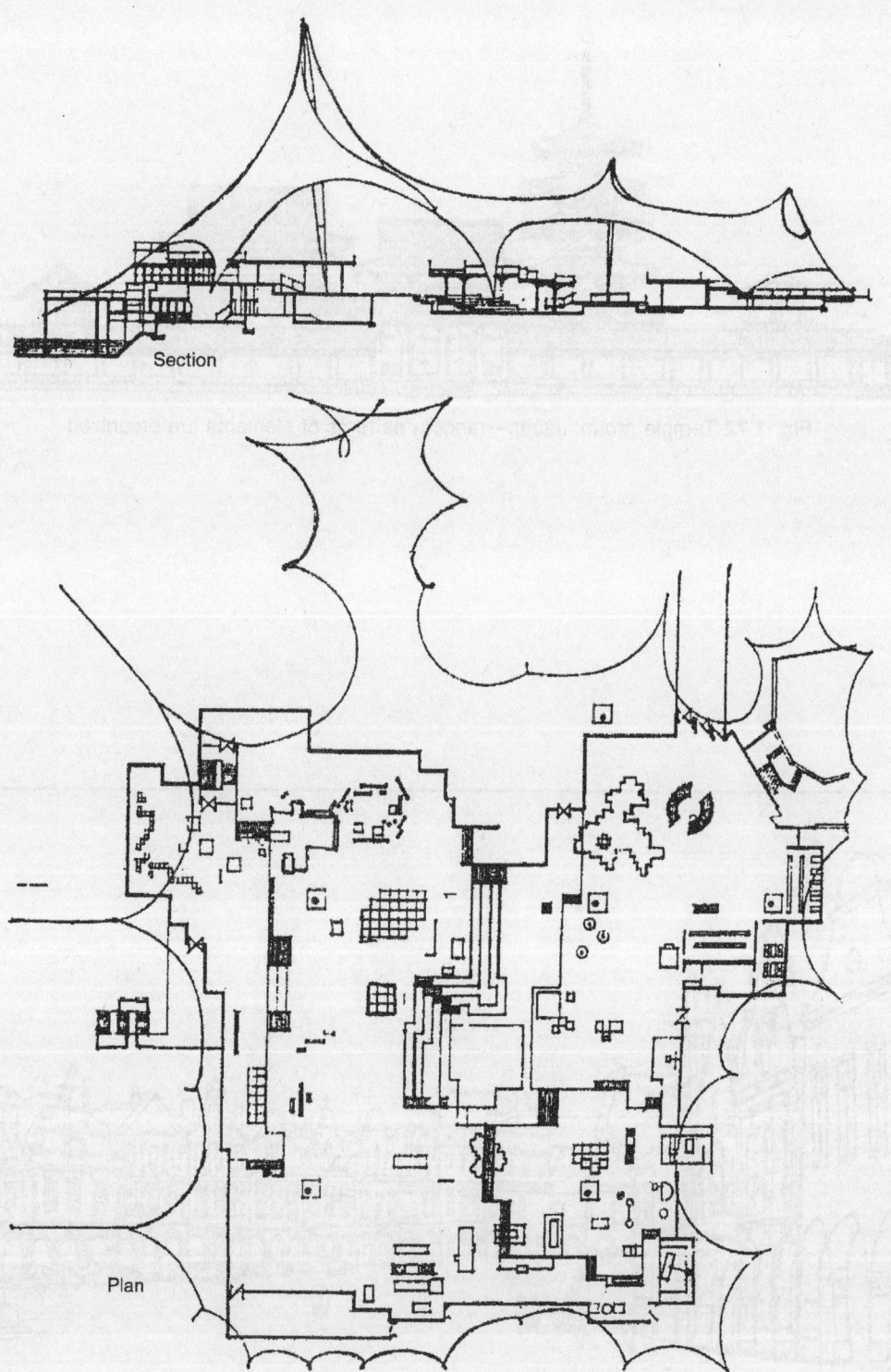

Section

Plan

Fig. 1.74 German pavilion in the World Exhibition, Montreal, Canada. The cable structure unifies the varied structures beneath it, through regularity, continuity, and constant pressure.

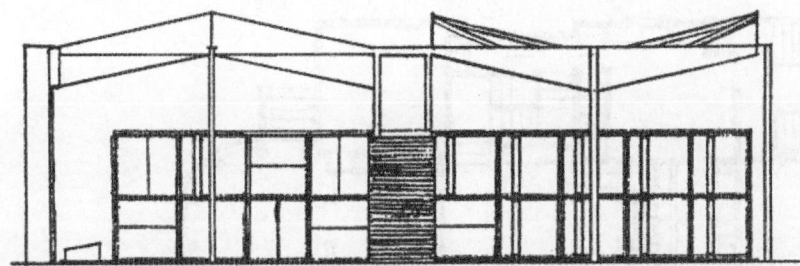

Fig. 1.75 Centre Le Corbusier, Zurich. The frames of columns and beams unify the total composition.

1.4.6 Transformation

As with any discipline, the study of architecture also involves the study of the past (the history). By studying the history, one learns about the past experiences and achievements of others. The principle of transformation is based on this study. This principle allows a designer to select a standard architectural model whose formal structure and ordering of elements can be transformed, through a series of changes (Fig. 1.76), to respond to the specific condition and context of the design, so that the original design concept can be strengthened and built upon, rather than destroyed. See Figs 1.77–1.80.

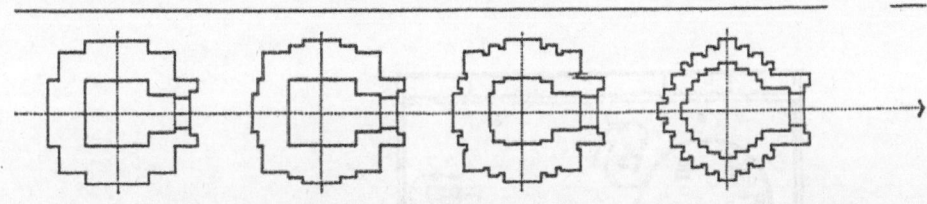

Fig. 1.76 Plan development of North Indian Cella. A formal structure is transformed through a series of changes.

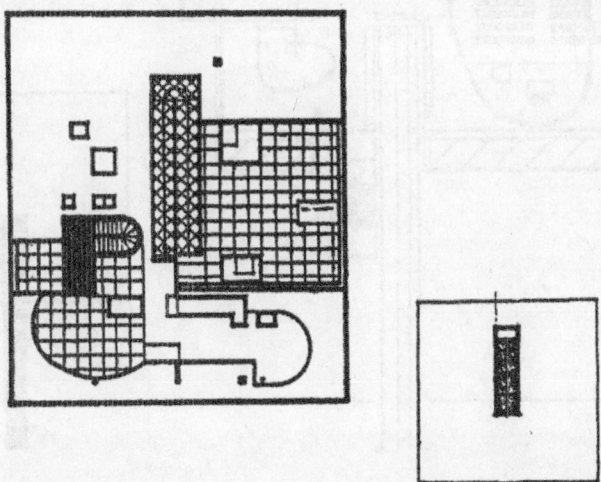

Fig. 1.77 Villa Savoye, France. Transformation of the ramp in a square, 'Free Plan' Organization by Le Corbusier.

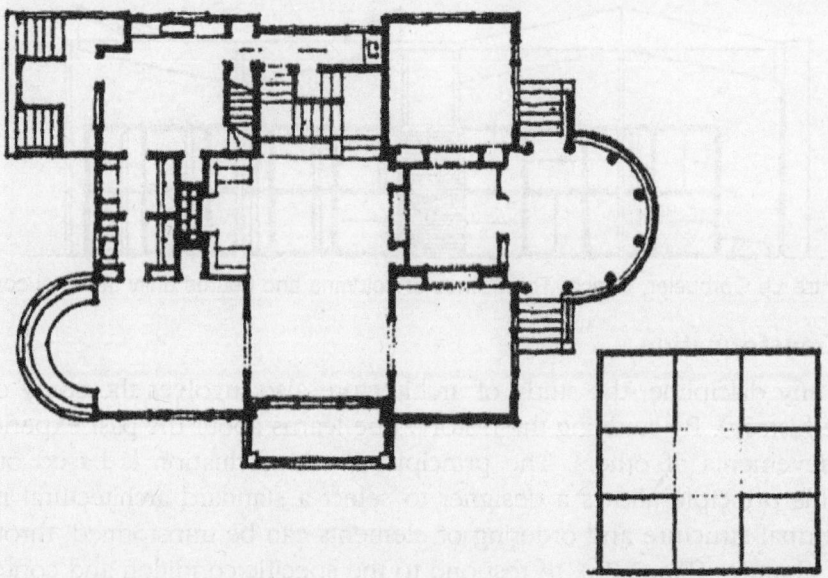

Fig. 1.78 George Blossom House, Chicago. Transformation of the Cruciform plan organization by F.L. Wright.

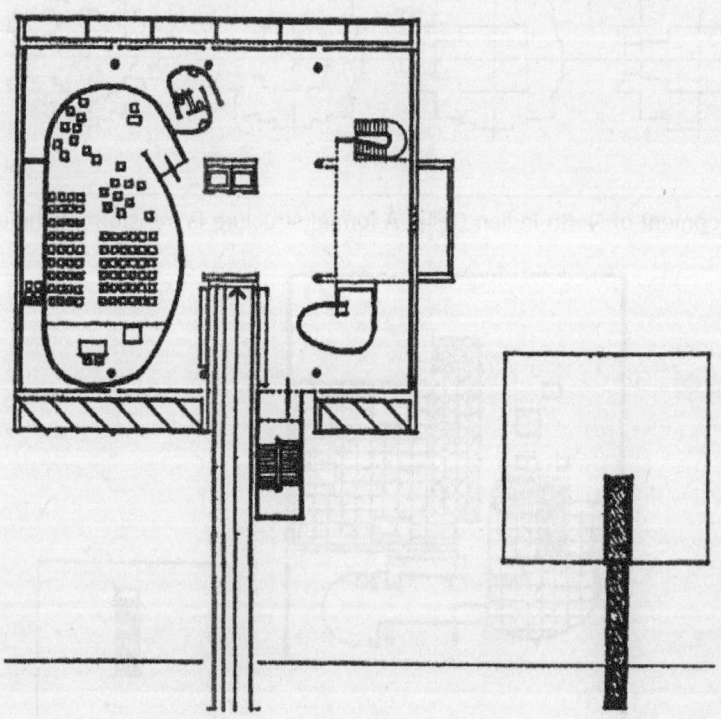

Fig. 1.79 Mill Owner's Association building, Ahmedabad. Transformation of ramp leading from the road to the elevated entrance of the building by Le Corbusier.

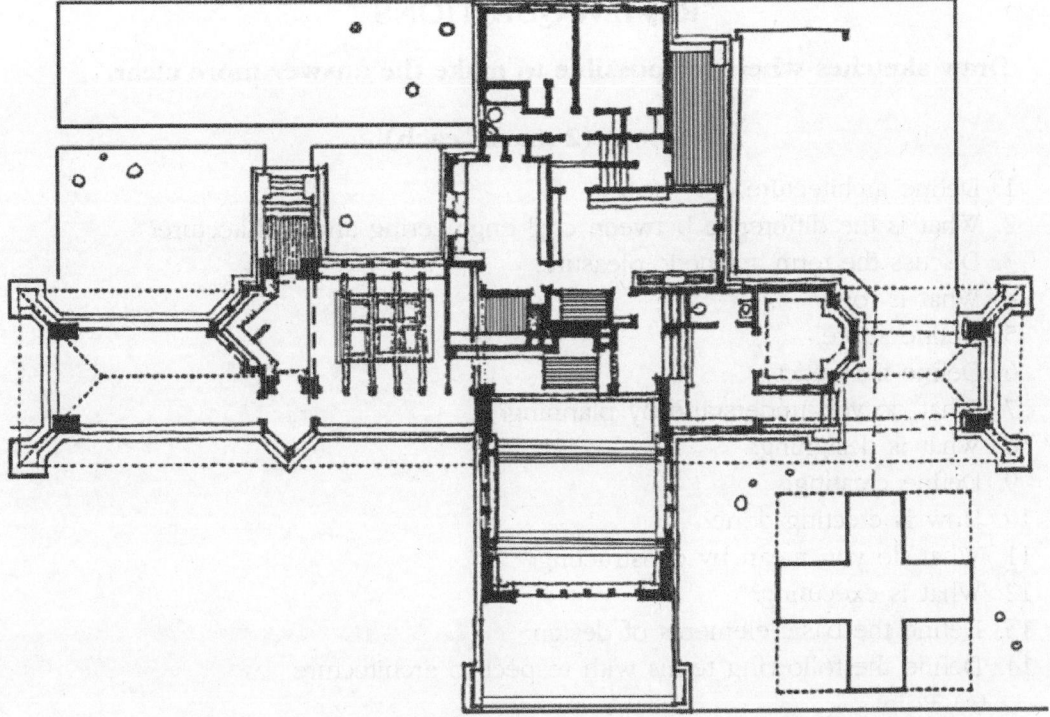

Fig. 1.80 Ward Willet's House, Illinois. Transformation of the Cruciform plan organization.

Summary

As civilizations evolved over millions of years, man felt the need not only to derive physical satisfaction from the basic requirements of life—food, shelter, clothing—but also to experience aesthetic pleasure.

Aesthetics can be defined as anything concerned with the appreciation of beauty. Architecture developed as man began to consciously incorporate aesthetic elements into the building activity. Architecture is, thus, essentially a design process, which results in functionally efficient, economically viable, and aesthetically satisfying building. The elements of aesthetics that play an important role in architecture are mass, space, proportion, symmetry, balance, contrast, pattern, and decoration.

The form or architectural design of a building is generated through conceptual visual elements—point, line, plane, and volume. We can better understand these concepts by studying examples from the history of architecture.

History also provides us with examples of the 'Ordering principles' of architecture—visual methods that allow the form and space of a building to co-exist and make a design complete. These are axis, symmetry, hierarchy, rhythm or repetition, datum, and transformation.

The next step is to know more about aesthetics, which plays a vital role in the study of architectural concepts. This is discussed in Chapter 2.

*REVIEW QUESTIONS

Draw sketches wherever possible to make the answer more clear.

Part—A (2 marks each)

1. Define architecture.
2. What is the difference between civil engineering and architecture?
3. Discuss the term aesthetic pleasure.
4. What is form?
5. Define space.
6. Define aesthetics.
7. What do you understand by planning?
8. What is designing?
9. Define creating.
10. How is erecting done?
11. What do you mean by constructing?
12. What is executing?
13. Define the basic elements of design.
14. Define the following terms with respect to architecture.
    (a) Point
    (b) Line
    (c) Plane
    (d) Volume
15. What are the principles of design?
16. What are the basic principles of architecture?
17. What is an axis?
18. What do you mean by symmetry?
19. What is the difference between radial and bilateral symmetry?
20. What is hierarchy?
21. Define rhythm or repetition.
22. What is datum?
23. What do you understand by transformation?

Part—B (16 marks each)

1. Write short notes on the following
   - Aesthetics
   - Planning
   - Designing
   - Erecting
   - Constructing
   - Executing

2. What do you understand by the basic elements of architecture (or) basic elements of design?
3. Describe in detail the principles used in architecture or in designing.

Note: According to the Anna University examination pattern, there are ten 2-mark questions, which are compulsory, and five 16-mark questions. Model question papers are given in the Appendix.

Aesthetic Components of Design

2.1 Aesthetic Qualities

We have familiarized ourselves with the term aesthetics in the previous chapter. Aesthetic qualities may be subdivided into a number of basic elements: unity, proportion, scale, balance, symmetry, and rhythm. All these qualities are capable of evoking responses only to the extent that they are seen, i.e., they belong to the realm of vision and it is principally through the organ of sight that we become aware of their collective impact. The act of seeing is, however, not a simple mechanical reproduction of external data, but is influenced by the nature/perception of the subject (person) viewing them. The act of seeing helps one to evaluate a structure on the basis of aesthetic qualities. In the following sections, we will discuss these elements of aesthetics one by one in detail.

2.2 Unity and Elements of Unity

The principle of unity deals with visual composition in design. Composition means the relationship between the visual elements. The brick work, timber, or concrete which we use as building materials for protection from weather or for structural support form the visual elements in an architectural composition. To get a good composition the elements of unity should be chosen carefully.

Unity therefore deals with the arrangement of the building materials and building parts (floor, wall, roof, column, beam, etc.) to create a good composition.

Consider materials such as stone, glass, and steel. They are available in a variety of colours, tones, textures, shapes, proportions, etc. Various combinations of these material properties are possible—the challenge lies in arriving at the most pleasing composition. The texture or colour of a single brick or wood panel will differ in effect when it forms part of a larger composition such as a brick wall or a door frame set in a wall as illustrated in Fig. 2.1.

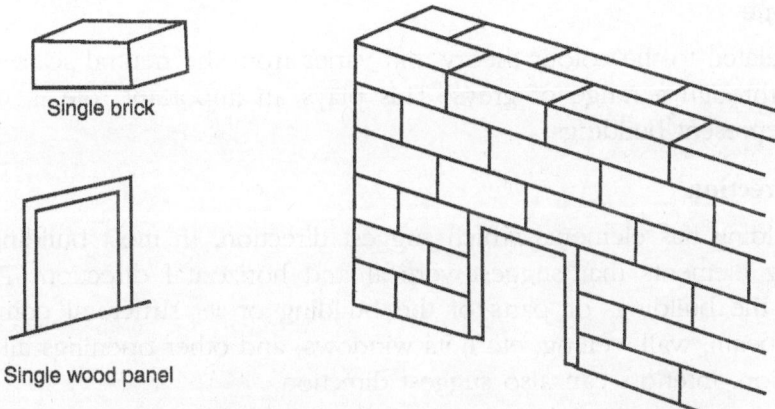

Fig. 2.1 Composite texture made of bricks and a wood panel

2.2.1 Texture

The word *texture* generally refers to the appearance and feel of a surface. However, it could also mean the physical composition or structure of something, especially with respect to the size, shape, and arrangement of its parts (see Fig. 2.2). In this figure, a variety of textures can be seen in the roofs, walls, and paving. Strongly identifiable shapes in roofs and battered walls are seen as repeated units, presenting a definite and distinguishable feel of the surface.

Fig. 2.2 A school in England

Individual dwellings within a group, as well as decoration and carvings in buildings, can create an effect of texture.

2.2.2 Colour

Here, colour refers to hue (colour range). Colour is one of the pronounced elements of aesthetics and its selection is very crucial to the overall effect it has on aesthetics. A variety of effects can be produced by varying colour luminance, fullness, and its transparency.

2.2.3 Tone

Tone is related to the colour theory and varies from the neutral scale of white to black through a range of greys. This plays an important role in drawings used to represent buildings.

2.2.4 Direction

Every building has elements which suggest direction. In most buildings there are strong elements that suggest vertical and horizontal direction. The total shape of the building, or parts of the building or its structural components (column, beam, wall, ceiling, etc.), its windows, and other openings all suggest the direction. Interiors can also suggest direction.

Fig. 2.3 The City Theatre, Helsinki, Finland

Figure 2.3 shows the strong dominance of horizontality given by the roof line and overhang with contrast from the columns. The directional emphasis is reinforced by the use of a strongly ribbed cladding tile, which can be seen running horizontally on the eaves soffit and the beams linking the column heads and the solid parts of the wall behind them.

A dominantly vertical composition is shown in Fig. 2.4 where the verticality is reinforced by the faceting of the envelope. Contrast is given by the generally horizontal emphasis of the fenestration. The curving round of the plan also affords more window space for the living accommodation on the southerly side relative to the service and circulation space on the north, i.e., stairs, lifts, passage access to flats.

2.2.5 Proportion

In this context, proportion is the geometric relationship of the sides of volumes (e.g., rectangles). It is also the ratio or comparitive size of individual parts of the composition. We cannot measure these relationships accurately by eye, but we can compare them and try to judge the relationship of one part to another on a proportional basis.

Buildings belonging to Classical and Gothic architecture have better proportional relationships than most buildings today. Figure 2.5 shows a building that illustrates the concept of unity through proportions.

Fig. 2.4 Flats in Bremen

Fig. 2.5 Ducal Palace, Urbino, Paris

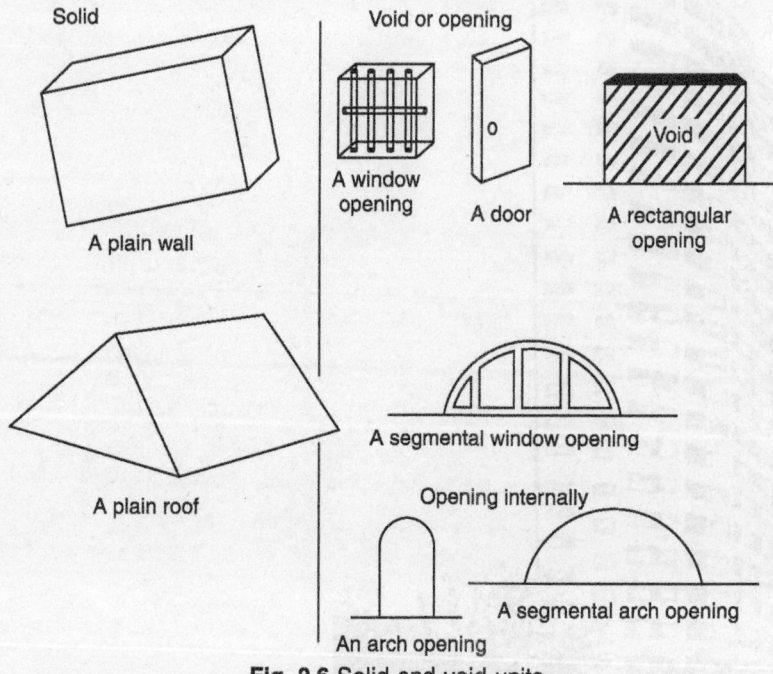

Solid

A plain wall

Void or opening

A window opening

A door

A rectangular opening

A plain roof

A segmental window opening

Opening internally

A segmental arch opening

An arch opening

Fig. 2.6 Solid and void units

2.2.6 Solid and void

The relationship between solid (walls, roofs, etc.) and void (windows and other openings) structural units is very important in order to form a good composition. Figures 2.6 and 2.7 show

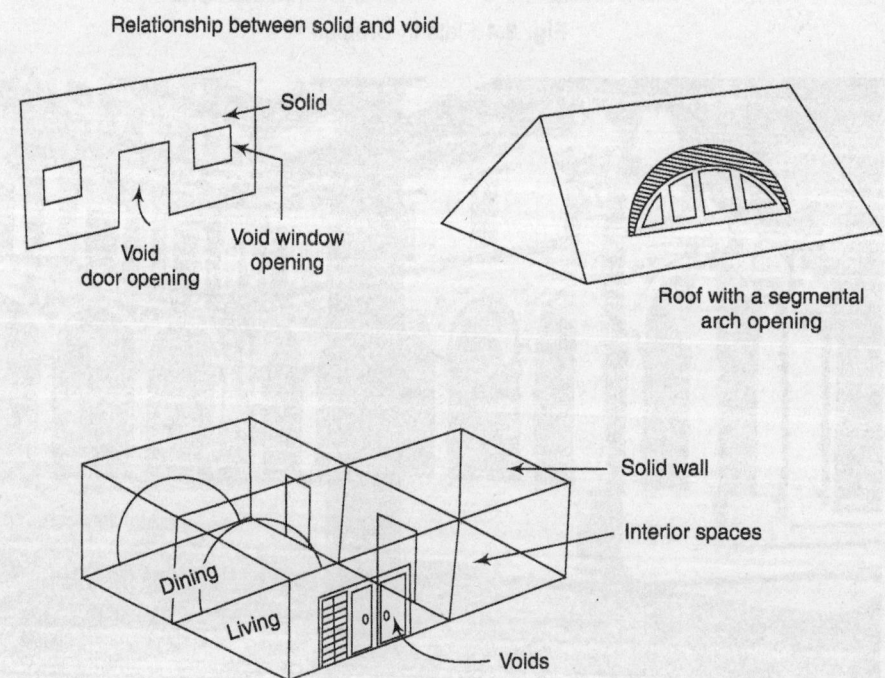

Relationship between solid and void

Solid

Void door opening

Void window opening

Roof with a segmental arch opening

Solid wall

Interior spaces

Dining

Living

Voids

Fig. 2.7 Interior spaces are formed by the arrangement of the solids and voids which surround them

Fig. 2.8 Cemetery Chapel, Turku, Finland

examples of some solid and void units. Figure 2.8 shows the relationship between solid and void structural units. In Fig. 2.8, the dominance of solid material contributes to the sense of enduring unity. The porch and opening above give some limited contrast, but the major source of vitality is the contrast between the building and the many natural elements of the setting.

2.2.7 Form and shape

Form and shape can be observed clearly in the overall arrangement of a building or in parts of a building (windows, doors, etc.) which have geometric shapes. Repetition or variation of a particular form can provide strong elements of composition as shown in Fig. 2.9.

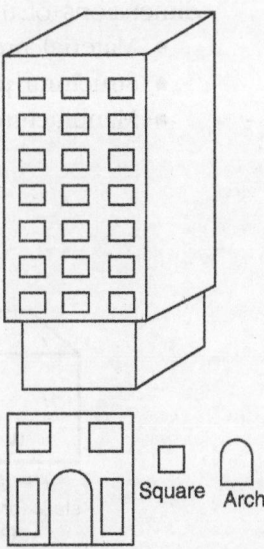

Square   Arch

Fig. 2.9 Housing arrangement: apartment blocks

Parts of the building, for example, windows, depicting form represent geometric shapes such as a square or a rectangle. Repetition of the window units forms a good element of composition.

Shape helps to identify different forms, for example, the pitch of a roof will provide a certain form which is easy to identify and also easy to relate to other roofs which have a similar pitch. When a roof is of a distinctly different shape, it will look strongly dissimilar.

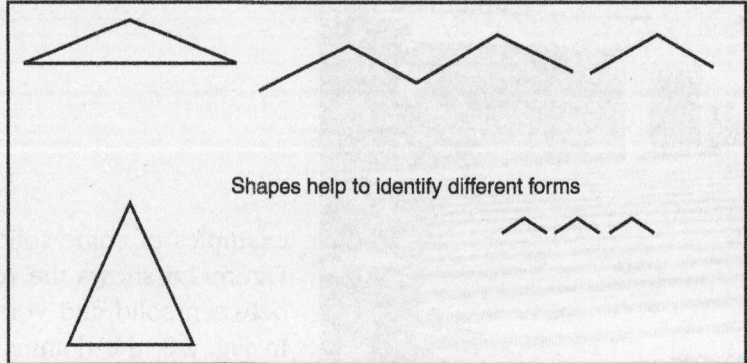

Fig. 2.10 Pitch of a roof

2.3 Proportion

Proportion is one of the main features of various masses or spaces. The primary properties of mass are height, width, and depth. Proportion is the visual relationship of these properties with respect to each other. The observer does not see these as individual, separate properties (height, width, depth) but as an entire mass. Any mass can be identified by the geometric shape and the dimensions of the form. There are three facets of proportion:

- Material proportion
- Structural proportion
- Manufactured proportion

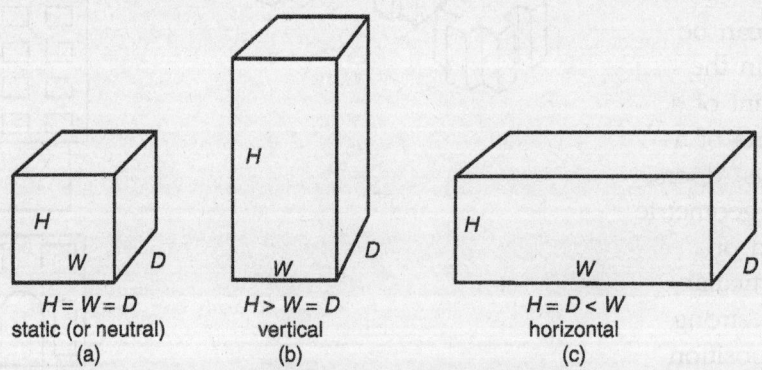

Fig. 2.11 Form and proportion

2.3.1 Material proportion

All building materials have distinct properties of stiffness, durability, and hardness. Also, all materials have an ultimate strength or breaking strength beyond which these cannot remain intact. For example, a stone slab 10 cm thick and 250 cm (2.5 m) long can support itself as a bridge as shown in Fig. 2.12. But if the size is increased four times, that is, it has thickness of 40 cm and length of 10 m, it would probably collapse under its own weight. All materials have rational proportions that are controlled by their in-built strength and weakness. For example, a brick of size 20 cm × 10 cm × 10 cm can take only a specific load before it crumbles. A load-bearing wall (wall supporting slabs and beams) should be 20 cm thick. For a partition wall, a half-brick wall, 10 cm thick, is enough for support. This is illustrated in Fig. 2.13. The concept of material proportion is brought out by the following example.

Fig. 2.12 2.5-m-long stone slab bridge

- The proportion of length and breadth of bricks (which are 20 cm and 10 cm, respectively) helps to construct a brick wall without continuous verti-cal joints to provide for bond (support).
- The proportion in concrete of cement, fine aggregate (sand), and coarse aggregate is 1:2:4. For a stronger or denser concrete mix, a proportion of 1:1.5:3 can be used.

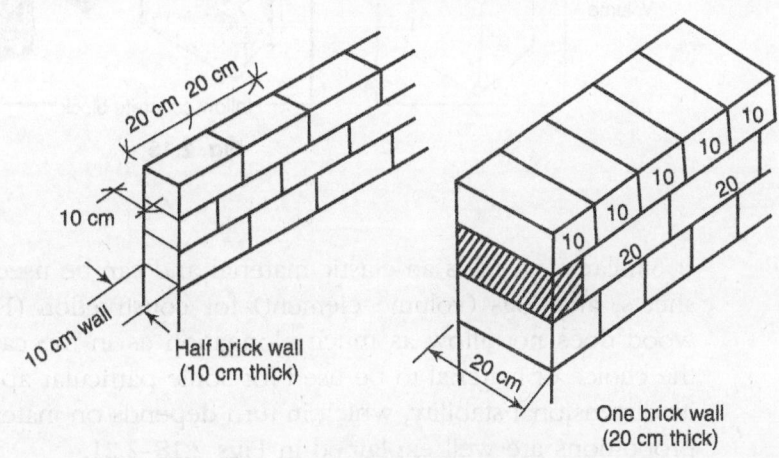

Half brick wall (10 cm thick)

One brick wall (20 cm thick)

Fig. 2.13

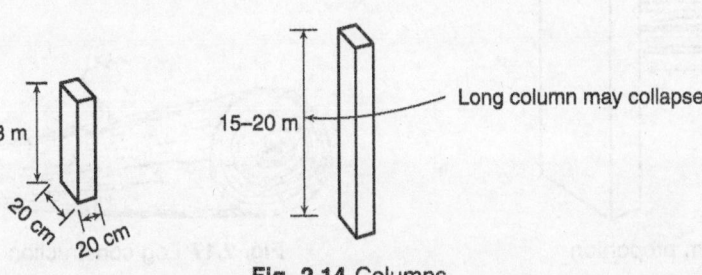

Long column may collapse

Fig. 2.14 Columns

- In the case of columns, the height and dimensions have a proportion, as shown in Fig. 2.14. For example, a 20 cm × 20 cm column can be used as a support for a single storey, but if this is extended to three or four storeys, it may collapse.

Materials such as steel are strong in both compression and tension and as such can be used for linear columns and beams and also as sheets (which are akin to planes) (Fig. 2.15). This means that there is possibility of greater elongation along any two axes if steel is used as compared to brick material.

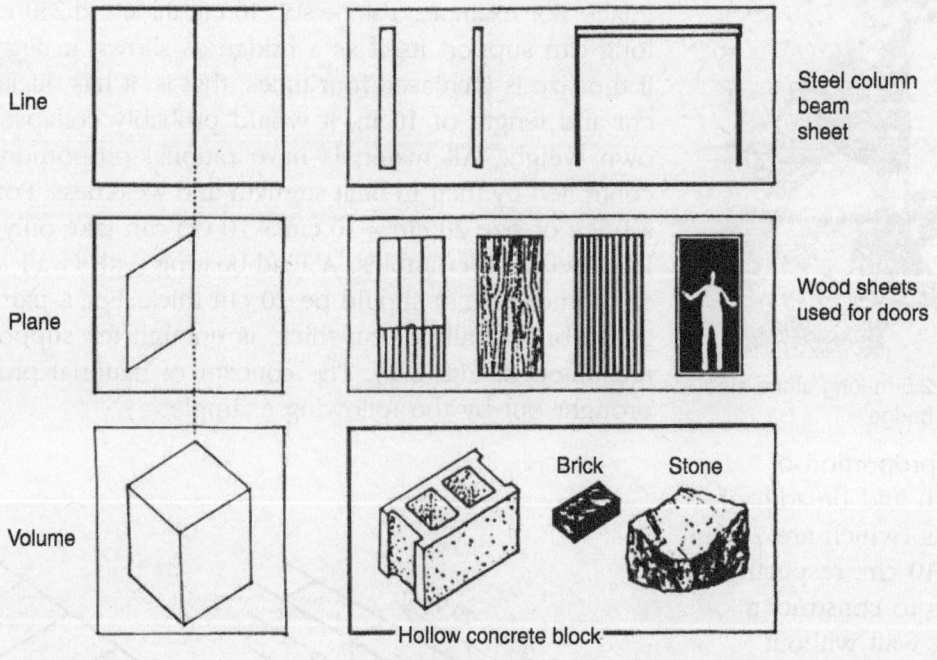

Fig. 2.15

Similarly, wood is an elastic material and can be used as linear posts, beams, sheets, and logs (volume element) for construction (Figs 2.16 and 2.17). But wood does not allow as much elongation as in the case of steel. To sum up, the choice of material to be used for some particular application is governed by its dimensional stability, which in turn depends on material dimensions. Material proportions are well explained in Figs 2.18–2.21.

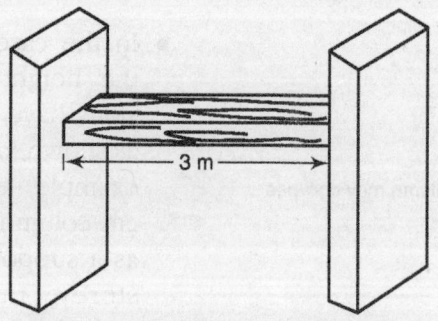

Fig. 2.16 Wooden beam proportion

Fig. 2.17 Log construction

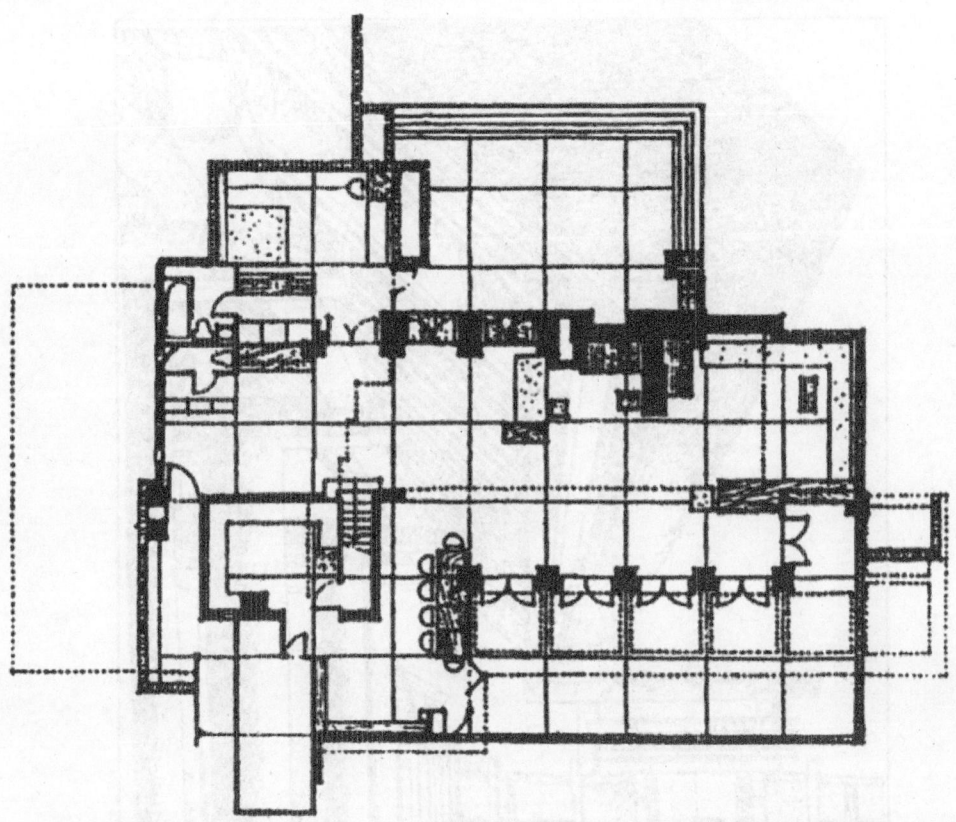

Fig. 2.18 Schwartz House, Wisconsin, 1939; architect: Frank Lloyd Wright; material proportion: wood and brick used (for smaller spans of beams, e.g., in residences, wood is used, and brick is used for walls)

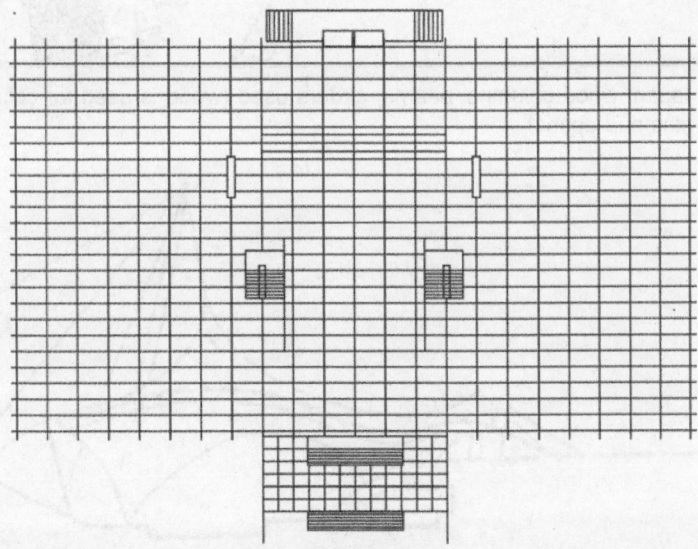

Fig. 2.19 Crown Hall, Illinois Institute of Technology, Chicago; architect: Mies van der Rohe; material proportion steel used (for halls, spans would be larger and steel beams used)

Fig. 2.20 A shrine in Japan; wood columns, beams, girders used (wood is used for columns, beams, and girders of small lengths and spans)

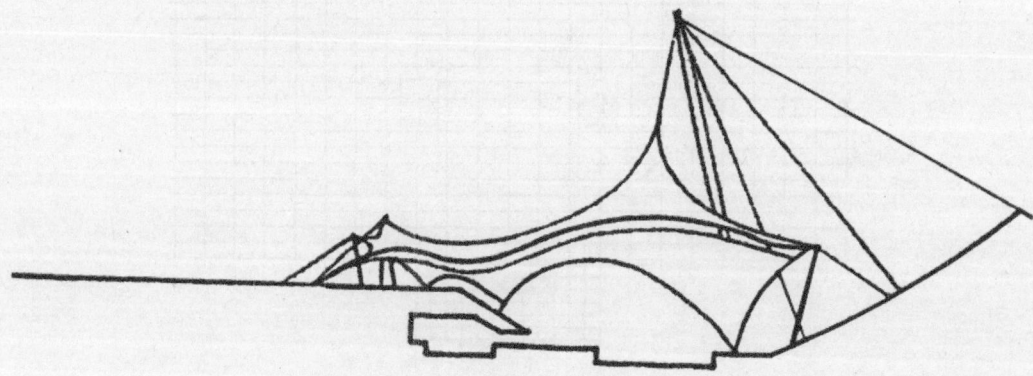

Fig. 2.21 Roof for swimming arena, Germany; material proportion: membrane used (a membrane is used for the roof of a large unsupported area)

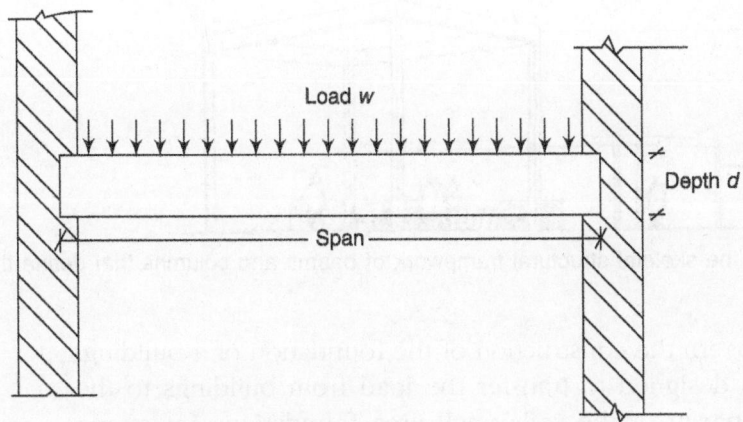

Fig. 2.22 Beam action; loads acting on a beam

2.3.2 Structural proportion

In the construction of buildings, structural elements such as beams and slabs are required to span over spaces and transmit their loads through vertical supports. (Fig. 2.22). Again, dimensional proportions play their role in restricting the height, width, and depth of these structural elements under a given load condition.

Checking the design of a beam

In a beam with load W, if the span is doubled, the stresses due to bending would be increased four times. If the depth of the beam is also doubled, the stress would remain the same. Thus there is a proportion between the span and depth of the beam for safe stresses.

Checking a beam design for deflection

The proportion of the span to the depth of a beam influences the deflection of the beam (Fig. 2.23). For pillars and columns, the ability to support a load depends on the height of the column.

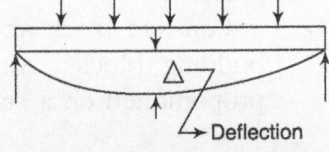

Fig. 2.23 Structural proportion: deflection of a beam due to bending

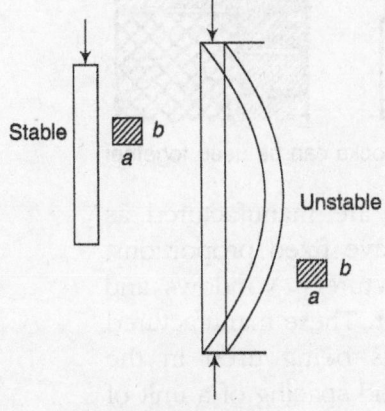

Fig. 2.24 A column

If a column is very tall and its cross-sectional dimensions are small, the column will buckle or bend sideways (Fig. 2.24). If the sectional dimensions are kept unchanged and the height is reduced, the column would be stable and safe. Thus, there is a proportion between the height of the column and dimension of the cross section of the column for safe construction of columns in buildings (Fig. 2.25).

Fig. 2.25 Structural proportions seen in the skeletal structural framework of beams and columns that define the modulus of space

Similarly, proportions govern the construction of the foundation of a building. The foundation should be designed to transfer the load from buildings to the soil within the bearing capacity of the soil: small area foundations for strong soil and large area foundations for weak soil.

2.3.3 Manufactured proportion

Many architectural elements have a size and proportion not only because of their structural properties but also as a result of the process by which they are manufactured. As these elements are mass produced in factories, they have standard sizes and proportions decided by the individual manufacturer or industry standards.

Concrete block and common brick, for example, are produced as modular building blocks. Although they differ from each other in size, they are proportioned on a similar basis. Figure 2.26 illustrates this.

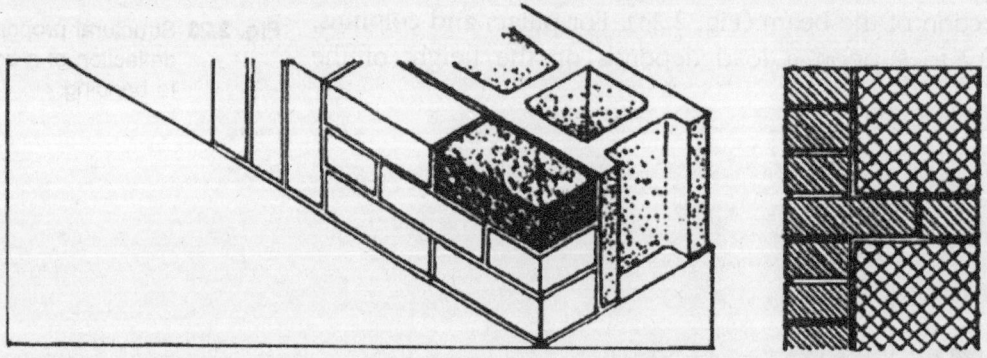

Fig. 2.26 Manufactured proportion showing how brick and hollow blocks can be used together

Plywood and covering material (sheathing material) are manufactured as modular units with a fixed proportion. Steel sections have fixed proportions, which are decided by ISO standards and steel manufacturers. Windows and doors have proportions which are set by the manufacturer. These manufactured materials should fit perfectly with the other materials being used in the construction, as they affect the overall size, proportion, and spacing of a unit of construction, such as a wall having doors and windows (Figs 2.27 and 2.28).

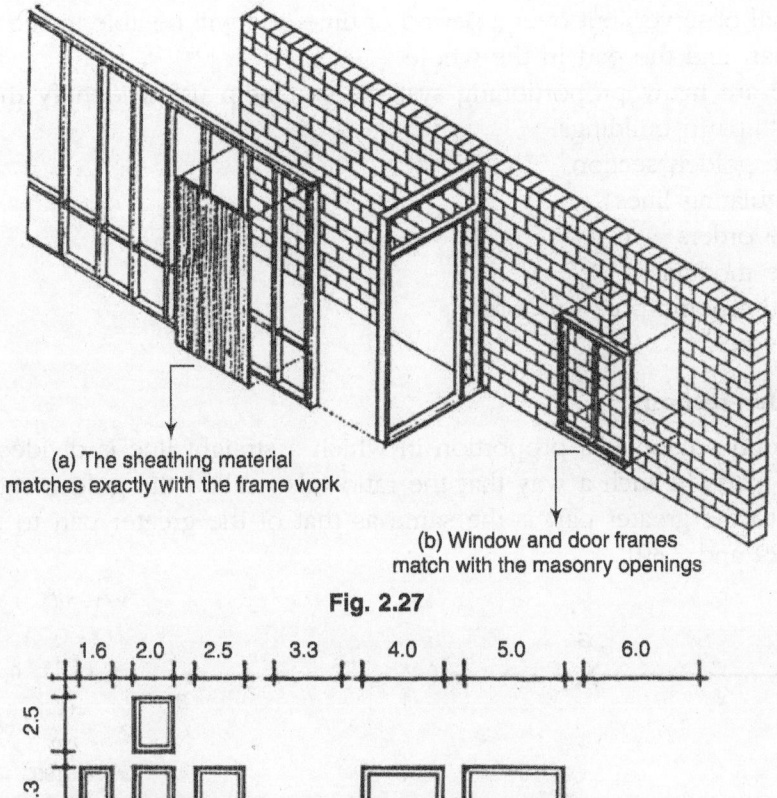

(a) The sheathing material matches exactly with the frame work

(b) Window and door frames match with the masonry openings

Fig. 2.27

Fig. 2.28 Basic element window unit (all dimensions in feet)

2.3.4 Proportioning systems

Though there are proportion restrictions imposed by the material, structural, and manufacturing processes, the designer's ability is the key to control the proportion of a building's form and space. If space of a required area is to be designed, the length, width, and the height ratios depend on the

- functionality of the space
- the nature of the activities to be performed in that space

There is a visual relationship between the parts of the building and the building as whole. These visual relationships cannot be seen immediately by the casual observer, but over a period of time one will be able to see the whole in the part and the part in the whole.

There are many proportioning systems that help us to identify these visual relationships in buildings.
- The golden section
- Regulating lines
- The orders
- The modular
- Anthropomorphic proportions
- The ken

The golden section

The golden section is a proportion in which a straight line is divided into two unequal parts in such a way that the ratio (also called the golden ratio) of the smaller to the greater part is the same as that of the greater part to the whole (Figs 2.29 and 2.30).

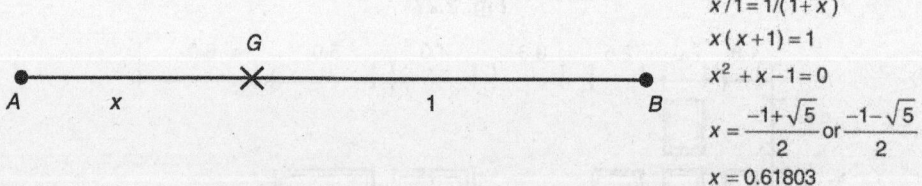

$$x/1 = 1/(1+x)$$
$$x(x+1) = 1$$
$$x^2 + x - 1 = 0$$
$$x = \frac{-1+\sqrt{5}}{2} \text{ or } \frac{-1-\sqrt{5}}{2}$$
$$x = 0.61803$$

Fig. 2.29 A straight line divided into two unequal parts

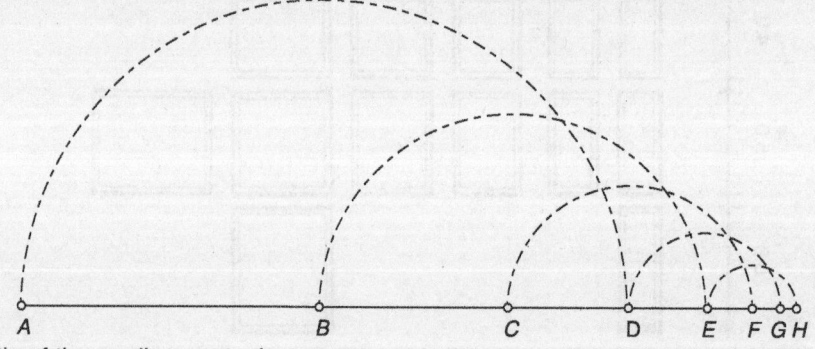

Fig. 2.30 The ratio of the smaller part to the greater part is the same as the ratio of the greater part to the whole

Likewise, this property of golden ratio can be applied to any form, be it circular, triangular, rectangular, or any other form (Fig. 2.31). A rectangle with sides in the proportion of 1:1.618 is called a *golden rectangle*. If a golden

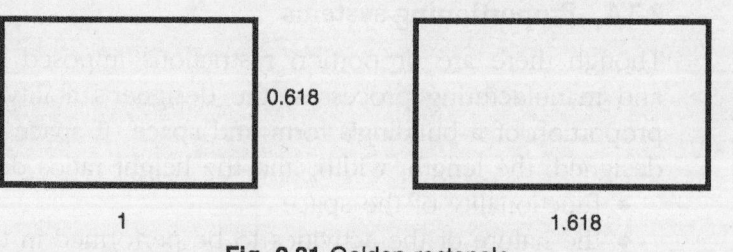

Fig. 2.31 Golden rectangle

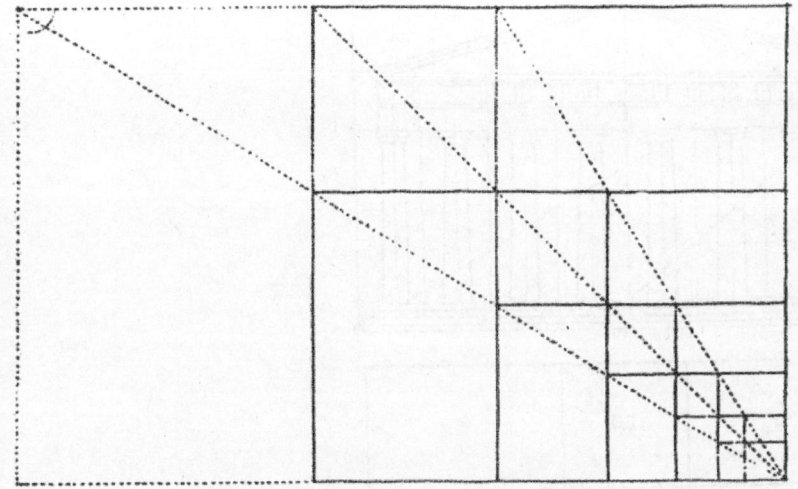

rectangle is subdivided as shown in Figs 2.32, 2.33, and 2.38, all the resulting smaller rectangles will have the golden ratio, i.e., smaller side : larger side = 0.618 : 1 or 1 : 1.618.

Fig. 2.32 Rectangle within rectangle

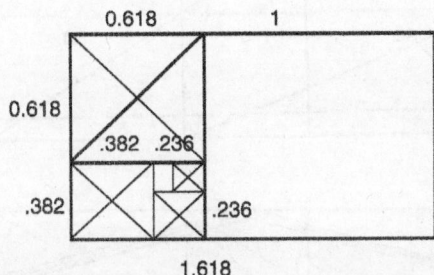

Fig. 2.33   Golden rectangle subdivided into rectangles

The ancient Greeks used this proportion of the golden rectangle in the design of the Parthenon Temples at Athens (see Figs 2.34 and 2.35).

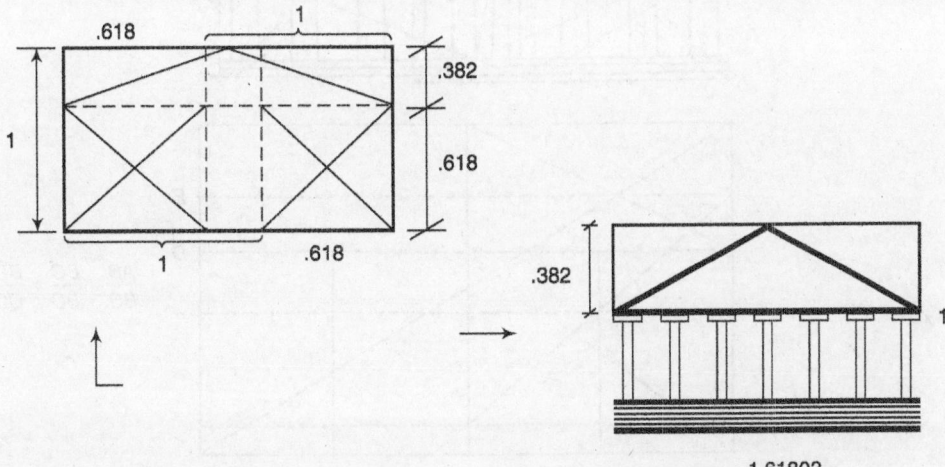

Fig. 2.34 Parthenon, Athens, Greece: facade divided in the golden proportion

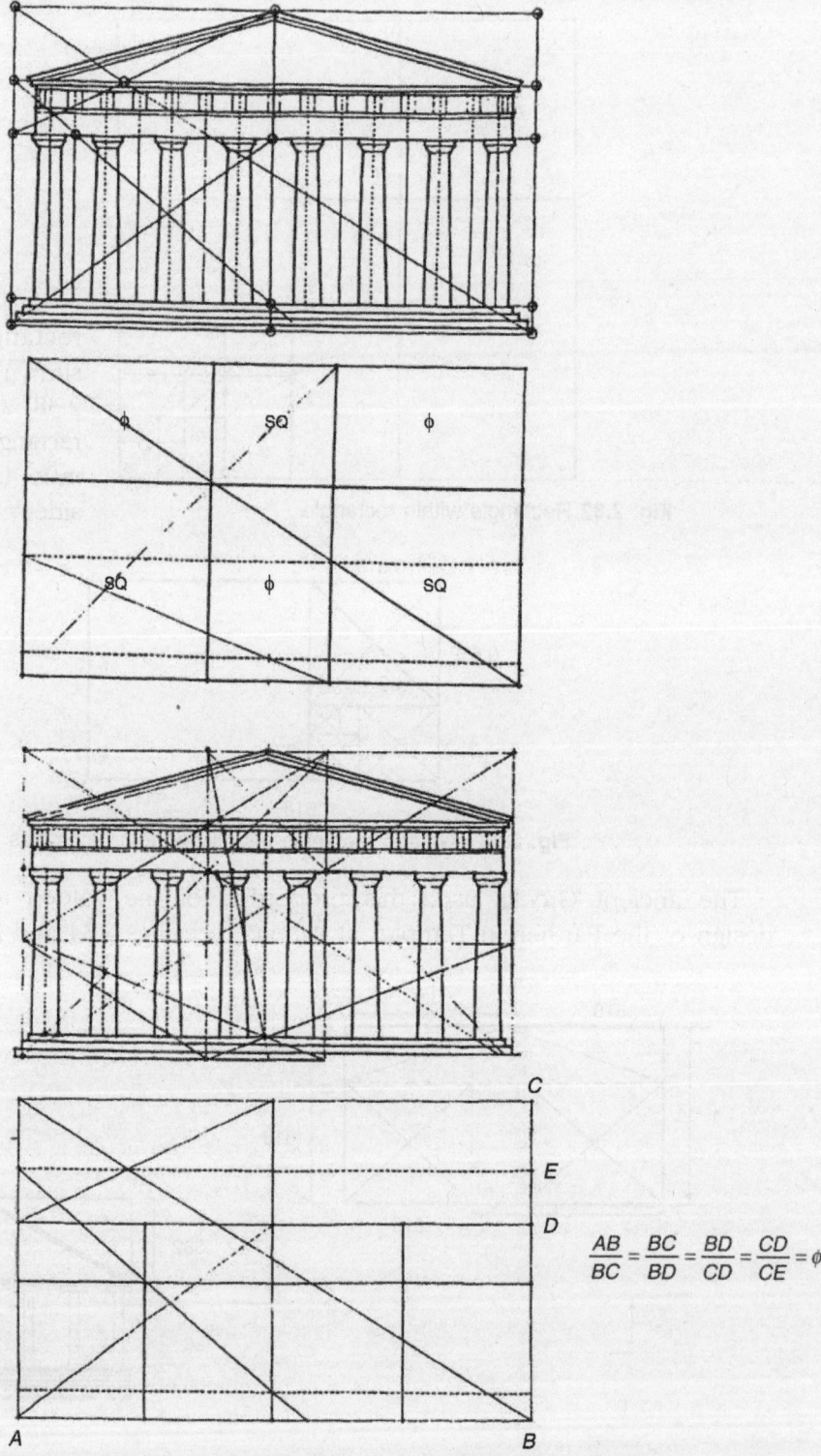

$$\frac{AB}{BC} = \frac{BC}{BD} = \frac{BD}{CD} = \frac{CD}{CE} = \phi$$

Fig. 2.35 The use of the golden section in the proportioning of the facade of the Parthenon, Athens

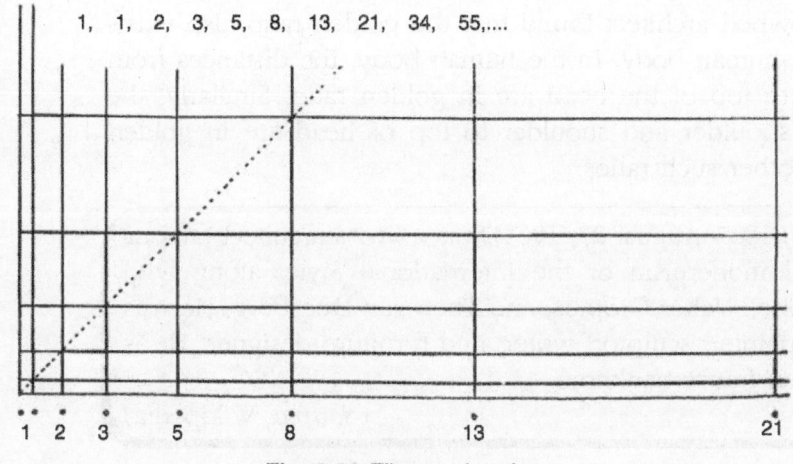

1, 1, 2, 3, 5, 8, 13, 21, 34, 55,....

Fig. 2.36 Fibonacci series

The golden ratio or golden mean or phi is referred to as 1.618033. The *Fibonacci series* is a progression of whole numbers where each term is the sum of the preceding two. The ratio between two consecutive terms approxi-mates the golden ratio (Fig. 2.36).

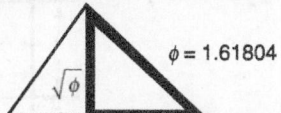

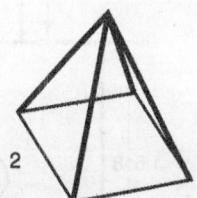

The ancient Egyptians were the first to use the golden ratio in the construction of pyramids. In a pyramid, if the base width is 2, the slant height of the pyramid should be 1.61804 (see Fig. 2.37).

Fig. 2.37 Egyptian pyramids

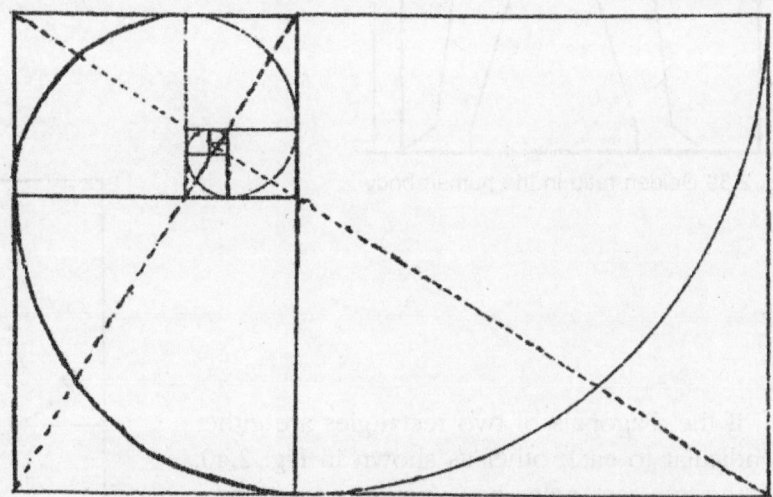

Fig. 2.38 A harmonic composition of golden rectangles

Le Corbusier, the renowned architect found that the golden ratio also exists in the proportions of the human body. In the human body, the distances from foot to navel and navel to top of the head are in golden ratio. Similarly, the distances from navel to shoulder and shoulder to top of head are in golden ratio. Figure 2.39 shows other such ratios.

Le Corbusier (October 6, 1887–August 27, 1965) was a Swiss architect famous for what is now called modernism or the International Style, along with Ludwig Mies van der Rohe, Walter Gropius, and Theo van Doesburg. He was also an urban planner, painter, sculptor, writer, and furniture designer. He is featured on the Swiss ten francs banknote.

(*Source:* Wikipedia)

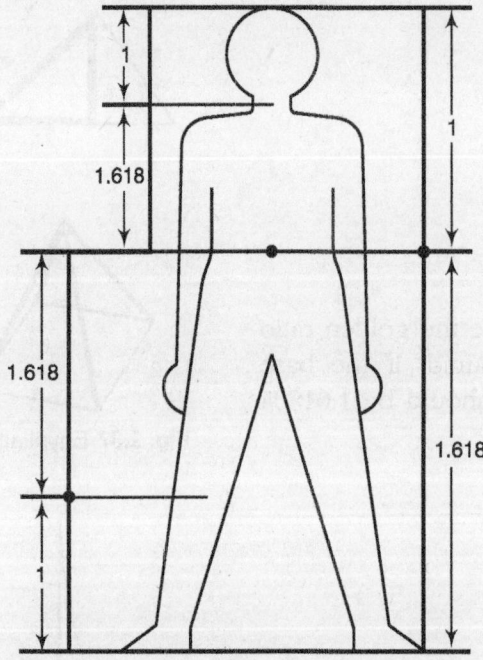

Fig. 2.39 Golden ratio in the human body

Regulating lines If the diagonals of two rectangles are either parallel or perpendicular to each other as shown in Fig. 2.40, they indicate that the two rectangles have similar proportions. These diagonals as well as the lines that indicate the alignment of elements with one another are called regulating lines.

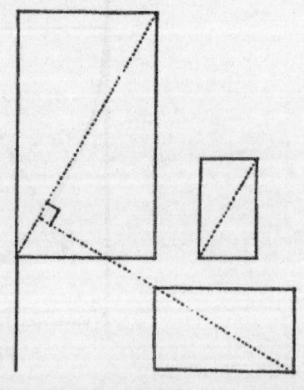

Fig. 2.40 Alignment of rectangles

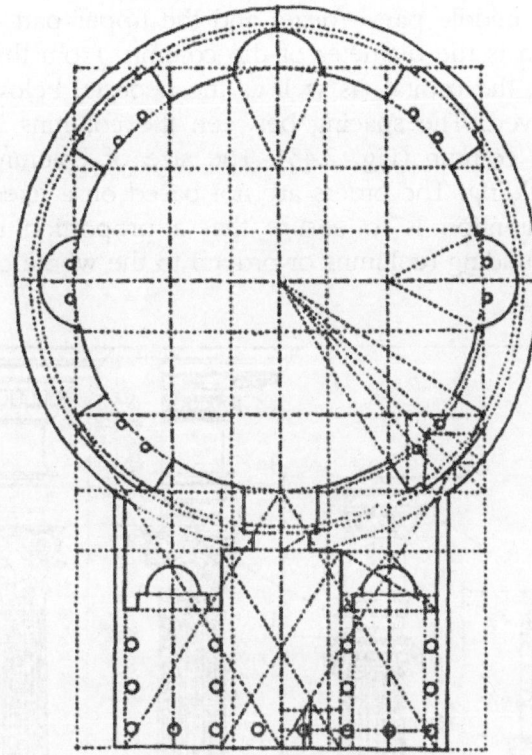

Fig. 2.41 The Pantheon, Rome

Regulating lines bring the form of mathematical order and help fix the fundamental geometry of the work (Figs 2.41 and 2.42).

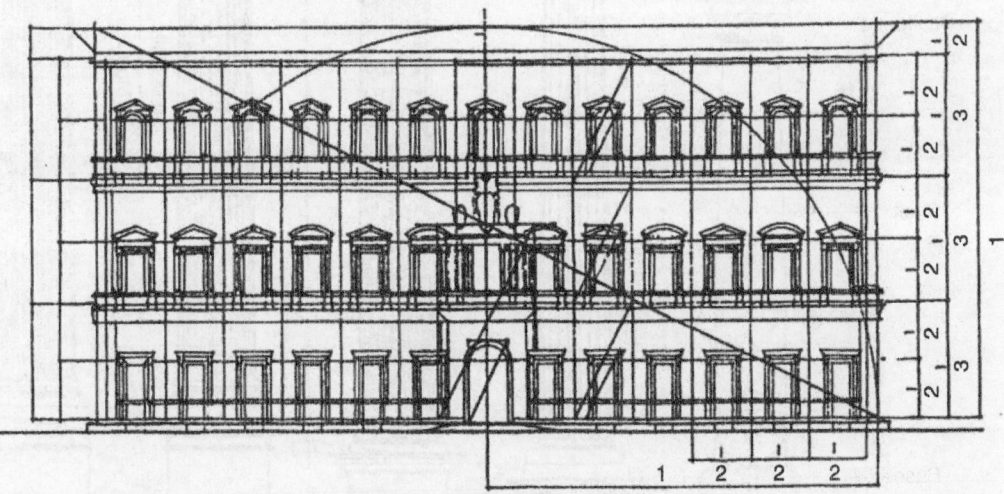

Fig. 2.42 Palazzo Farnese, Rome

The orders   The Greeks and Romans developed the orders. The orders represent the proportioning of the elements and the perfect expression of beauty and harmony. An order consists of an upright column or support, which includes the entablature, capital, shaft, base, and the pedestal. The entablature is divided into the lower part—architrave, the middle part—frieze, and the upper part—cornice. The basic unit of dimension is the diameter of the column. From this module the dimensions of the shaft, the capital, as well as the pedestal below and the entablature above are derived. The spacing between the columns is also based on the diameter of the column (Fig. 2.43). The size of columns varies according to the size of a building. The orders are not based on a fixed unit of measurement. The main intention is to ensure that a proportion is maintained between the parts of a building (columns or orders) to the whole of a building and vice versa.

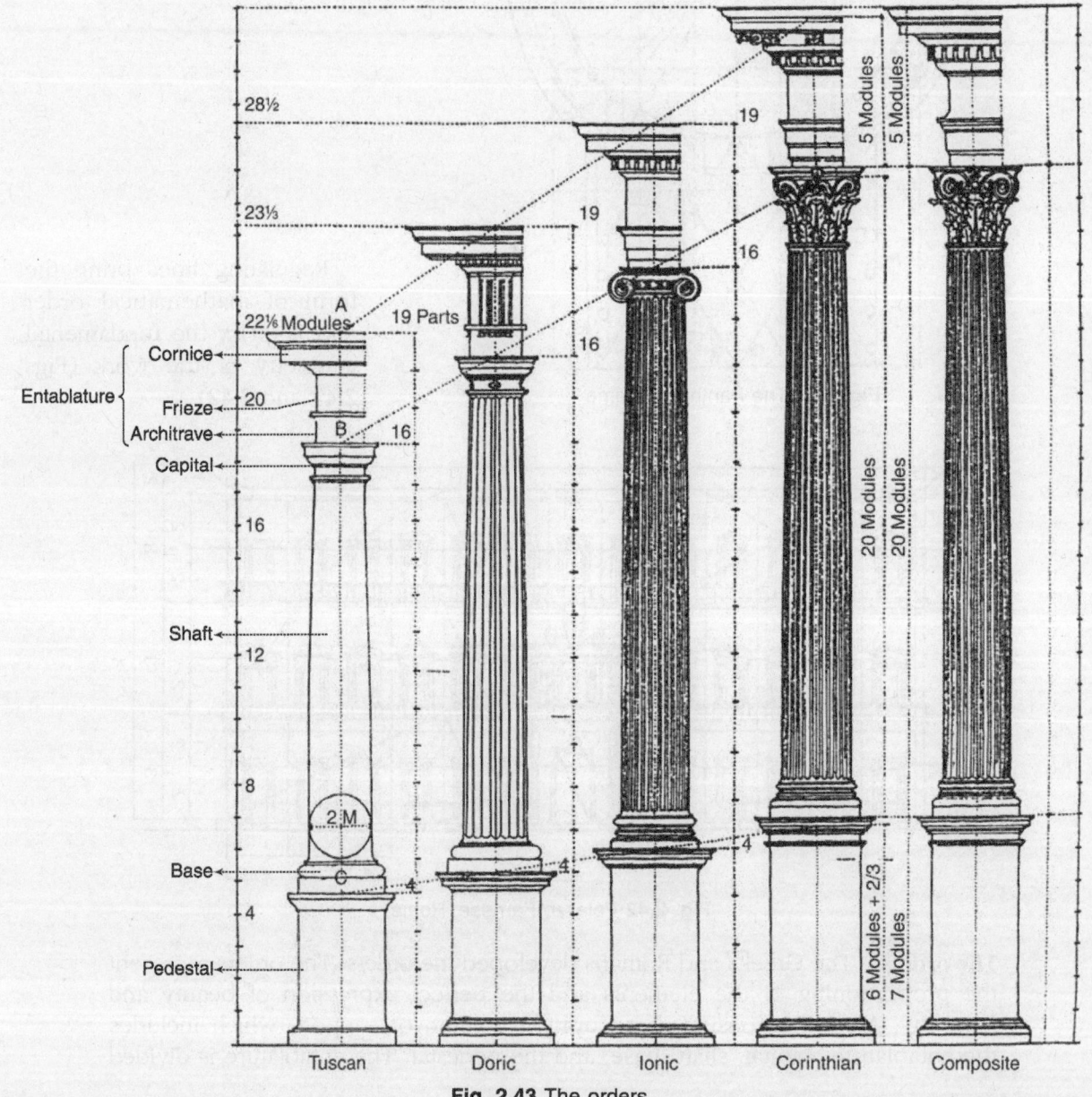

Fig. 2.43 The orders

The two original orders of architecture developed by the Greeks were Doric and Ionic. Then came the Corinthian. The Etruscans developed the Tuscan order and the Romans developed the Composite order. These were the five orders of architecture of the classical times.

Figure 2.44 shows an Etruscan temple illustrating the Tuscan order.

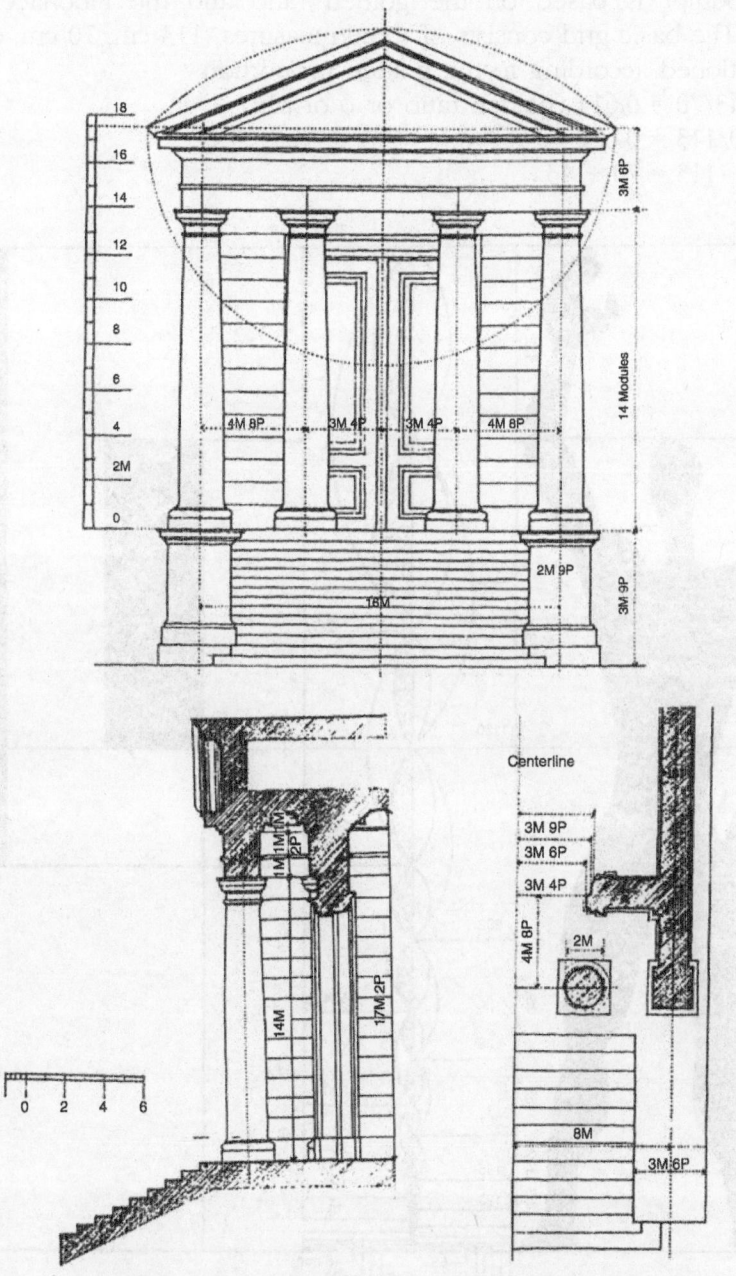

Fig. 2.44 Temple front in the Tuscan order

The Modulor

Le Corbusier studied the modulation of the human figure. He developed a proportioning system, called 'the Modulor', based on the mathematics and proportioning of the human body. Le Corbusier hoped to describe a set of universal truths with regard to the architectural application of the Modulor in the development of form, space, and surface.

The Modulor is based on the golden ratio and the Fibonacci series of numbers. The basic grid consists of three measures, 113 cm, 70 cm, and 43 cm, all proportioned according to the golden proportion:

$$43/70 = 0.614 \text{ (golden ratio or } \phi \text{ or phi)}$$
$$70/113 = 0.629$$
$$113 = 70 + 43$$

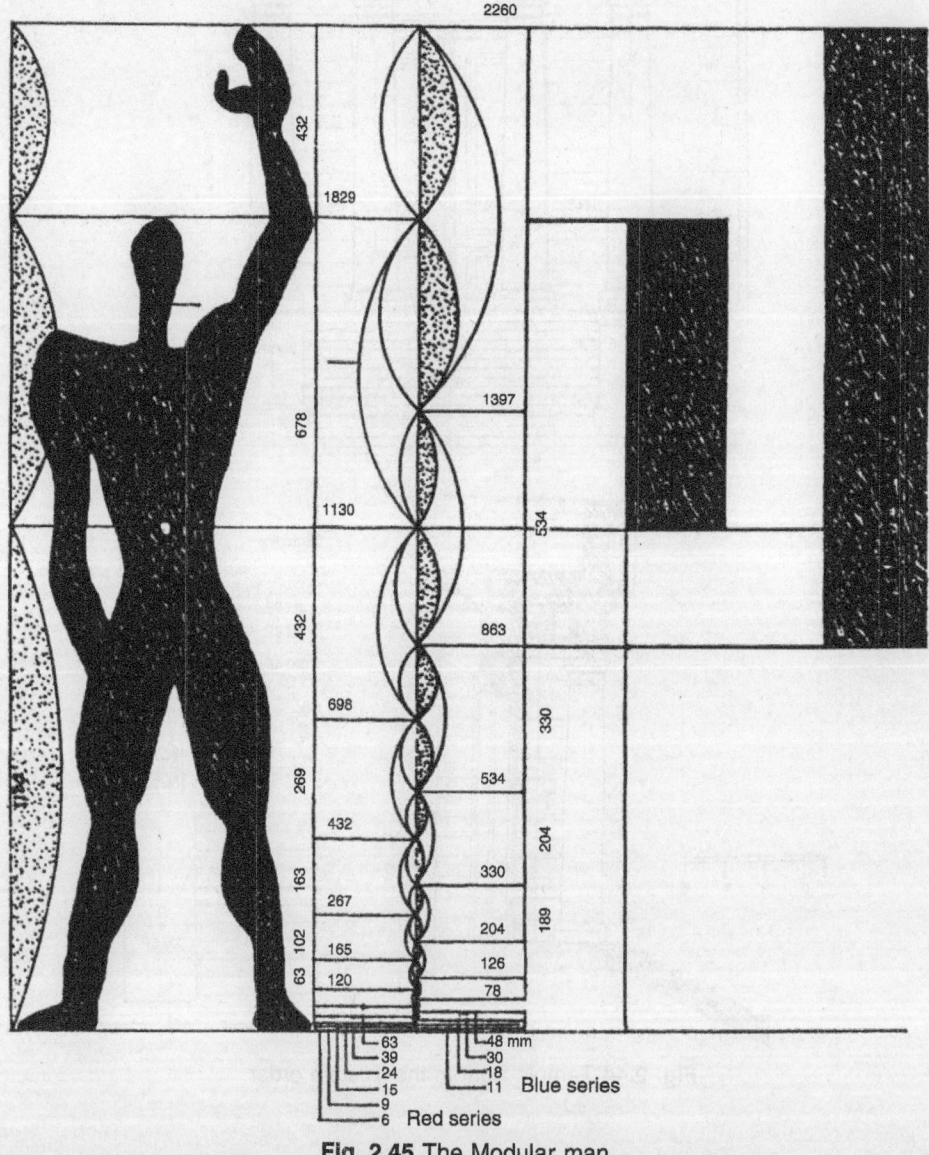

Fig. 2.45 The Modular man

Le Corbusier believed this proportion to be evident in the human body and that 'Art involves mathematics'. This is a system of measurement that decides on the dimensions of surfaces and volumes and maintains a human scale everywhere. It is based on man's height (of 1830 mm) and his height with hands stretched up (2260 mm) (see Fig. 2.45).

Le Corbusier suggested that the Modulor would give harmonious proportions to everything, from the size of cabinets and door handles to buildings and urban spaces. In a world with an increasing need for mass production, the Modulor was supposed to provide a model for standardization.

Anthropomorphic proportions

This is based on the dimensions and proportions of the human body. There is an assumption in architecture that form (building) and space are decided by the dimensions of human body (see Figs 2.46 and 2.47).

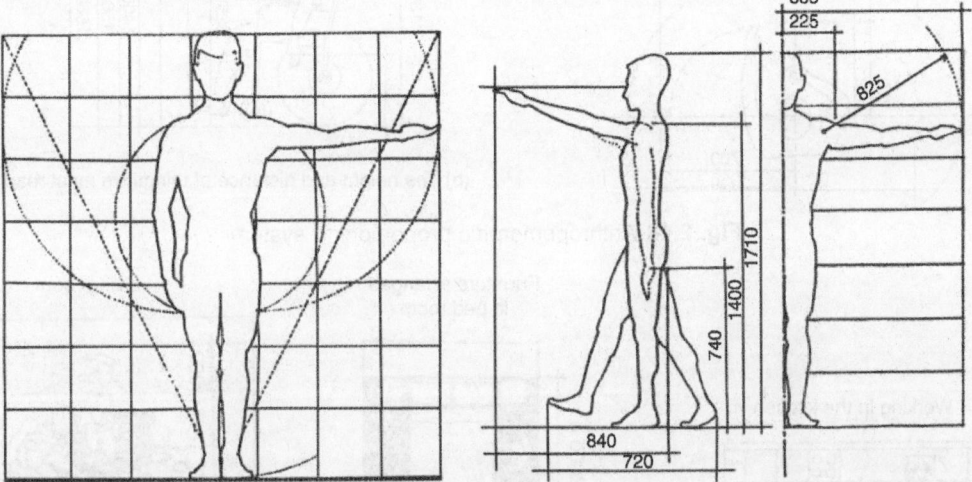

Fig. 2.46 The dimensions and proportions of human body affect the heights and distances of things we must reach

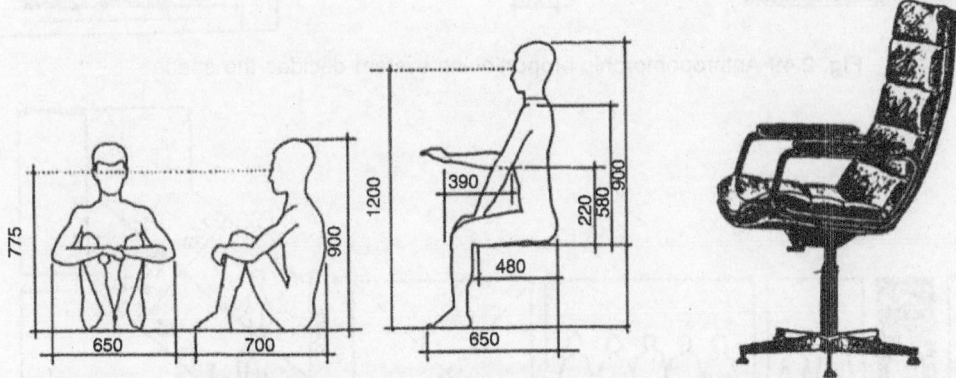

Fig. 2.47 Dimensions of the human body decide the space

The dimensions and proportions of the human body affect the heights and distances of things we must reach.

The dimensions and proportions of the human body affect the following:
- The proportion of things we handle [Fig. 2.48(a)].
- The height and distance of things we must reach [Figs 2.48(a) and (b)].
- The dimensions of the furniture we use [Fig. 2.48(b)].
- The volume or space required for movement, activity, and rest (Figs 2.49 and 2.50).

(a) The proportion of things we handle

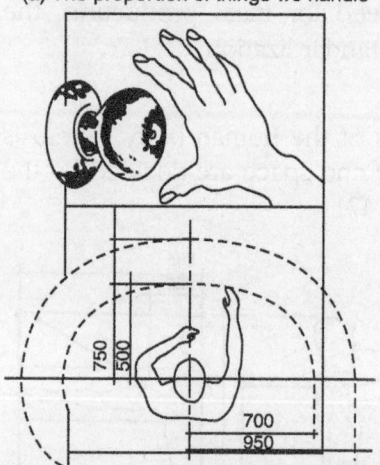

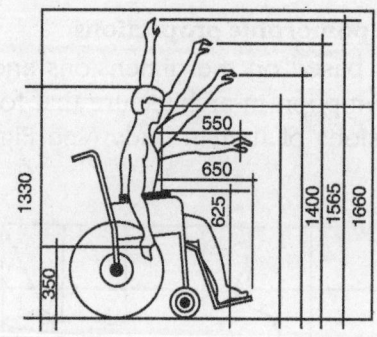

(b) The height and distance of things we must reach

Fig. 2.48 Anthropomorphic proportioning system

Working in the kitchen

Furniture arranged in bed room

Living room

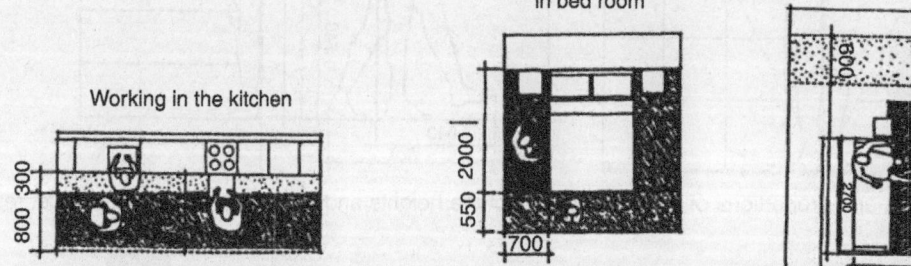

Fig. 2.49 Anthropomorphic proportioning system decides the space

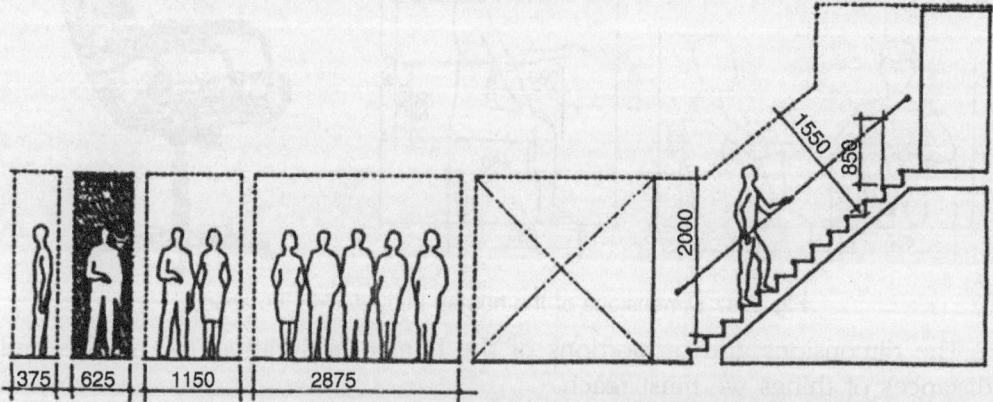

Fig. 2.50 The volume of space required for movement, activity, and rest

The ken

The traditional Japanese unit of measure was *shaku*, which is almost equivalent to the English foot and divisible into decimal units. Later, one more unit, called *ken*, was introduced. It was used to measure the interval between two columns.

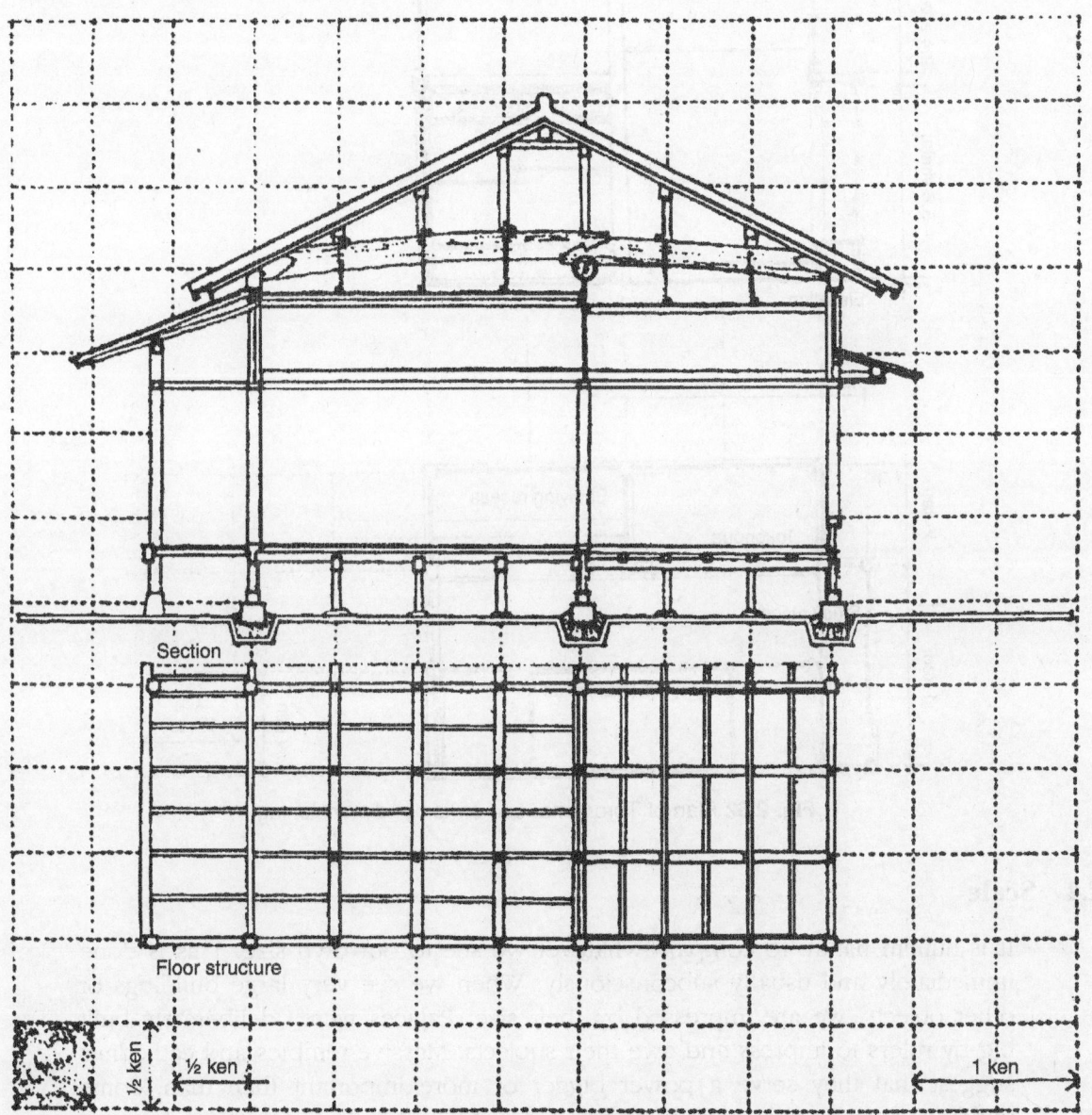

Fig. 2.51 Typical Japanese house

The ken was standardized for residential buildings (Fig. 2.51). It was an absolute measurement. The ken was not only used as a measurement for the construction of buildings, it was also made into an aesthetic module that decided the structure, materials, and space of Japanese architecture

Figure 2.52 shows the ken modular grids (6 shaku) which decided the centre-to-centre spacing of columns. The room dimensions were decided based on the ken module. The ceiling height was also decided accordingly.

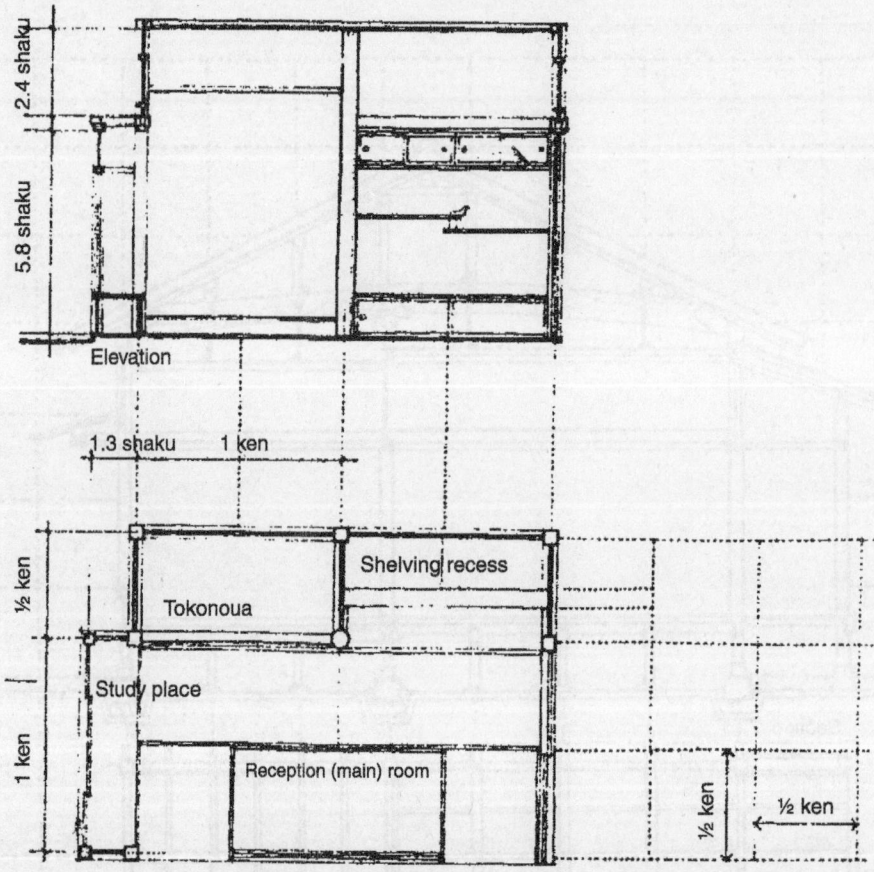

Fig. 2.52 Plan of Toronoma: use of ken in dimensioning

2.4 Scale

It is human nature to compare whatever we see to our own size. This is done immediately and usually subconsciously. When we see very large buildings or other objects, we are impressed by their size. Palaces were deliberately built big by rulers to impress and awe their subjects. Massive temples and cathedrals suggest that they serve a power bigger or more important than man. Some buildings such as railway stations and large machine halls are massive because of their purpose.

In modern times, we have seen a great increase in the number of large buildings. This is as a result of the developments in structural engineering and high land values. Some buildings are made large for prestige purposes because the name of the company is associated with the building, even though the company offices occupy only a part of the building.

New, tall buildings are very exciting and impressive. However, the concentration of big buildings without any open spaces results in a cold and impersonal environment. (Fig. 2.53). As against these, low-rise buildings with open spaces are more human in scale. People may not feel like living, working, and spending their time in surroundings that dwarf them.

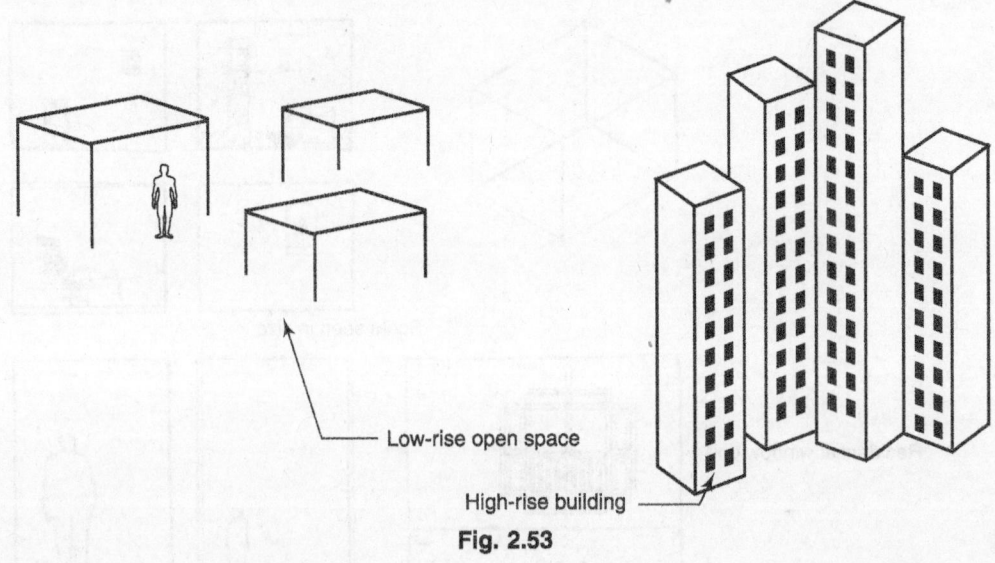

Low-rise open space

High-rise building

Fig. 2.53

The sense of scale and changes in height and space can provide visual contrast, giving vitality to cities. Narrow spaces leading to open areas give vitality and a distinctive character to cities such as Florence and Rome.

So, scale can be defined as how one views the size of a building element or space relative to other forms. When we want to measure the size of an element, we use other elements of known size. These elements are known as scale-giving elements. There are two kinds of scale-giving elements.

Generic scale: Building elements whose size and characteristics are known.

Human scale: The size of a building element or space related to the dimensions and proportions of the human body.

Overall dimensions of the facade. *b* denotes the space between windows

Fig. 2.54

2.4.1 Generic scale

All building elements have a certain size determined by the manufacturer or the designer from a range of choices. Additionally, the size of each element of a building is decided upon with respect to the size of other elements around it. For example, the size and proportion of windows in a building façade are visually related to one another, to the space between them, and also to the overall dimensions of the facade (Fig. 2.54). If the windows are all of the same size and shape, then a scale relates to the size of the façade (Fig. 2.55).

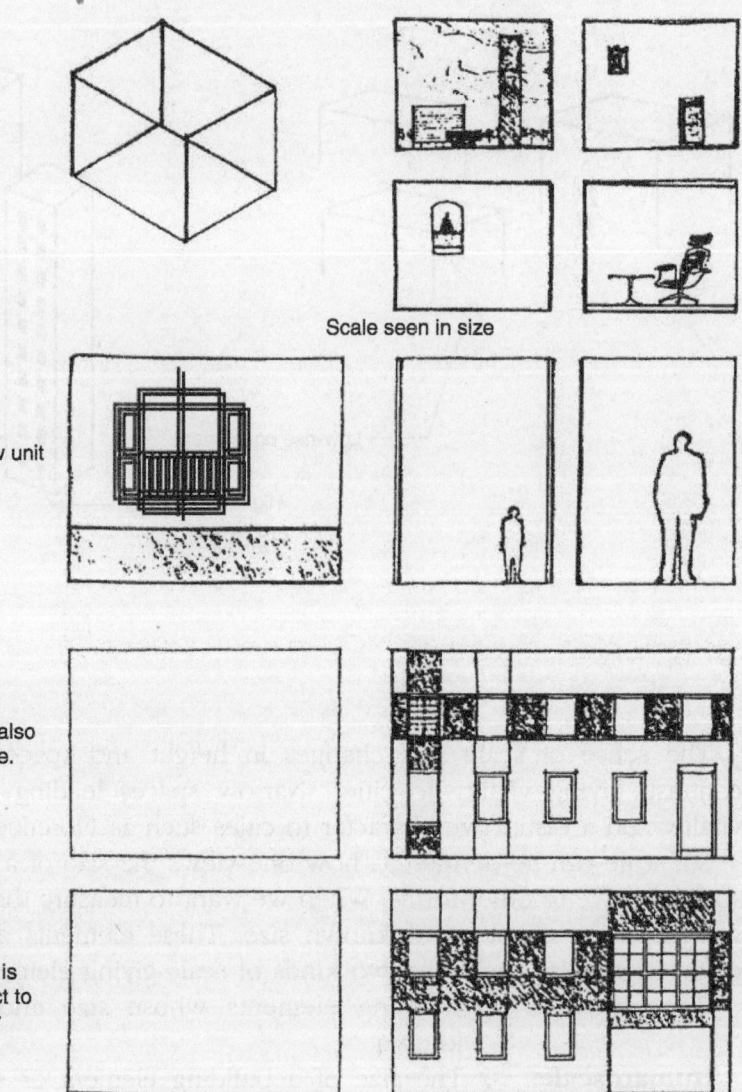

Scale seen in size

Residential window unit

Door, window elements which signify the scale and also relate to the overall facade.

Size of each element is understood with respect to the size of the other elements around it

Fig. 2.55 Scale

If one of the windows is larger than the others, it would create another scale with respect to the composition of the façade. This big size of the window could suggest the importance of the space behind the window (Fig. 2.56).

Scale of window units

Scale of opening size

Fig. 2.56

Many building elements' sizes are known or familiar to us and these elements help us to determine/guess the sizes of other elements around them. For example, residential window units and doorways give an idea as to how small or large the building is or how many storeys it has. Stairs and handrails also help us to measure the scale of a space.

Some buildings and spaces have two scales operating simultaneously. The entrance portico of the Rotunda at the University of Virginia is scaled to the overall building form. The doorway and windows behind it are scaled to the size of the spaces within the building (Fig. 2.57). In the Reims Cathedral, France, shown in Fig. 2.58, the recessed entrance is scaled to the overall dimensions of the façade and from a distance it appears like the building's entrance. As one comes closer to the building, it can be seen that the entrances are really simple doors scaled to the human scale.

Fig. 2.57 Rotunda, University of Virginia

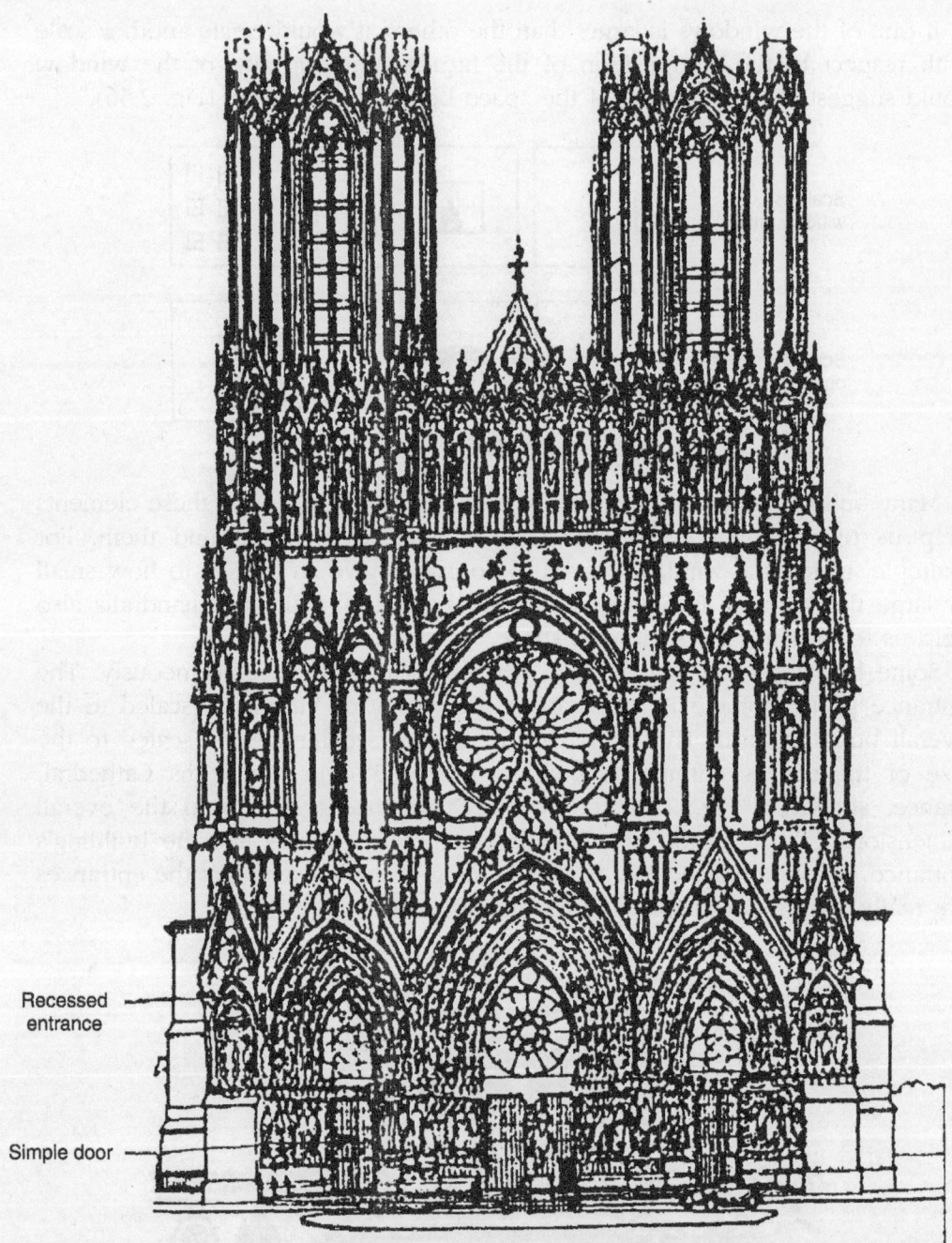

Recessed
entrance

Simple door

Fig. 2.58 Reims Cathedral, France

2.4.2 Human scale

The human scale is based on the dimensions and proportions of the human body. We can measure a space whose width is such that we can stretch out our arms and touch the walls. Similarly the height can be measured by reaching up and touching the ceiling. This is known as anthropomorphic proportioning.

The Vitruvian Man is a famous drawing with accompanying notes by Leonardo da Vinci made around the year 1490 in one of his journals. It depicts a naked male figure in two superimposed positions with his arms apart and simultaneously inscribed in a circle and square. The drawing and text are sometimes called the Canon of Proportions or, less often, Proportions of Man.

(*Source*: Wikipedia)

Leonardo da Vinci's Vitruvian Man

Vitruvius, an ancient Roman architect, used the proportions and measurements of the human body to describe architectural proportions. He wrote in his famous treatise *De Architectura*: 'For without symmetry and proportion no temple can have a regular plan; that is, it must have an exact proportion worked out after the fashion of the members of a finely shaped human body.'

'Good architecture' was described by Vitruvius through a careful analysis of both the proportions and measurement of the human figure. He found that each separate part of a human body was a simple fraction of the whole. For example, the head measured from forehead to the chin was exactly one-tenth of the total height, the outstretched arms were always as wide as the body was tall, a palm was the width of four fingers, a foot was the width of four palms, a cubit was the width of six palms, a man's height is four cubits, etc. Vitruvius believed that the same principles should be used when designing a building.

The objects used by human beings such as a table, a sofa, a chair, stairs, a window, or a door help to decide the size of a space and also give it a human scale (Fig. 2.59). For example, a close arrangement of chairs and tables in a hotel lobby will convey an impression of spaciousness. Similarly, a staircase leading to a second-storey balcony will give an idea of the vertical dimension of a room as shown in Fig. 2.59. A window in a blank wall suggests

Fig. 2.59 Objects used by human beings such as tables, sofas, chairs, and windows help to decide the size of the space

that the space behind could be a living room or a bedroom. A ventilator will indicate that the space behind is a toilet. All these elements convey the presence of humans.

2.4.3 The effect of vertical height on scale

Of the three dimensions, height has a greater effect on scale than width or length. The walls of a room provide enclosure and the roof decides the quality of the shelter (a big building or a small one, a thatch or a tile roof) and intimacy. For example, raising the ceiling height of a 3.5 m × 5 m room from

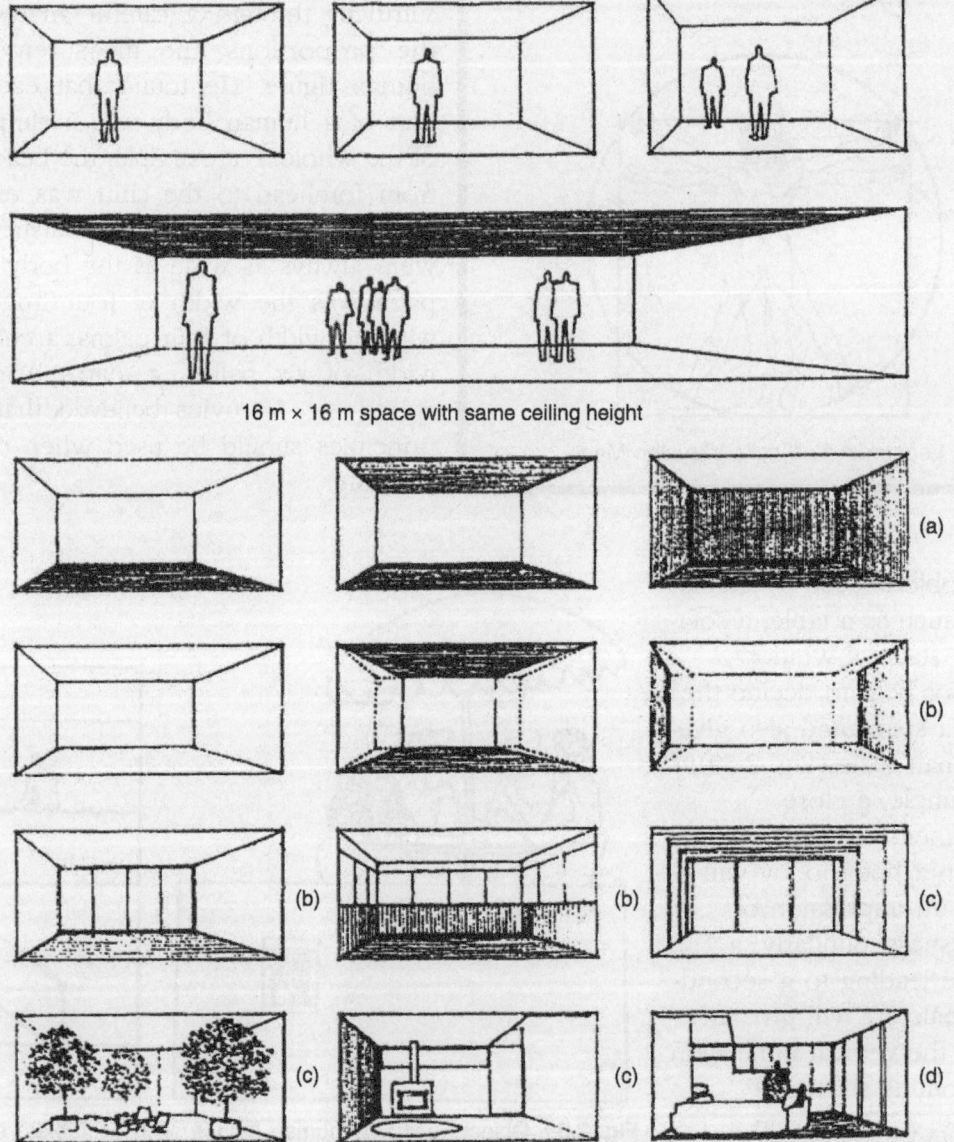

16 m × 16 m space with same ceiling height

Fig. 2.60 (a, b) Shape colour and pattern of the enclosing space, (c) the shape and position of enclosing space, (d) the nature and the scale of the elements placed within it

2.75 m to 3 m will be more noticed and affect the scale more than if the width was increased to 4 m and length to 5.5 m. While the 3.5 m × 5 m room with a 2.75 m ceiling is comfortable to most people, a 15 m × 15 m space with the same ceiling height will create a feeling of distress.

In addition to the vertical dimension of a space, the other factors (Fig. 2.60) that affect scale are:
- the shape, colour, and pattern of the enclosing surface.
- the shape and position of openings.
- the nature and the scale of the elements placed within it.

2.5 Balance

Balance in architecture is used to denote visual equilibrium. Here equilibrium does not mean equal elements distributed symmetrically about a centre of gravity, but unequal elements distributed asymmetrically and still remaining balanced. In this sense, it is the exact opposite of symmetry. This can be understood from Fig. 2.61.

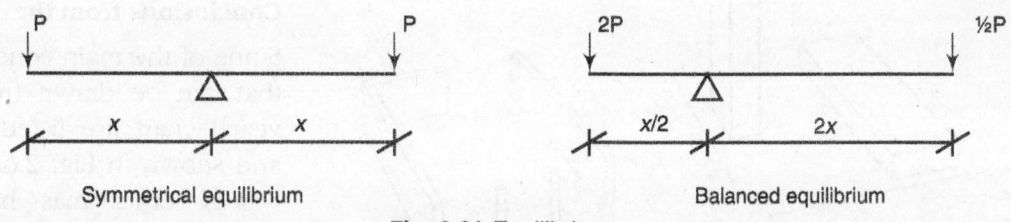

Fig. 2.61 Equilibrium

In order to balance dissimilar elements, every element is given some kind of visual weightage. The equilibrium of various weightages about a point of reference leads to balance.

2.5.1 Understanding balance

The weightage given to various architectural elements needs to be understood. Imagine a rectangular surface divided equally into two halves by a vertical line with one half painted yellow and the other blue. As yellow is a much stronger colour than blue, its weightage is also greater. This means the two surfaces are not in balance. Now if we shift the dividing line towards the yellow surface so that it progressively decreases and correspondingly increases the area of the blue surface, a stage will be reached when the two surfaces will appear to have an equal visual impact, in balance with each other. The two are no longer symmetrical, but they are balanced. So, it is the eye that decides the parameters for balance. There can be balance within each of the architectural elements of mass, space, line, surface, point, etc. Balance should be within elements of the same class, for example, between one mass and another, one surface and another.

2.5.2 Chart of visual weightage

It is the perception of the designer that determines visual balance. So the visual qualities of an architectural object can be compared by studying the chart of visual weightage shown below. For example, the first relationship in the chart shows how a vertical mass has a higher weightage than a horizontal or neutral mass. The '>' symbol indicates greater visual weightage.

- Vertical > Horizontal > Neutral
- Pyramidal > Vertical > Rectanguloid > Horizontal
- Rectanguloid > Cube > Cylinder > Sphere
- Points > Lines > Surfaces > Openings
- Solid > Perforated
- Projecting > Receding
- Convex > Concave
- Height > Dark
- Red > Yellow > White > Green > Blue > Black

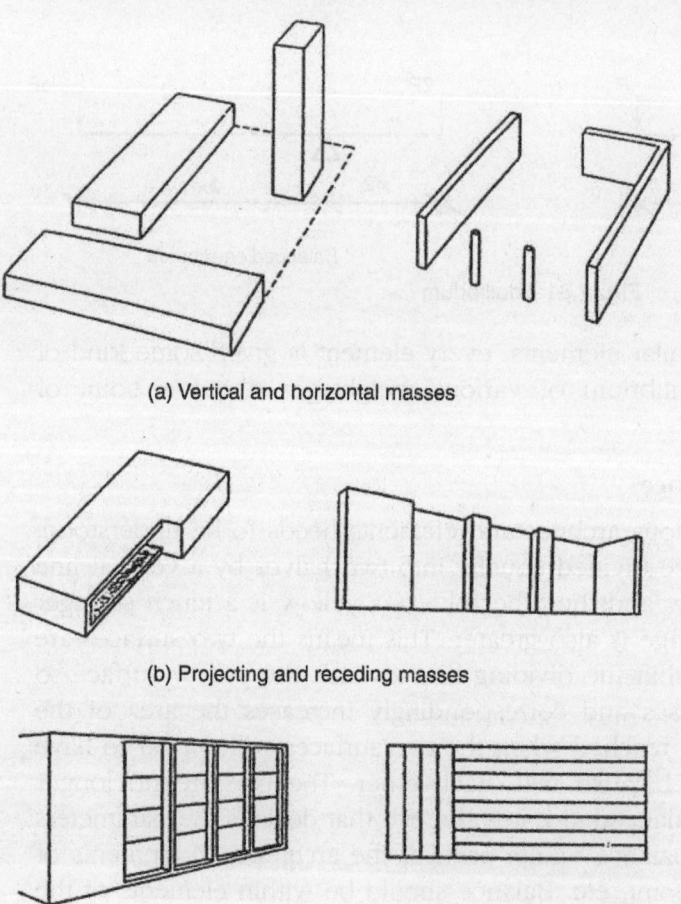

(a) Vertical and horizontal masses

(b) Projecting and receding masses

(c) Solid and perforated   (d) Vertical and horizontal

Fig. 2.62 Visual weightage

Conclusions from the chart

Some of the main conclusions that can be drawn from the visual chart are listed below and shown in Fig. 2.62.

- A vertical mass having a maximum weightage may appear visually weaker if it is fully perforated, is painted a dull colour, or is covered with horizontal bands [Fig. 2.62(c)].
- A horizontal mass can be made to look more vertical if it is painted with vertical stripes and its visual weightage is increased [Fig. 2.62(d)].
- Many combinations need to be tried because achieving balance is a more difficult job than achieving equilibrium through symmetry.

2.5.3 Balance occurs horizontally, not vertically

Consider cases (c) and (d) of Fig. 2.62. The two elements where the balance is to be maintained are located laterally to each other. In part (c) the solid mass is alongside the perforated one and in (d) the vertical pattern lies alongside the horizontal. In no case does one element lie above or below its neighbour. If arranged vertically, there would be no balance. In Fig. 2.63 also, the forces creating the equilibrium are horizontally arranged. Thus balance, whether visual or physical in nature, is only the horizontal arrangement of forces in a vertical plane.

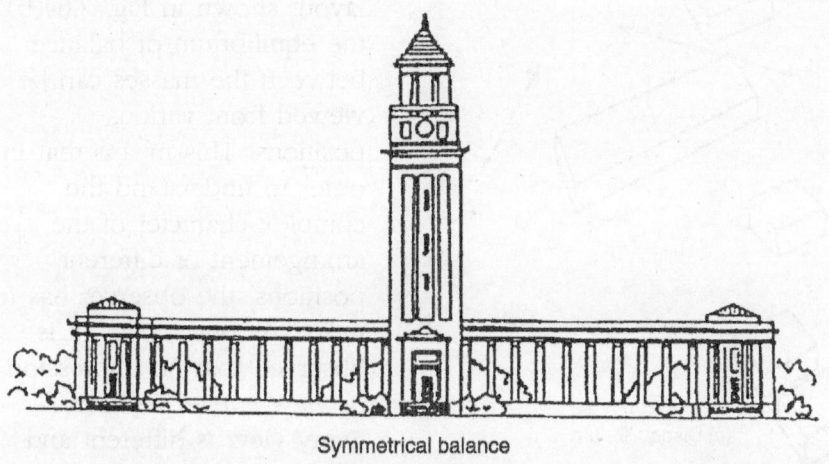

Symmetrical balance

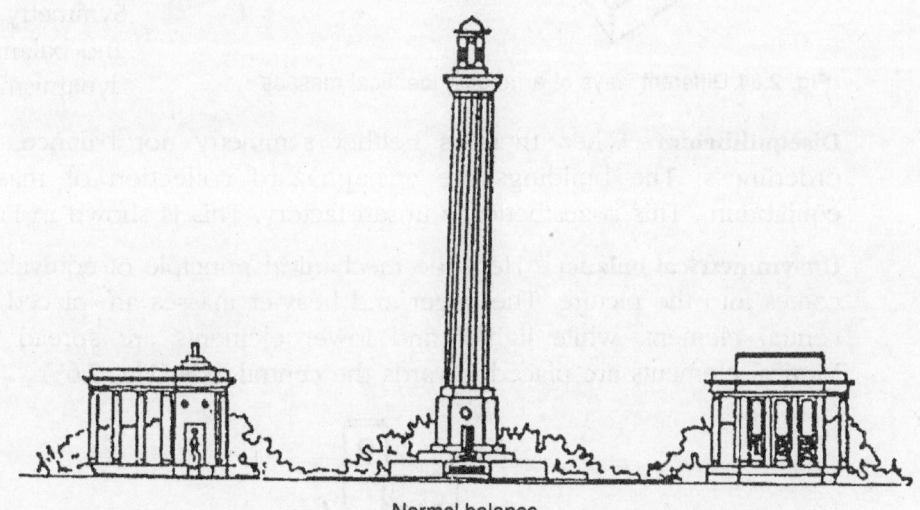

Normal balance

Fig. 2.63 Balance

2.5.4 Emotional effect of balance

Identical masses are arranged in four different ways, as shown in Figs 2.64 and 2.65. We will discuss these ways one by one.

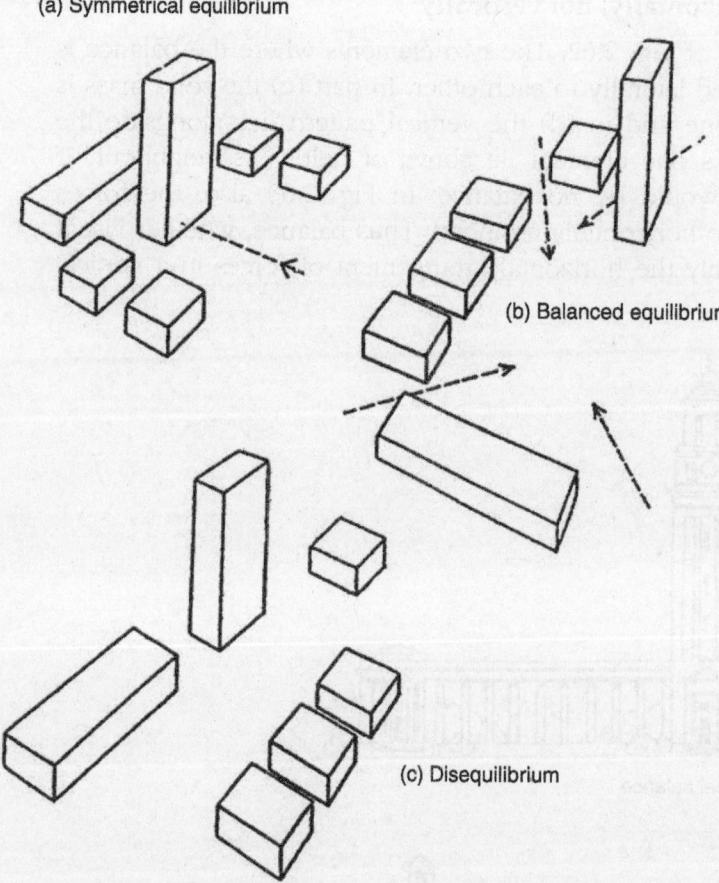

(a) Symmetrical equilibrium

(b) Balanced equilibrium

(c) Disequilibrium

Fig. 2.64 Different ways of arranging identical masses

Symmetrical equilibrium In Fig. 2.64(a) the equilibrium is maintained only along the axis. The layout will appear to be in equilibrium only from one location. From other locations, disequilibrium is seen.

Balanced equilibrium In this layout shown in Fig. 2.64(b) the equilibrium or balance between the masses can be viewed from various positions. This means that in order to understand the complex character of the arrangement of different positions, the observer has to move. As one moves, it is observed that no impression or view is ever repeated. Every view is different and cannot be predicted. Symmetry causes stability and balance causes dynamism.

Disequilibrium When there is neither symmetry nor balance, there is no orderliness. The buildings are a haphazard collection of masses lacking equilibrium. This is aesthetically unsatisfactory. This is shown in Fig. 2.64(c).

Unsymmetrical balance Here the mechanical principle of equivalent moments comes into the picture. The larger and heavier masses are placed close to the central element, while lighter and lower elements are spread horizontally. Vertical elements are placed towards the central axis (Fig. 2.65).

Fig. 2.65 Unsymmetrical balance

2.6 Symmetry

Humans are able to appreciate symmetry because they themselves are symmetrically constituted. The human body is symmetrical about an imaginary plane bisecting it vertically and passing between the eyes. Every object or group of objects in nature that can be similarly bisected by an imaginary vertical plane into two identical halves will appear symmetrical to the observer.

Consider Figs 2.66(a) and (b). If we rotate the two configurations by 90° to the vertical plane, we will get Fig. 2.67. It will be seen in Fig. 2.67 that configuration (a) assumes a symmetrical appearance and configuration (b) becomes asymmetrical. The objects themselves have not changed, yet the quality of symmetry has obviously changed due to the change in their positions with respect to our eyes. Thus, we can conclude that symmetry is not a quality of the object itself but is related to an imaginary vertical plane, which in turn depends on the position of the observer. When an object is continually seen from different angles, it may appear symmetrical from one side and asymmetrical from another.

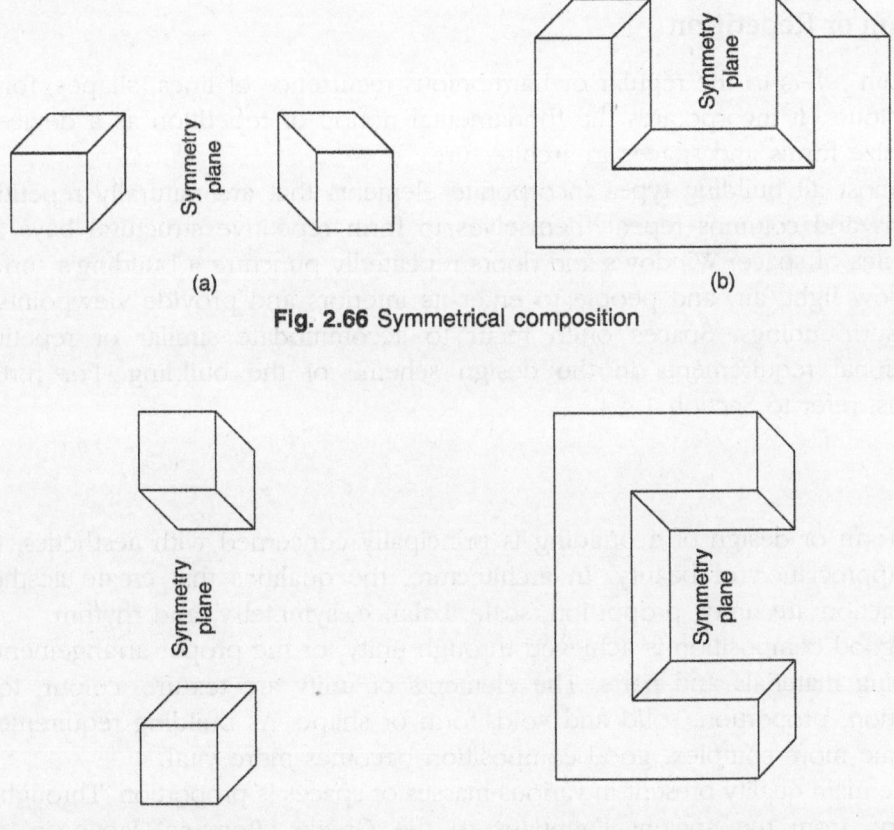

Fig. 2.66 Symmetrical composition

Fig. 2.67 Symmetry and asymmetry

Other than symmetry of masses, a structure could display symmetry of surfaces or of openings (Fig. 2.68). Symmetry has various visual and psychological effects. It is one of the modes of orderliness. In primitive societies, every man-made object or arrangement, such as a village layout or shelter, was symmetrical. A symmetrical object is visually more stable and better organized and hence easy to capture mentally. Excessive symmetry and orderliness can also appear monotonous. An architectural object must combine orderliness with interest. An example of this can be seen in the medieval Gothic cathedral shown in Fig. 2.58, in which the two entrance towers are symmetrically arranged and the location of a large circular opening between these two towers, set within a multitude of decorative elements, is also visually interesting. For further details, refer to Section 1.4.2.

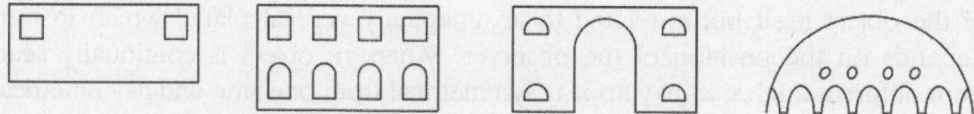

Fig. 2.68 Symmetry of surfaces and openings

2.7 Rhythm or Repetition

Rhythm refers to the regular or harmonious recurrence of lines, shapes, forms, or colours. It incorporates the fundamental notion of repetition as a device to organize forms and spaces in architecture.

Almost all building types incorporate elements that are naturally repetitive. Beams and columns repeat themselves to form repetitive structural bays and modules of space. Windows and doors repeatedly puncture a building's surface to allow light, air, and people to enter its interiors and provide viewpoints to the surroundings. Spaces often recur to accommodate similar or repetitive functional requirements in the design scheme of the building. For further details, refer to Section 1.4.4.

Summary

The form or design of a building is principally concerned with aesthetics, i.e., the appreciation of beauty. In architecture, the qualities that create aesthetic satisfaction are unity, proportion, scale, balance, symmetry, and rhythm.

A good composition is achieved through unity, or the proper arrangement of building materials and parts. The elements of unity are texture, colour, tone, direction, proportion, solid and void, form or shape. As building requirements become more complex, good composition becomes more vital.

The main quality present in various masses or spaces is proportion. Throughout history, from the ancient Egyptians to the Greeks, Romans, Japanese, and twentieth-century Europe, architects have developed 'proportioning systems'— the golden section, regulating lines, the orders, the modular, anthropomorphic system, and the ken. These help the architect to retain control over the proportion of a building's form and spaces, overcoming restrictions imposed by material, structural, and manufacturing processes.

Scale is used in architecture to study the size of a building relative to other measures. There are two types of scale—generic and human. This concept can be studied through the examples of many existing buildings.

This chapter also discusses the concept of balance in architecture, which is used to denote visual equilibrium, and how balance is achieved through various combinations of elements. Symmetry and rhythm are also elements that contribute to the achievement of an aesthetic composition.

The study of these helps to understand how aesthetics define the style and character of buildings.

REVIEW QUESTIONS

Part—A (2 marks each)

1. What are the aesthetic components of design?
2. Define unity in architecture.
3. What do you mean by proportion?
4. Define golden ratio.
5. Write short notes on the golden rectangle.
6. What is modular?
7. What do you understand by Le Modular?
8. What do you understand by ken?
9. What do you understand by 'The Order'?
10. What is anthropomorphic proportioning system?
11. What are regulating lines?
12. Mention the various proportioning systems.
13. What is Doric order? (Draw a neat sketch of the order and mark the parts)
14. What is Ionic order?
15. What is Tuscan order?
16. Describe the Corinthian order.
17. What is Composite order?
18. Mention the various orders.
19. What do you understand by scale in architecture?
20. What is balance in architecture?
21. Define symmetry.
22. What do you understand by rhythm?

Part—B (16 marks each)

1. Giving examples from history, explain the concept of unity.
2. Explain how proportion is brought out in a building, with examples from history.

3. In the design of a school building, explain with sketches the concept of proportion.
4. Explain scale in architecture with examples from history.
5. Explain the concept of balance in architecture with examples from history.
6. Symmetry in architecture—explain the concept with examples from history.
7. How is rhythm brought out in an architectural composition? Explain the phenomenon with examples from history.
8. When you are asked to design an institution block (college building) how will you bring out the aspects of unity, proportion, scale, balance, symmetry, and rhythm?

Aesthetic Relationships, Character, and Style in Buildings

3.1 Character and Style in Buildings

Every building, in its final form, depicts a particular character and style of its own. The character of a building brings out its exact theme. The style of a building is its character expressed with definite concepts. The true style of a building reflects the particular period/era in which it has been built. There are styles that represent the traditional architecture of the Greek, Roman, and Medieval periods and the modern architecture of the industrial era, twentieth century, and the Art Nouveau movements. Architects have made remarkable contributions to these styles from earlier periods to the present day. In India, we have varied styles of architecture ranging from Hindu temple architecture (which took different forms during the various periods of Hindu rulers—Cholas, Pandyas, Guptas, etc.) and Islamic architecture to Buddhist architecture, etc. For ease, character and style have been classified under traditional and modern architecture.

3.1.1 Character of a building

The character of a building depends upon its capacity to express a particular function and status. In the historical styles of architecture, the function and status of a building were successfully expressed in a variety of styles. *No building can have a neutral character.*

Just like an individual's personality is made up of unique qualities resulting from a combination of genetic factors, environment, and upbringing, which distinguish one person from another, the character of a building comprises aspects on the basis of which one can distinguish one building from another. If a building can by itself reveal the use for which it was designed and built, then it has the right character. A great number of buildings in a city's commercial centre are now office buildings which have the same kind of functions. For example, the Reserve Bank building is often distinguished by its massive appearance, which is symbolic of the country's financial stability. The character of civic buildings reflects the form of the building and the purpose for which it was built (to serve the public).

In its appropriate setting, a building, through its character and style, is capable of evoking abstract values such as truth, beauty, grace, life, sacrifice, power, and obedience—the seven lamps of architecture as enumerated by the English poet John Ruskin. All the aesthetic components, such as unity, composition, contrast, and scale, together make up the character of a building. Character also brings out the utility and purpose of the building. It expresses the idea with which the architect has designed the building. Whether it is a bank, a church, or a library, a building should possess appropriate character. In short, a building has a real architectural character when its exterior expresses its internal functions. Architectural character can be divided into three categories as described in the following sections.

Functional character

The purpose of a building decides the arrangement of the various parts of the building. This arrangement affects its external appearance through which its character is expressed. For example, externally,

(a) skylighting instead of windows can indicate a museum
(b) a long wall having numerous windows in a particular order can also indicate a museum
(c) very tall and wide windows suggest a library
(d) symmetrical arrangement with a central block having a wide entrance to express balance and rhythm indicates public buildings

Fig. 3.1 Functional character: A public building—the exterior of the building depicts the interior function

Associated character

The character of some buildings is expressed through elements associated with certain influences. Such influences often go thousands of years back in history

and culture. We recognize a building by its features, which we associate with a particular structural style. A spire on a tower by the side of a spacious building, as shown in Fig. 3.2, indicates a church. A *gopuram* depicts a Hindu temple, whereas a crescent on a white dome is the symbol of a mosque.

Fig. 3.2 Associated character: the impression created from past experience helps to identify the building (here a church)

Personal character

Character in architecture is comparable with the attributes of an individual. If a building is designed in the proper spirit, personal characteristics such as grace, dignity, and vitality can be expressed as an integral part of the structure (Fig. 3.3).

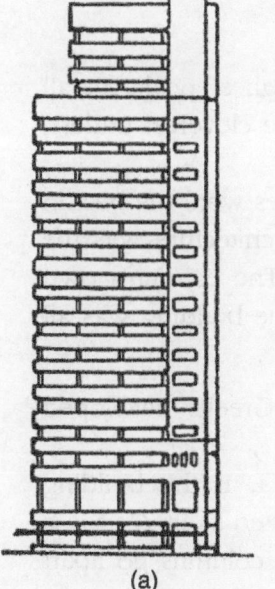

(a)   (b)

Fig. 3.3 Every building has a personal character

3.1.2 Style in buildings

Style can be defined as a distinctive or characteristic mode of presentation, construction, or execution in any art form, e.g., costume styles. It is a quality which gives a particular identity to artistic expression, e.g., the Gothic style of architecture. 'A gentleman is always known by his style', the popular expression goes.

Style can also be defined as a particular mode or form of skilled construction, execution, or production. Style in architecture is the manner in which a building is constructed in a particular region (or area), during a particular period, which defines certain special characteristics of general design or construction and ornamentation.

A building may not conform to any particular style of architecture, historical or modern, but still have a perfect composition. Style is character expressed with definite ideas. A religious style is a true style of building, which reflects the ideas of sacredness, ritual rightness, and stability. One such style is the architecture of Hindu temples. The 'modern styles' of Christian churches in Europe and America are also examples of style in architecture.

3.1.3 Realization of character and style in traditional architecture

Character and style can be studied under the various periods of traditional architecture. They can be classified under the following periods.
 (a) Greek architecture (3000 BC to 30 BC)
 (b) Roman architecture (300 BC to 365 AD)
 (c) The medieval period (300 AD to 1300 AD)

Greek architecture (3000 BC to 30 BC)

The Greeks developed a civilization which reached the highest peaks in all walks of life. They were the greatest producers of goods, the cleverest traders, and the bravest warriors during their time.

Of all the buildings constructed by the Greeks, their temples were the best in terms of architectural aspects. Their main contribution to architecture was the development of the column as a main structural member. The column's own proportions and its proportions with respect to the rest of the building was all standardized.

Greeks developed optical corrections in architecture The Greeks developed the use of optical illusion in architecture, as follows.
 (a) Consider the diagrams E, F, and G shown in Fig. 3.4. If the building elevation has to look as in E, then it should be constructed as in G. If it is constructed as in E, then it will appear as in F, as the columns go apart towards the top.

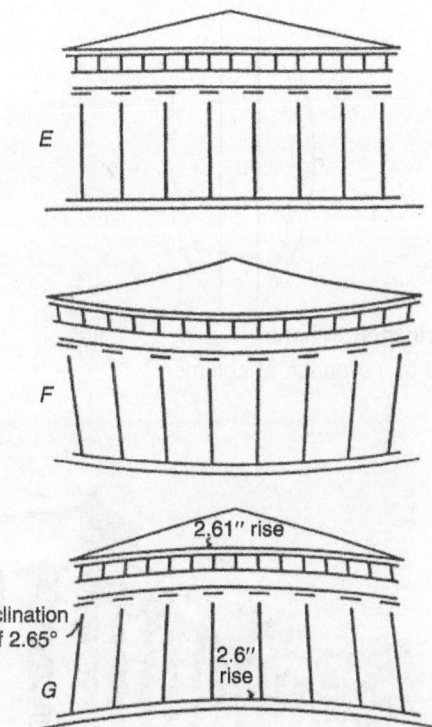

E

F

2.61" rise

Inclination
of 2.65°

2.6"
rise

G

Fig. 3.4 The Parthenon, Athens (east front): optical correction applied in elevation

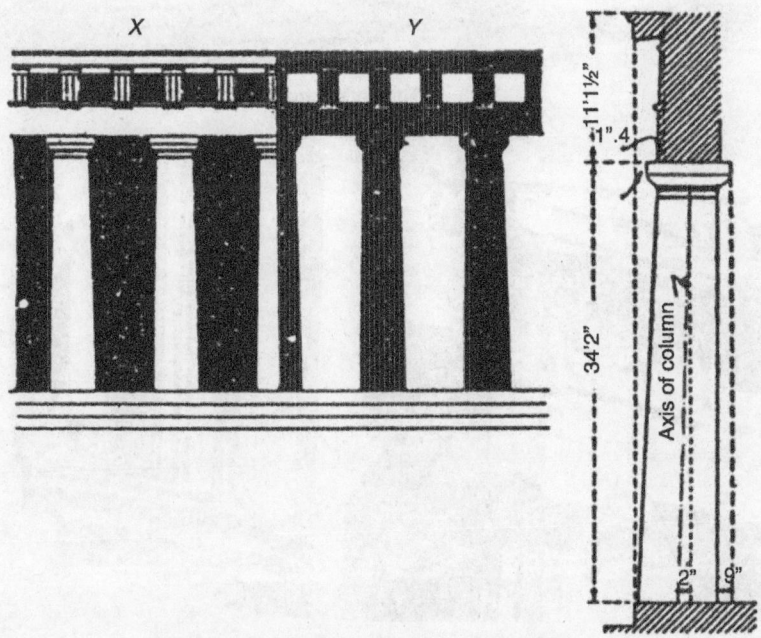

X

Y

11'1½"

1".4

34'2"

Axis of column

2" 9"

Fig. 3.5 The effect of colour on proportions

(b) The effect of colour on
proportions: Consider X
and Y (Fig. 3.5), the
columns look sturdier
(in X) if the
background and the
entablature is dark-
ened. In Y the columns
look slender.

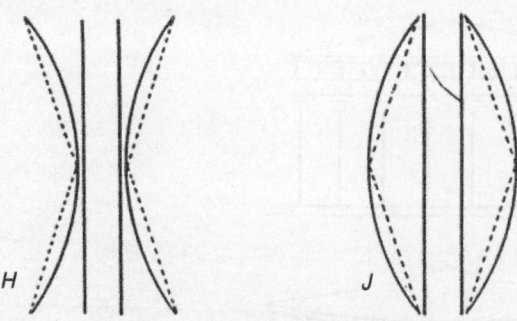

Fig. 3.6 The Parthenon, Athens
Optical correction applied to inclination of colums

(c) In *H* (Fig. 3.6) the curved lines make the column look wide. In *J* the column seems to be narrower at the centre.

Greeks developed columns
The Greeks perfected and developed three types of columns: Doric, Ionic, and Corinthian. The best example of the Doric column is the Parthenon at Athens (Figs 3.7 and 3.8). The best example of the Ionic column is the Erechtheion at Athens (Figs 3.9 and 3.10). The Monument of Lysicartes at Athens is the best example of the Corinthian column (Figs 3.11 and 3.12). The Doric column looks sturdy. The Ionic and Corinthian columns look more slender and elegant.

Greek architecture is characterized by mouldings, and their main contribution is the column and the pediment (the base of the column).

The entire structure

The entablature   The south passage
Fig. 3.7 The Parthenon, Athens: the Doric column

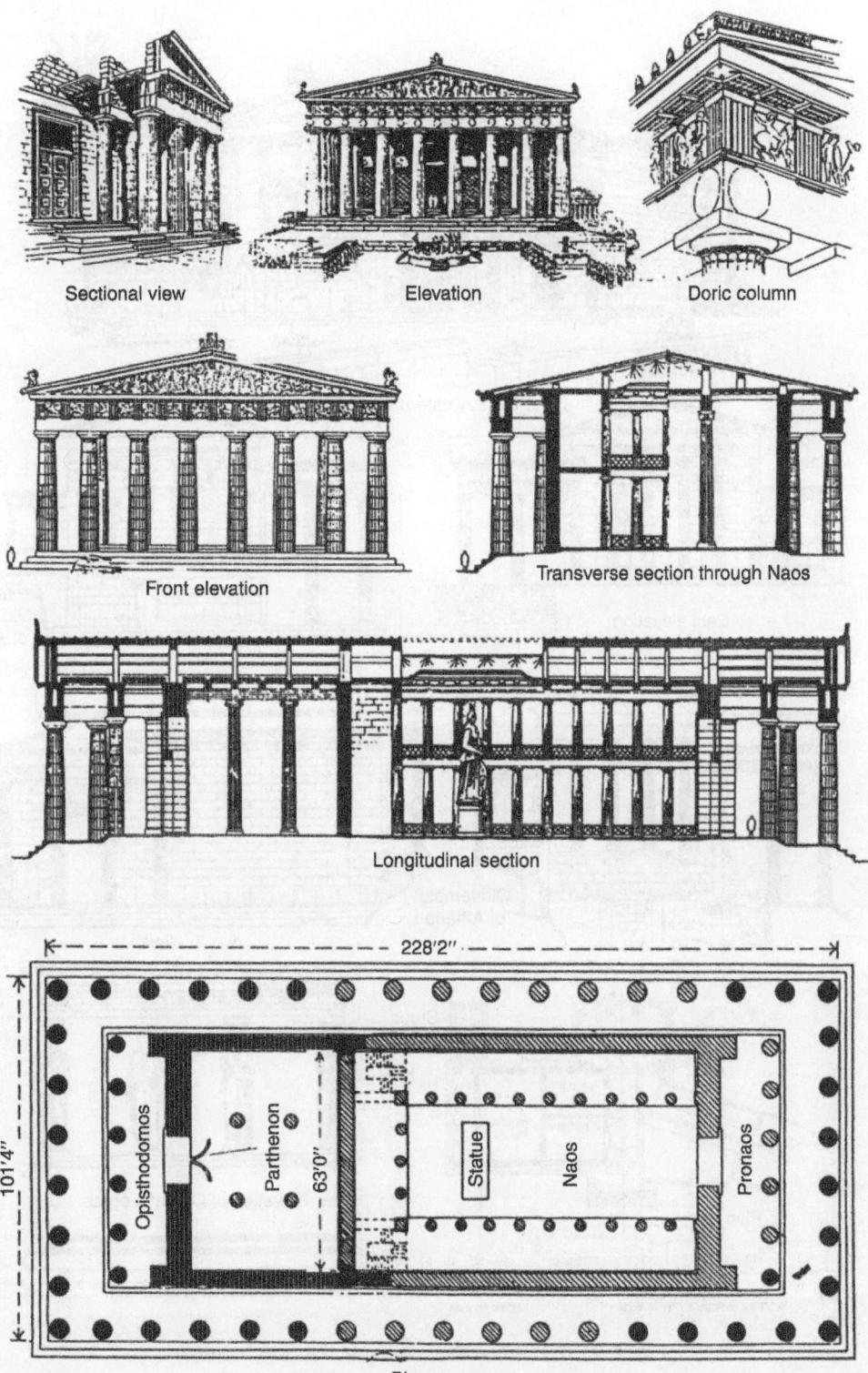

Sectional view Elevation Doric column

Front elevation Transverse section through Naos

Longitudinal section

Plan

Fig. 3.8 The Parthenon, Athens

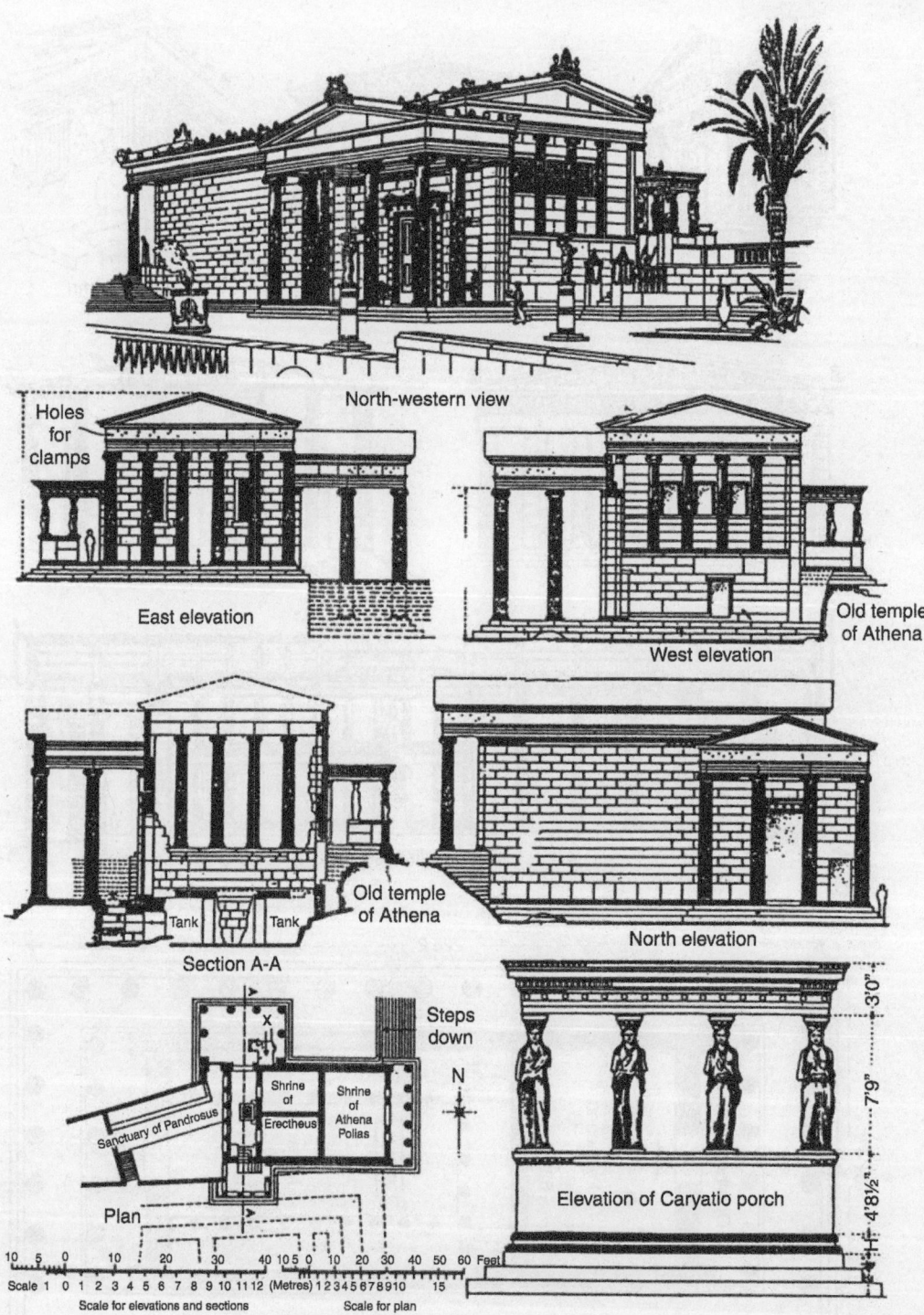

Fig. 3.9 The Erechtheion, Athens: the Ionic column

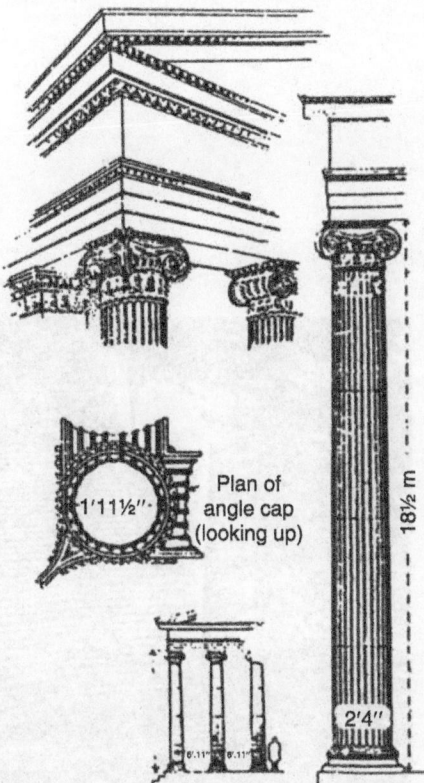

Fig. 3.10 The Erechtheion, Athens: the Ionic column details

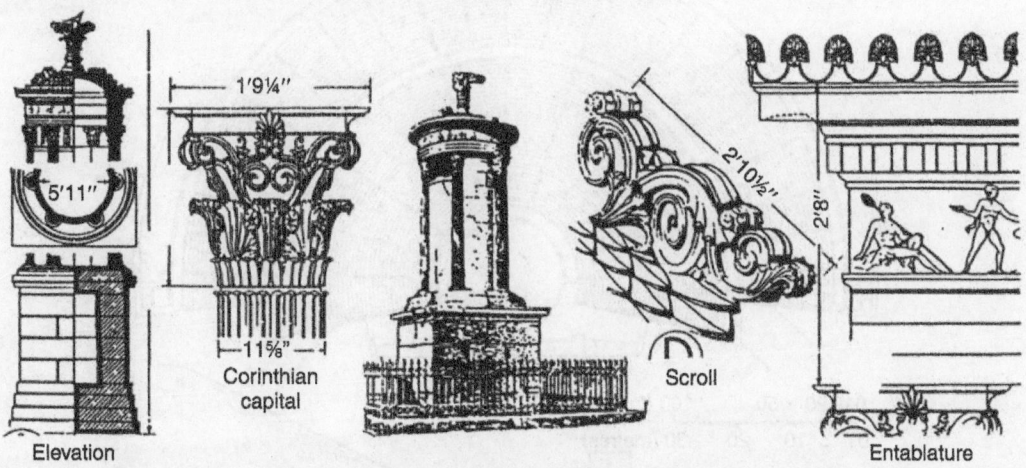

Fig. 3.11 Monument of Lysicrates at Athens: Corinthian order

Fig. 3.12 Corinthian capital

Greeks constructed theatres The Greeks were extremely fond of outdoor and spectator sports, for which they developed handsome theatres. The best example is Theatre Epidauros, Greece [Fig. 3.13(a)]. It has a fan-shaped [Fig. 3.13(b)] plan; the seats are arranged along the fan. The stage is circular in shape. There is also provision for an orchestra and dressing rooms. Our present-day theatres have their origins in Greek theatres.

(a)

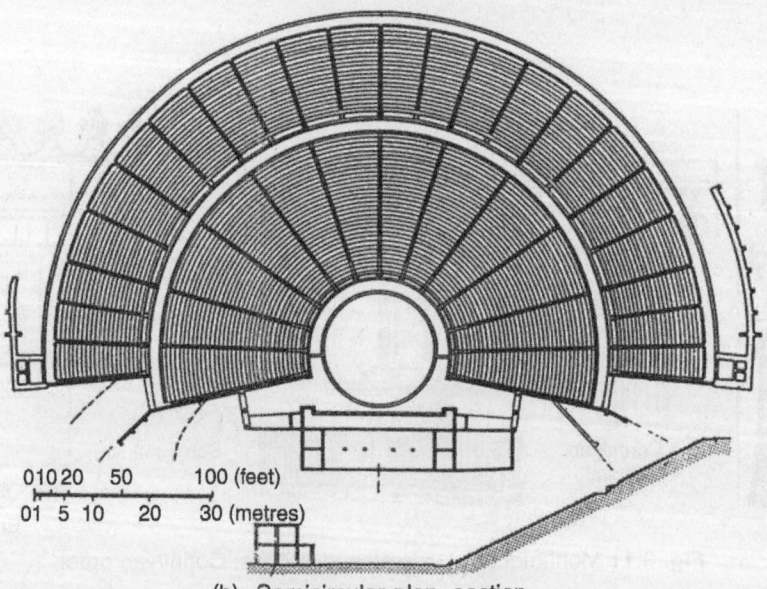

0 10 20   50   100 (feet)

0 1 5 10   20   30 (metres)

(b) Semicircular plan, section

Fig. 3.13 Theatre Epidauros, Greece

Greeks constructed houses The pattern of a typical Greek house was such that it did not have street windows. There was a central courtyard and there was a foyer, which led to the 'Triclinium' (present day drawing/living room).

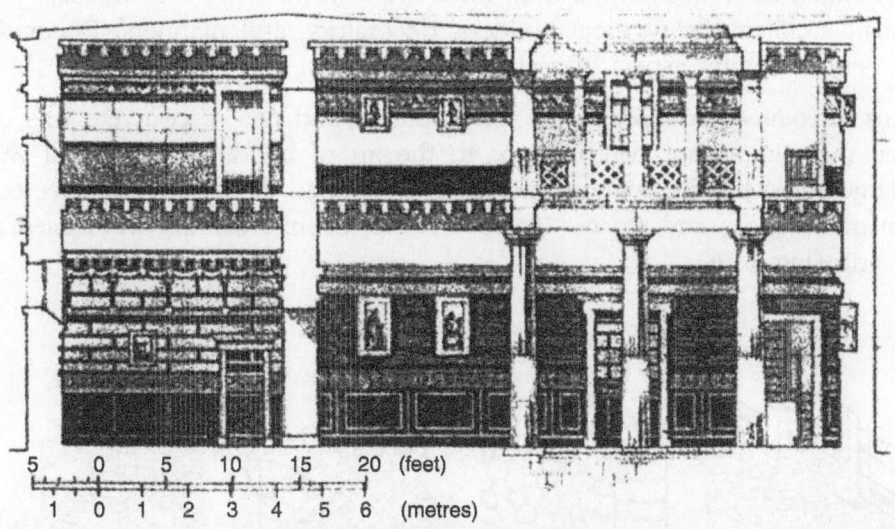

Fig. 3.14 Greek house (two-storey) section

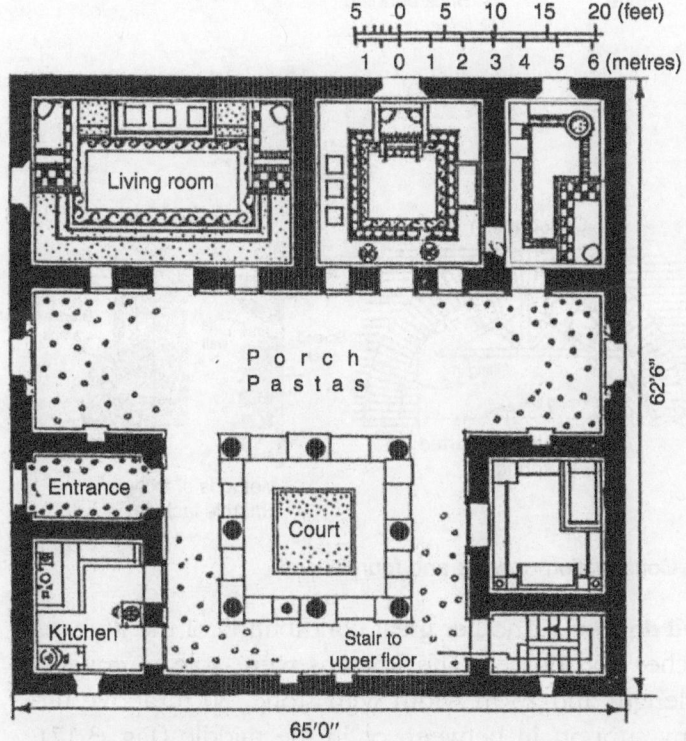

Fig. 3.15 Dalos, green house: two-storey apartments

Greek houses were double-storeyed (Fig. 3.14). They also developed pastas (Fig. 3.15) (long, shallow rooms, crossing the house from one side to the other).

Roman architecture (300 BC to 365 AD)

The Romans spent a great deal of time and effort in improving the living conditions in their empire. They were not only great warriors, but good administrators as well. Many of their ideas were inspired by the Greeks.

Roman architects were great builders, decorators, and planners. Decoration was a very important aspect of architecture for them.

Romans introduced concrete   The development and use of concrete was one of their most important contributions to the art of building. They built walls with large stone blocks, without mortar, and finished the surface or covered it with marble, brick, and stucco. To fix marble, bronze clamps were used for better grip (Fig. 3.16).

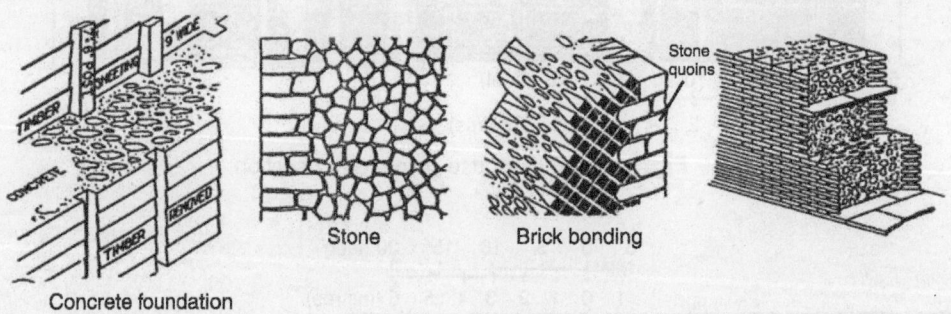

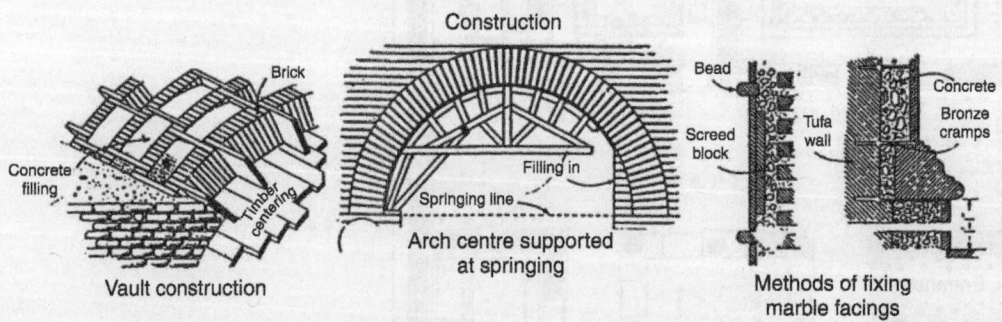

Fig. 3.16 Construction of walls and foundations

Romans developed vaults and domes   Another great contribution of the Romans was the development of arches and domes. The Romans were able to cover a large space of up to 60 m length and 25 m width with stone, 30 m above the floor without columns or any support in between or in the middle (Fig. 3.17). They introduced a new concept of space.

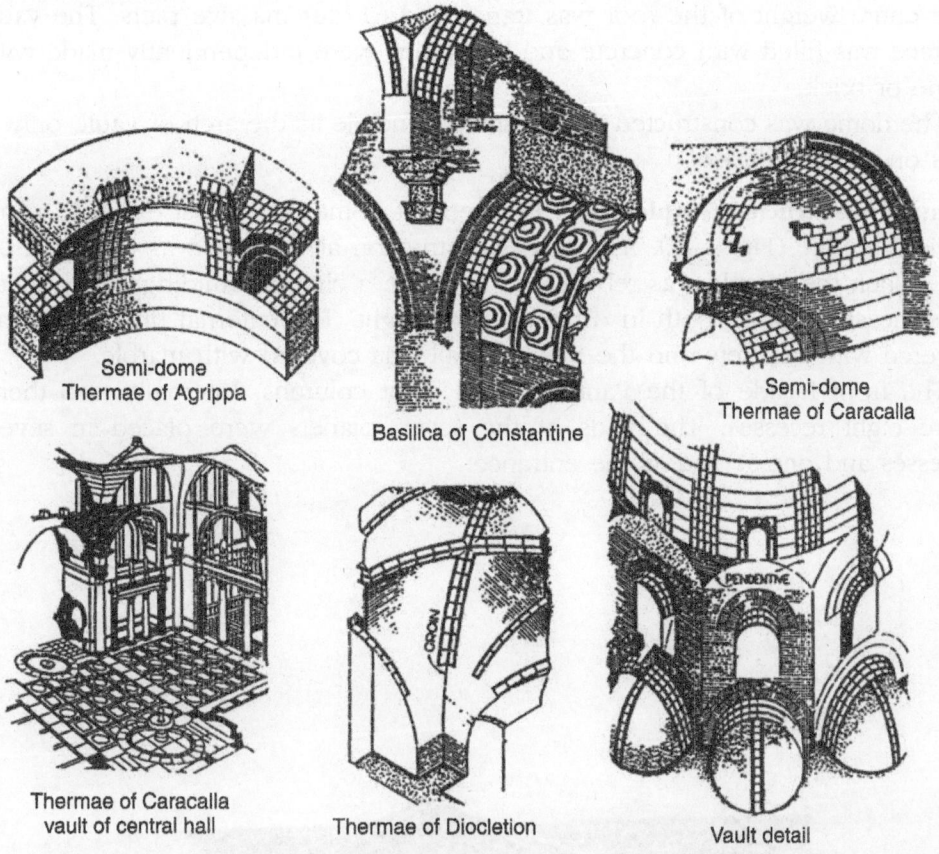

Semi-dome
Thermae of Agrippa

Basilica of Constantine

Semi-dome
Thermae of Caracalla

Thermae of Caracalla
vault of central hall

Thermae of Diocletion

Vault detail

Fig. 3.17 Construction of vaults and domes

The Roman engineering genius created the groined vault by the intersection of a barrel vault with a cross vault (Fig. 3.18).

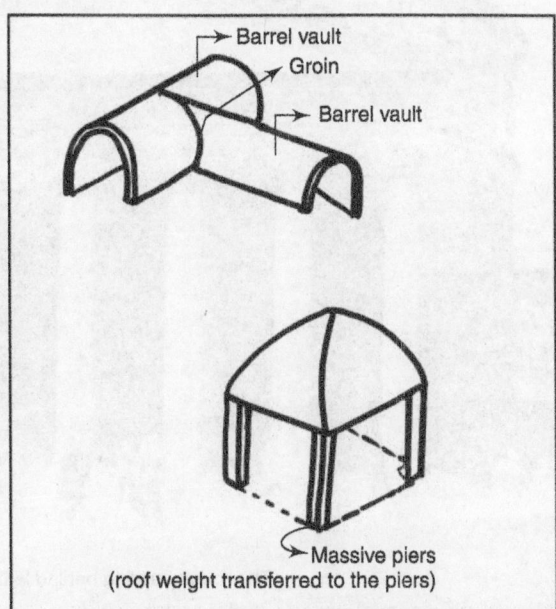

Fig. 3.18 Intersection of the barrel vault

The entire weight of the roof was transferred to four massive piers. The vault surface was filled with concrete and the groins were independently made with stone or brick.

The dome was constructed on the same principle as the arch or vault, only it was on a circular base.

Romans constructed temples The Pantheon at Rome is the best example of an ancient temple (Fig. 3.19). An earlier construction at this site was damaged by fire. When the temple was rebuilt, the Rotunda, a circular building, was added. This measured 43 m both in diameter and height. The external brick wall was covered with concrete and the internal wall was covered with marble.

The front façade of the Pantheon had eight columns. Internally also there were eight recesses. The gods of the seven planets were placed in seven recesses and one served as the entrance.

The round building behind is the Rotunda

Fig. 3.19 The Pantheon, Rome (Roman temple): portico in the front

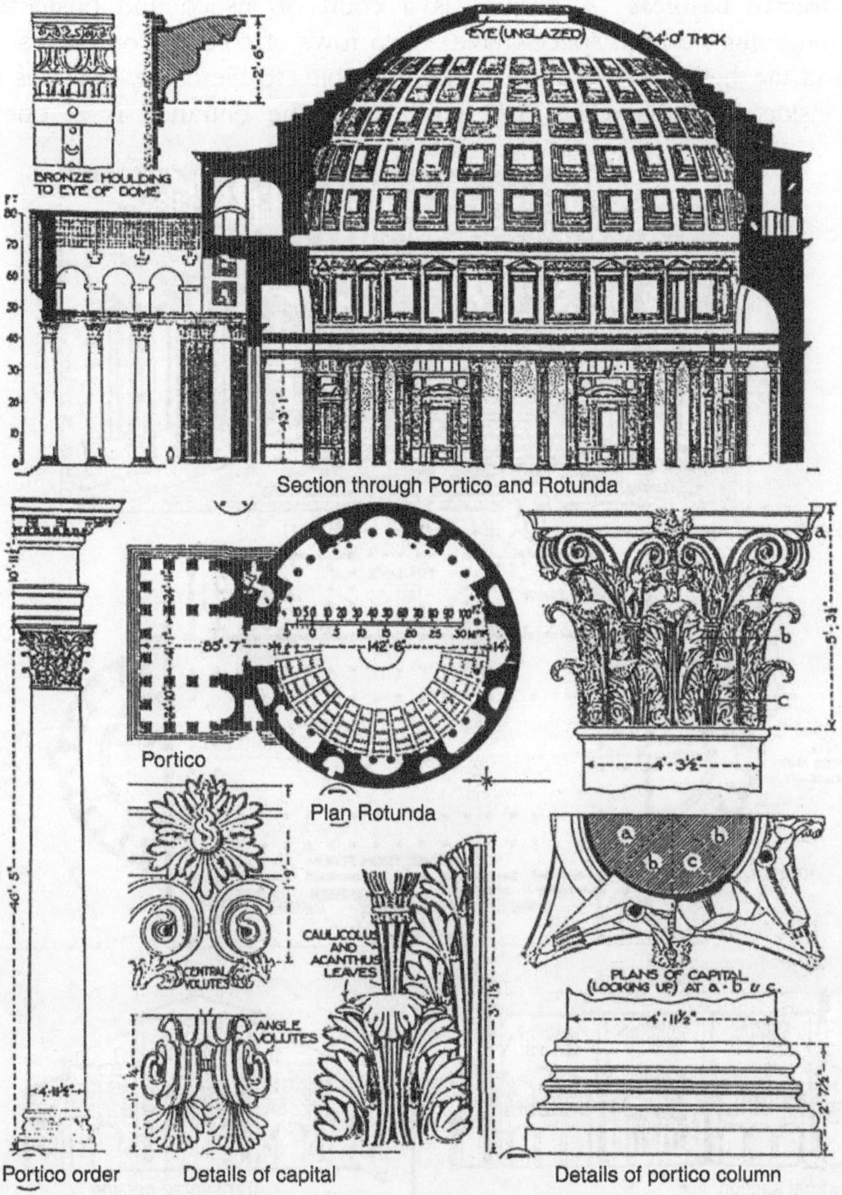

EYE (UNGLAZED)   **4'-0" THICK**

BRONZE MOULDING TO EYE OF DOME

Section through Portico and Rotunda

Portico

Plan Rotunda

CAULICOLUS AND ACANTHUS LEAVES

CENTRAL VOLUTES

ANGLE VOLUTES

PLANS OF CAPITAL (LOOKING UP) AT a · b · & c

Portico order   Details of capital   Details of portico column

Fig. 3.20 The Pantheon, Rome

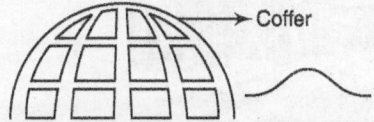
Coffer

Fig. 3.21 Section of a single coffer

The inner part of the dome was coffered (Figs 3.20 and 3.21). This coffer was provided not only for aesthetic reasons but also to reduce the weight of the dome. Appropriate lighting was also provided.

Romans constructed basilicas   A basilica is a court of justice and business exchange. It consists of a long, high, central space (nave). Two rows of columns on each side separate the aisle space, and the height of the nave is considerable, so clerestory space is created, which admits light inside the nave (Figs 3.22 and 3.23). The entrance is at one end and the

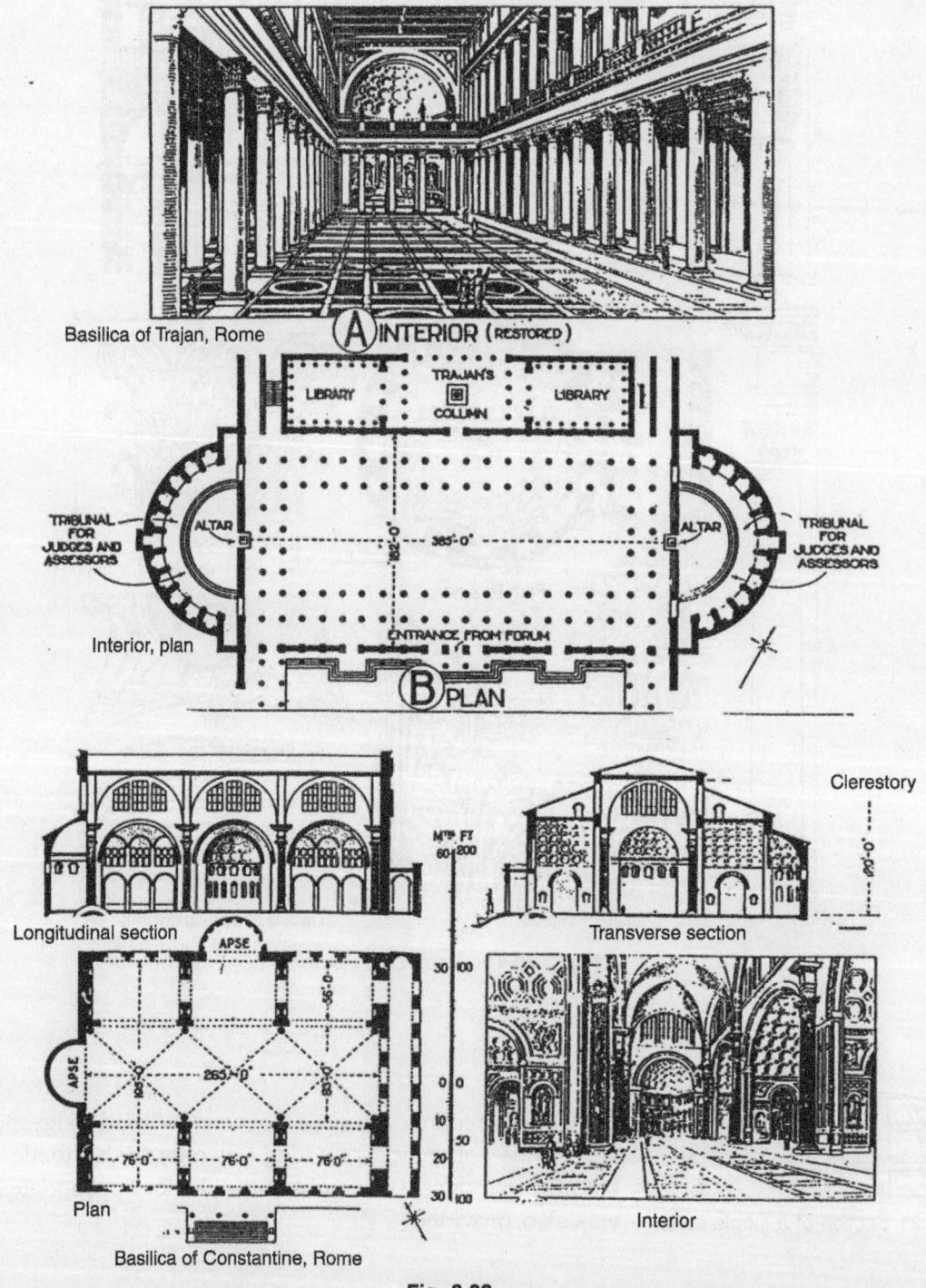

Basilica of Trajan, Rome

Interior, plan

Longitudinal section

Clerestory

Transverse section

Plan

Interior

Basilica of Constantine, Rome

Fig. 3.22

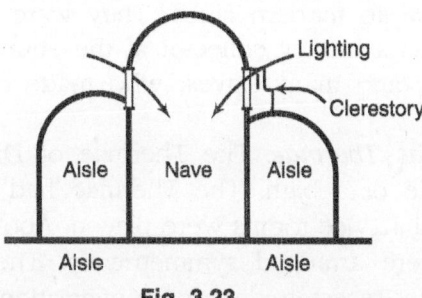

Fig. 3.23

semicircular recess or the apse lies at the opposite end, in which the seat for the magistrate is provided. An altar is provided for offering sacrifices before starting business.

Romans constructed baths   The Romans constructed palatial public baths. These not only served the purpose of bathing, but were also a place for socializing.

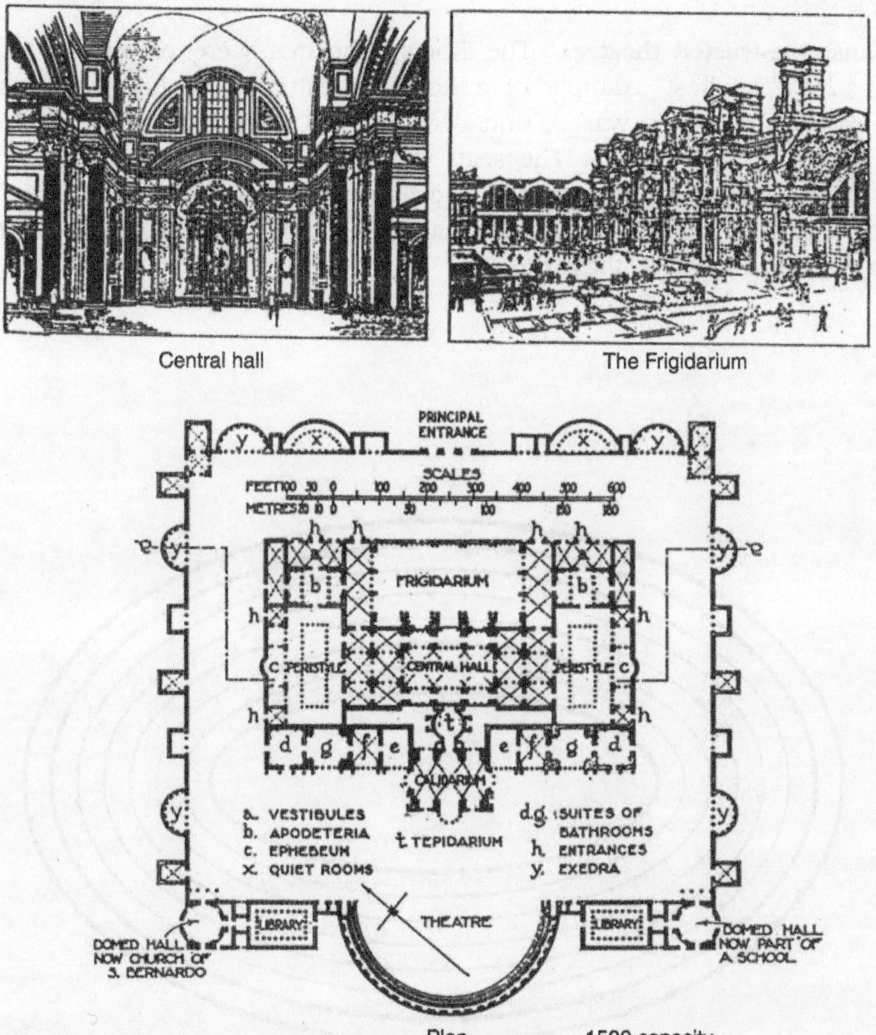

Central hall

The Frigidarium

PRINCIPAL ENTRANCE

SCALES

a. VESTIBULES
b. APODETERIA
c. EPHEBEUM
x. QUIET ROOMS

t TEPIDARIUM

d.g. SUITES OF BATHROOMS
h. ENTRANCES
y. EXEDRA

DOMED HALL NOW CHURCH OF S. BERNARDO

LIBRARY   THEATRE   LIBRARY

DOMED HALL NOW PART OF A SCHOOL

Plan   1500 capacity

Fig. 3.24 Thermae (bath) of diocletion, Rome

The baths were, in that sense, similar to modern clubs. They were used for lectures and athletic sports. There was a money collector at the entrance, and manicurists, barbers, bath attendants, and many slaves, who made bathing a luxurious experience.

The Romans used to call the baths *Thermae*. The Thermae of Diocletian, Rome (Fig. 3.24), is a good example of a bath. The Thermae had a raised platform below which the furnace and service rooms were placed. Above it was a large hall about which rooms were arranged symmetrically. There were lecture halls, libraries, shops, and quiet recesses ideal for conversation.

The main building had a warm hall (Tepidarium), a hot room (Calidarium), and a cold room (Frigidarium), which was a cool swimming pool. Surrounding these were private bathrooms and massaging rooms. The entire area was well landscaped.

Romans constructed theatres The Roman theatres were elliptically planned (Fig. 3.25). The best example of a Roman theatre is the Roman Colosseum (Fig. 3.26). The façade was colonnaded and had a capacity of 50,000 people with thrones for the rulers. The seats for citizens were separated by a huge wall. There were four storeys and each storey had a different order—Doric, Ionic, Corinthian. The fourth storey had Corinthian pilasters. Martial combats were displayed in these theatres.

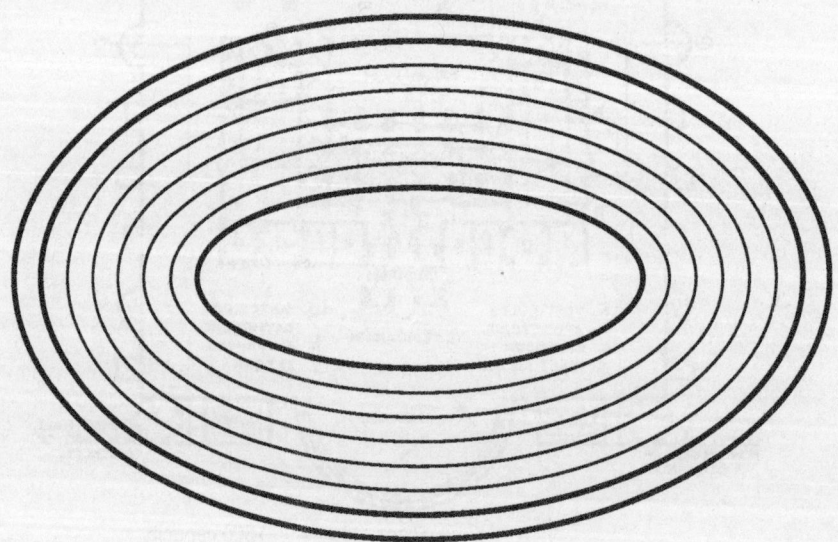

Fig. 3.25 Roman elliptical theatre

(a) External view

(b) View of the arena and auditorium

Fig. 3.26 The Colosseum, Rome

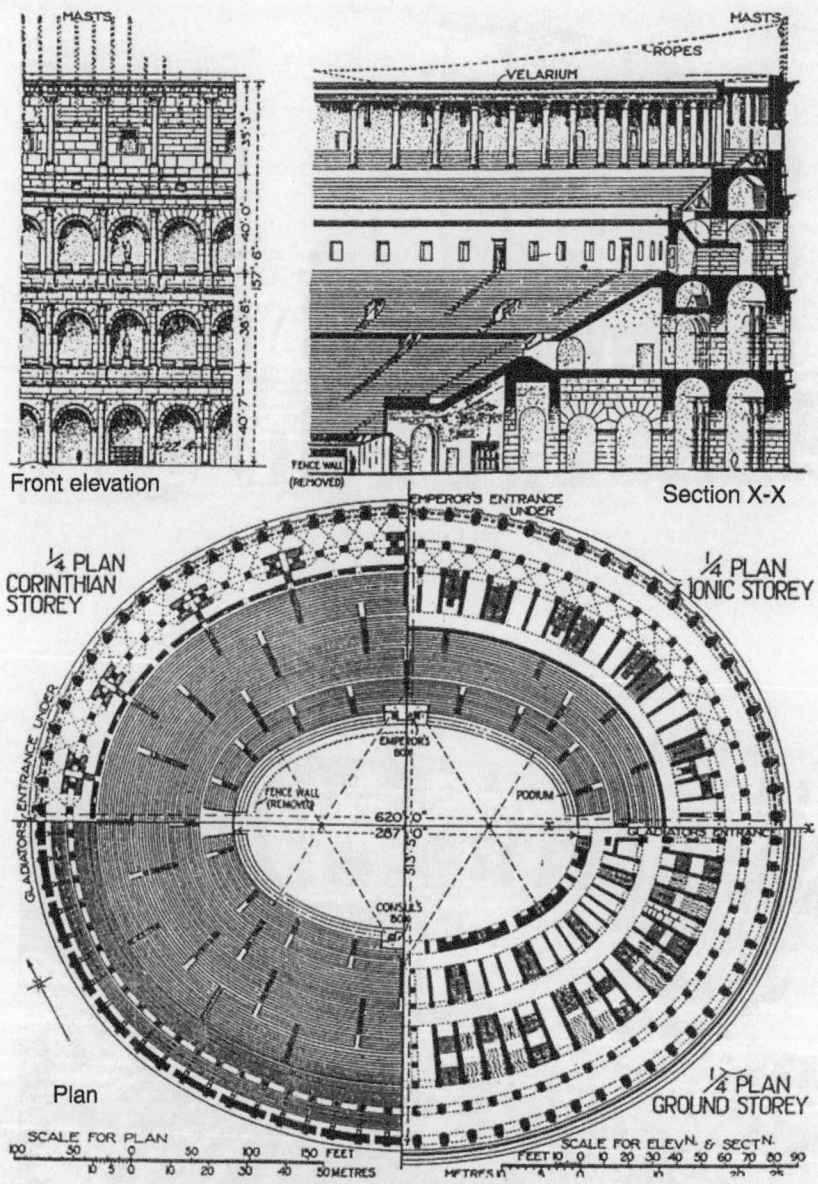

Fig. 3.27 Elliptical plan of the Colosseum, Rome

Romans constructed villas/houses The Romans constructed huge villas (bungalows) for the rich. They also made apartment blocks because of space constraints. Homes were made of brick-faced concrete. Continuous balconies were made in concrete or timber. Large windows surrounded the external façade and were also present internally around a landscape court. Window glass was rare, so hanging cloth was mostly used. Ground-floor flats were occupied by shops. The block of flats at Ostia (Fig. 3.28) is a good example.

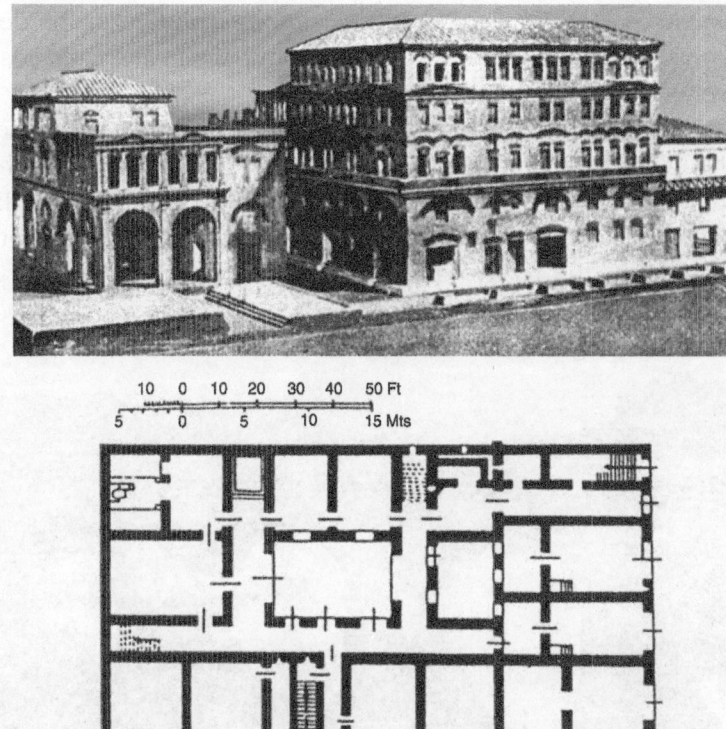

Fig. 3.28 Roman apartment house—block of flats at Ostia

Romans constructed aqueducts   The construction of aqueducts throughout the empire show the importance given to water supply. Large quantities of water were required for the Thermae and the public fountains. The supply of water required for domestic use was much less when compared to the other uses. About 350 million gallons of water were daily poured into Rome through 11 great aqueducts. The Romans followed the simple hydro-static law for the distribution of water through lead pipes. They were unable to cast iron pipes to withstand the great pressure of water. So they constructed aqueducts, and as labour was abundant, it was practical to build tiers of stone or concrete arches about 30 m high over valleys and low-lying areas (Fig. 3.29). These conveyed water from their sources (springs or rivers) to the reservoir, from which distribution began. The Acqua Claudia at Rome with arches over 30 m high is the finest example of an aqueduct. It brings water to Rome from a distance of 72 km.

Fig. 3.29 Roman aqueduct

Romans constructed bridges Bridges were simple, solid, easy to design, and practical to construct. They were designed to offer well-calculated resistance to the rush of water.

The early bridges were made of timber. The later and the finest bridges were made of stone. Very large spans were achieved where necessary. The arch of the Augustan Bridge near Aosta (Fig. 3.30) was 36 m across. In the Roman bridge Alcantara, the larger arches are 27.43 m wide. This romantic bridge with its gigantic arch spans the rocky valley of Tagus.

Fig. 3.30 Roman bridge of Augustus, Rimini

The medieval period (300 AD to 1300 AD)

The medieval period can be classified into
- Early Christian architecture
- Byzantine architecture
- Romanesque architecture
- Gothic architecture
- Renaissance architecture

Each of these eras made a special contribution to the development of the medieval period architecture. The birth of Christ and the spread of Christianity not only transformed the religious beliefs of the time but also influenced the formation of new cities and empires. The Roman empire under Constantine became Christian and its capital, Constantinople, became a great city.

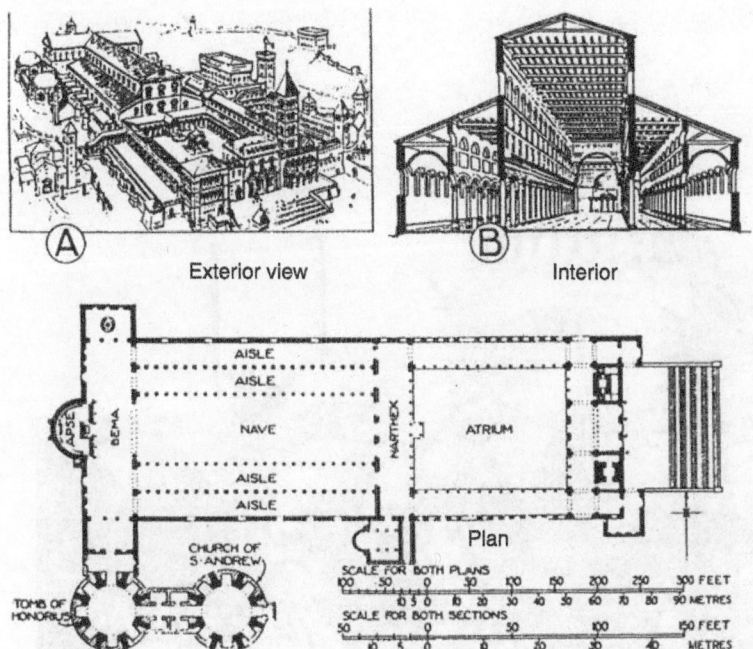

Exterior view   Interior

Plan

Fig. 3.31 Early Christian architecture: basilican church of St. Peter

Early Christian architecture
When Emperor Constantine declared Christianity as the state religion, the Roman temples, baths, and basilicas were all converted into churches (Fig. 3.31).

The basilica (Fig. 3.32) was converted into a church (Fig. 3.33) by modifying the aisles and converting the apse into the altar.

Two important churches were built, one over the basilica of St. Peter and the other over the basilica of St. Paul.

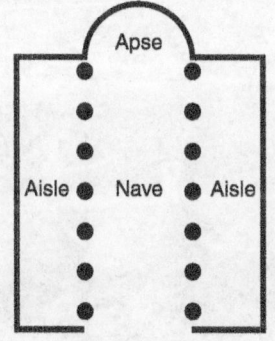

Fig. 3.32 Basilica

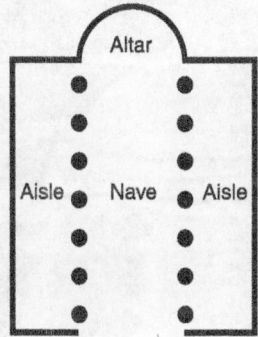

Fig. 3.33 Church

Byzantine architecture The most important contribution of the Byzantine builders was the dome. The problem of covering a square building was easily solved using a dome. This was best achieved in the great church of Sophia in Constantinople (Fig. 3.34). The church, built by Justinian in 532 AD, was both structurally and aesthetically satisfying.

Their next important contribution was the use of marble chips and bits of precious stones in mosaic-like designs. This technique was used to decorate interior walls and domes. They were made of glittering, bright colours. The dome, however, was still the dominant feature.

Fig. 3.34 Byzantine architecture: great church of Sophia, Constantinople; the minarets are a Turkish addition

Romanesque architecture Though the Church of Rome was very powerful and united the entire western world, the churches built all over the empire were dark, with small windows. Architecture seemed to have lost its glory.

In Italy there was a movement to revive and improve the design of churches. Following this, a baptistry and a campanile were added to the existing church plan.

Pisa Cathedral

(A) Pisan group

Plan

SCALE FOR PLAN

Interior

Bird's eye view from campanile

Transverse section

(F) LONGITUDINAL SECTION

Fig. 3.35 Romanesque architecture, Italy

The baptistry was a circular building covered with a conical dome for baptism. The campanile was a bell tower.

The best example was the Pisa Cathedral plan (see Figs 3.35 and 3.36), in which a baptistry and a campanile (the famous leaning tower of Pisa) were added to the existing Cathedral.

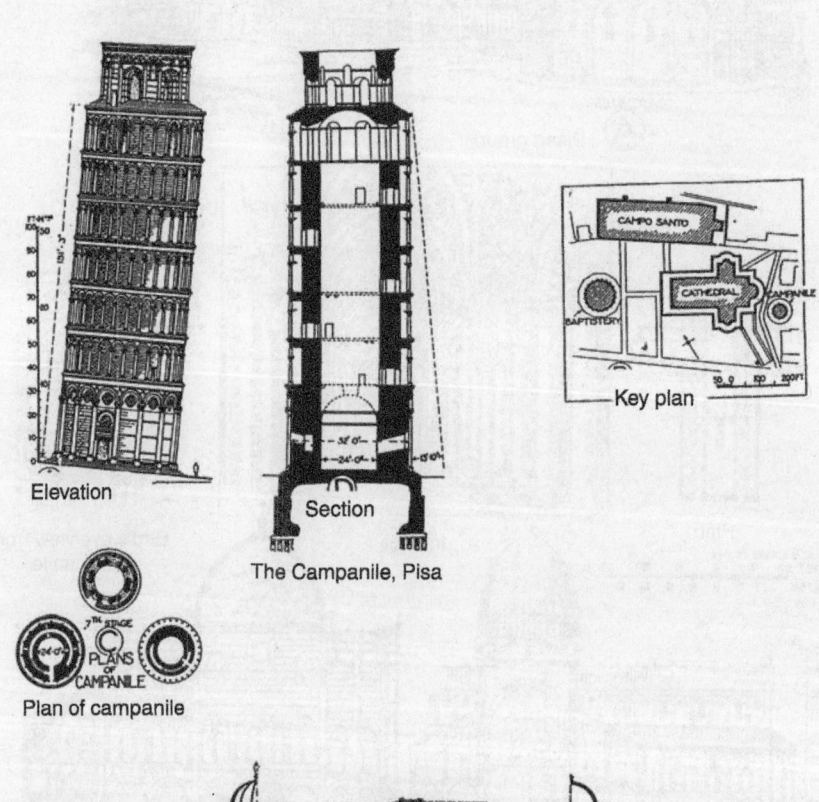

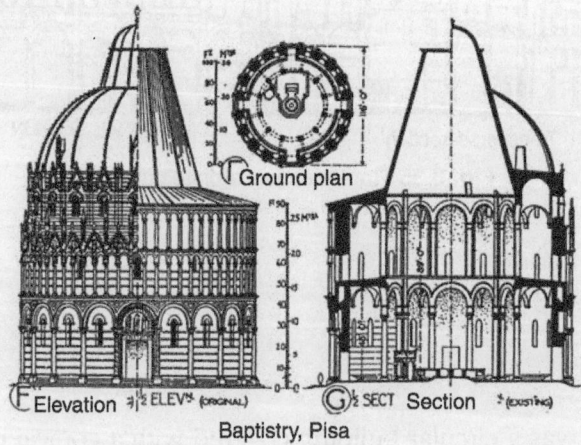

Fig. 3.36 Pisa Cathedral

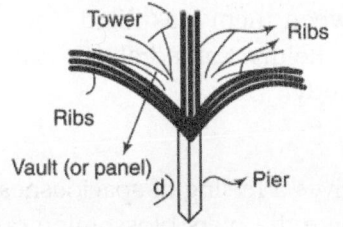

Fig. 3.37 Romanesque vault system

The next important development was the improvement of the vaulting system, in which ribs were directly supported on piers and constructed from one pier to another (Fig. 3.37). Cross ribs and diagonal ribs were also constructed (Fig. 3.38). The rib and panel construction was used. The arch framework was semicircular.

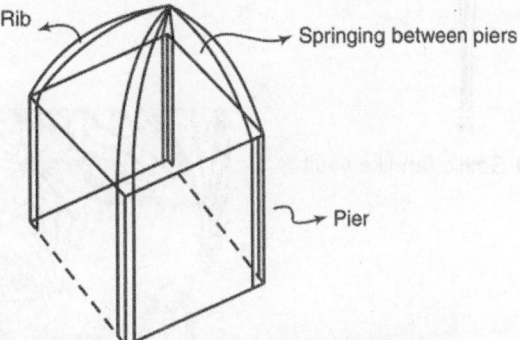

Fig. 3.38 Cross ribs and diagonal ribs

Gothic architecture   The Goths contributed many new architectural ideas. They developed the following elements.

(i) Pointed arch

The main feature of Gothic architecture is a vaulting framework of intersecting, pointed stone arch ribs, which support thin stone panels. The ribs were

Fig. 3.39 Gothic architecture: Reims Cathedral, France

constructed as a permanent 'formwork' and the space between them was filled with the pointed arch. The best known example is the Reims Cathedral in France (Fig. 3.39). Gothic vaulting makes for fascinating study.

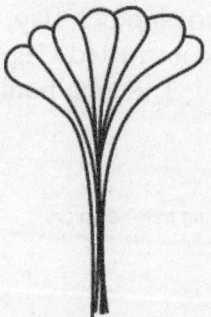

(ii) Fan-like vault

The interior of a Gothic cathedral gives a feeling of spaciousness and great height. One can appreciate the piers blossoming out into a fan-like vault (Fig. 3.40). The structural elements acted as ornaments (Fig. 3.41).

Fig. 3.40 Gothic fan-like vault

Fig. 3.41 Gothic architecture: vaulted staircase of Christ church, Oxford

(iii) Flying buttress

The evolution of the exterior is an outcome of the interior, and also the method of construction. In the vaulting system used, the pressure due to the vault and the arch was transferred to the ground by buttresses, which were called *flying buttresses*, thrusted on top by a pinnacle. Flying buttresses were also used as structural elements. Notre Dame, Paris, is a good example (Fig. 3.42).

Gothic architecture was an intricate system of construction, which was practiced by generations of experts consisting of masons, artists, and supervisors, who were both engineers and architects. They were the *master builders*.

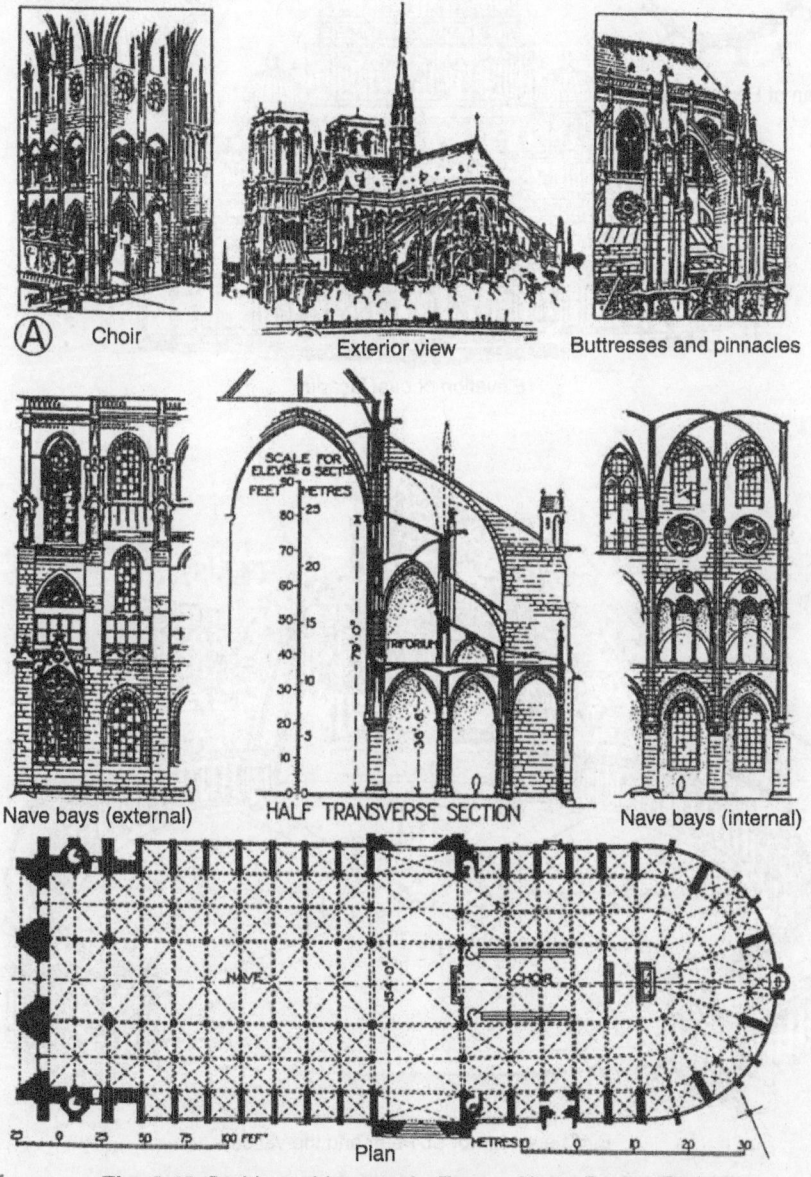

Choir

Exterior view

Buttresses and pinnacles

Nave bays (external)

HALF TRANSVERSE SECTION

Nave bays (internal)

Plan

Fig. 3.42 Gothic architecture in France Notre Dame, Paris

Renaissance architecture In the Renaissance period, individual master builders such as Brunelleschi, Bramante, and Michelangelo rose to the level of architects and later landscape architects and town planners; Brunelleschi invented perspective. The Renaissance movement spread through Italy, France, and

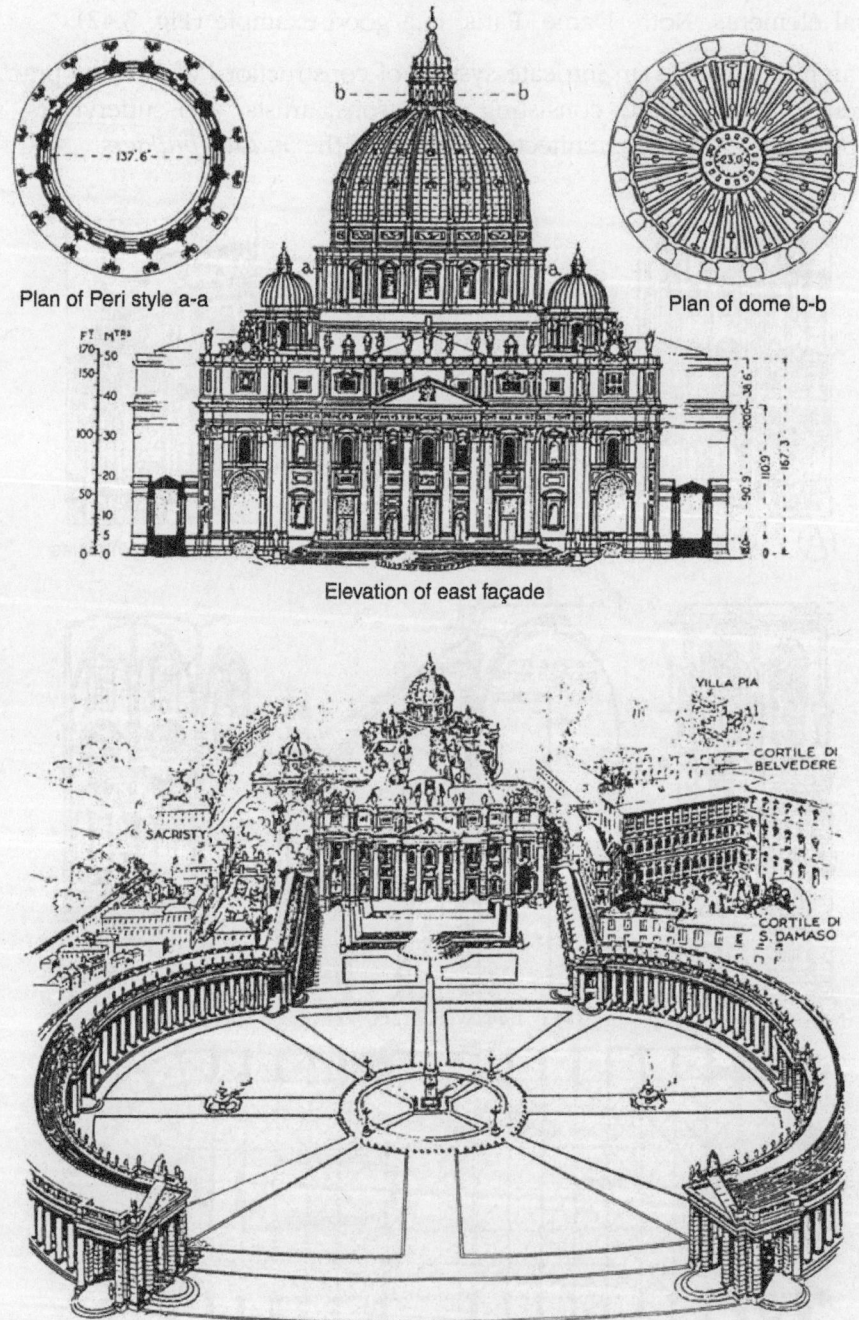

Plan of Peri style a-a

Plan of dome b-b

Elevation of east façade

Bird's eye view of St. Peter and the Vatican

Fig. 3.43 Renaissance architecture: St. Peter's Church, Rome

England. Public works such as water supplies, drainage systems, and machines for defence purposes were all developed by these builders.

St. Peter's Church in Rome (Fig. 3.43) is a famous example of Renaissance architecture in Italy. It was of vast proportions—an external length of 212 m and an internal length of 182 m. The internal diameter of the dome was 48 m. The perfect proportion of the dome was an important feature of the construction.

St. Paul's Cathedral, constructed in England, is a good example of English Renaissance architecture. One of the salient characteristics of Renaissance architecture was the application of the classic Roman orders.

3.1.4 Realization of character and style in modern architecture

Modern architecture can be classified under the following three periods.
 (a) The industrial era
 (b) Twentieth century architecture
 (c) The Art Nouveau movement

The industrial era

The industrial revolution took place both in Europe and America to overcome the problems of shortage and suitability of housing. The new towns that grew during the industrial era were trade and commerce centres. Architecture was no longer dominated by religious buildings. In Europe, housing, to meet the demands of an industrial society, became the most important architectural, structural, and planning problem. Later, mass housing developments came into being, because of which community facilities for education, health care, commerce, and recreation gained importance over palaces, churches, and cathedrals. Architecture became a profession in which buildings were constructed with great speed to meet the functional requirements of a growing family. In England, row or terrace housing and flats were built. Salt box housing became very popular. In this plan, the houses were square or rectangular with two, four, or six rooms and a main central chimney. Architecturally these buildings had low, hipped roofs.

Twentieth century architecture

Contemporary architecture was influenced by two important aspects:
 (a) The use of three building materials—steel, concrete, and glass
 (b) A series of architectural philosophies or thoughts

The availability of contemporary building materials led to many new constructions such as railway stations, bridges and aqueducts, stock exchanges, banks, and industrial buildings such as mills, warehouses, recreational clubs, and departmental stores. As cities developed with more and more facilities, people began to migrate to them in great numbers. The resulting housing shortages led to the development of multi-storeyed buildings. Building activity grew very fast. There was demand for an enormous amount of quick construction. Also, the buildings had unusual sizes in terms of height, strength, and complexity. This needed professionals or technologists and, thus, the 'civil engineer' came into being.

With buildings becoming more complex, greater accuracy and more calculations were required, for which 'quantity surveyors' came into existence. Then came the 'general contractors', who had to carry out speedy and quick construction as competition also grew. With all these developments, the architect's work was limited to purely consultative and creative work. Institutions, associations, and schools of architecture were established to train students in the art of building design.

The Art Nouveau movement of Europe (1910)

This movement gave importance to the ornamental values of buildings. The architectural philosophy of this period was to integrate art with social life.

The Chicago school of thought in Mid-East America (1900) This school of thought emphasized the need for a metal frame as the basic structural system and the expression of details on the exterior of a building in a modern way.

Bauhaus school of thought in Germany (1920) This school of thought believed that machines should be made subservient (subordinate) to the creative designer. The hand tool of the craftsman was valued greatly. It was also believed that buildings should express themselves.

3.2 Aesthetic Impact

Aesthetic impact concerns beauty or things of beauty and the theory or theories of beauty. It is a science which deals with visual composition. The overall composition of a building is affected by various aesthetic aspects such as harmony, contrast, dominance, punctuation, and climax. These affect the overall building form and space.

3.2.1 Harmony

Harmony can be defined as the pleasing arrangement of the various parts or a combination of parts to make a good composition. The principle of harmony is to carefully select elements that share common traits (characteristics) such as shape, colour, texture, or material, or even variety in a single characteristic.

Sharing a common trait or characteristic

Elements can share common characteristics (Fig. 3.44) such as a common
- size
- shape
or similar
- orientation

- colours and values
- materials
- detail characteristics

When harmony is over-emphasized in buildings it will result in a unified but uninteresting composition. Variety, when used too much to create interest, may result in visual chaos or confusion. The careful way to create interest is by creating a proper balance between order and disorder.

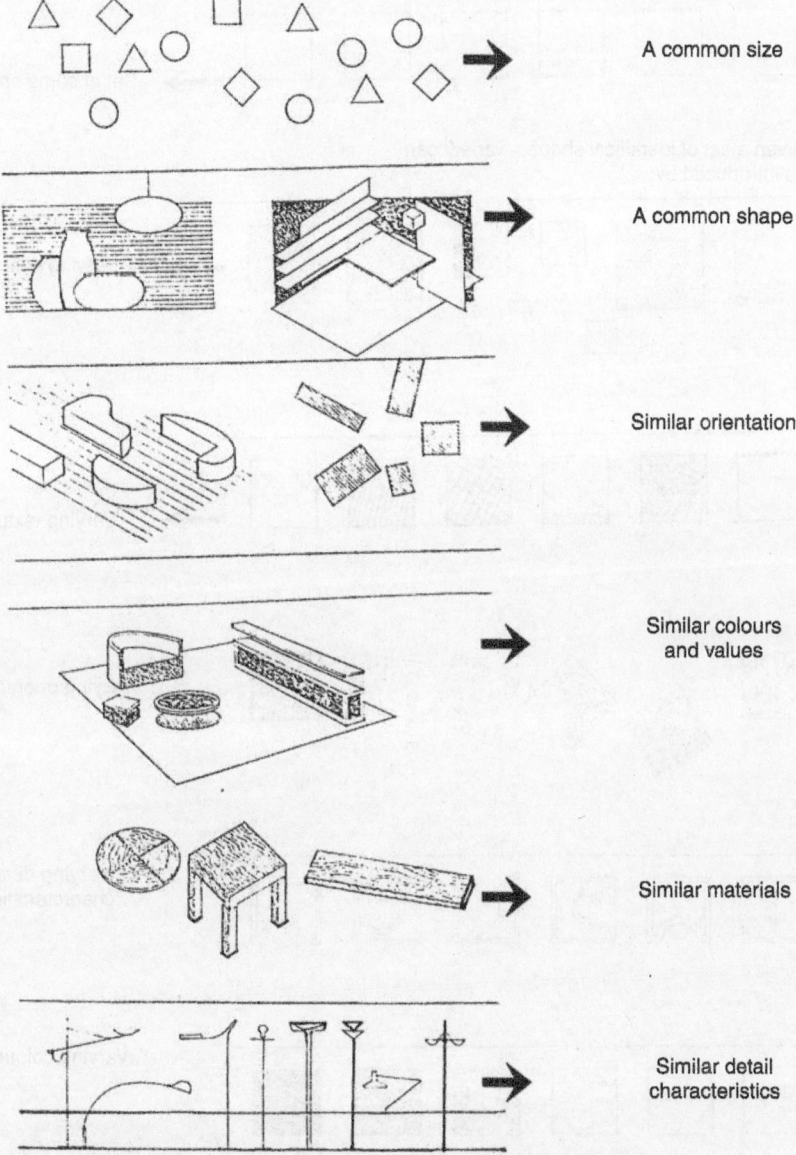

A common size

A common shape

Similar orientation

Similar colours and values

Similar materials

Similar detail characteristics

Fig. 3.44 Sharing a common trait for characteristics

Introducing variety by various means

Given a set of identical shapes, variety can be introduced (Fig. 3.45) by varying the

- size
- texture
- orientation
- detail characteristics
- colour

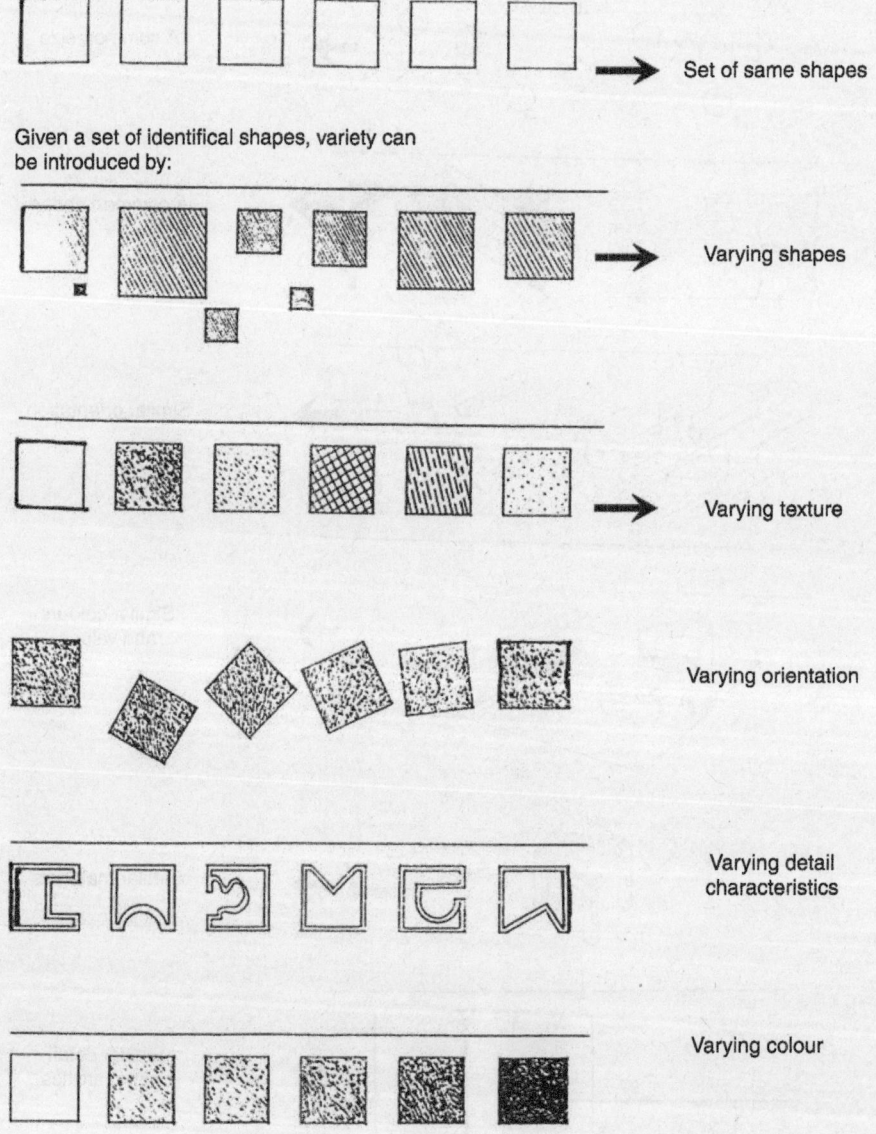

Fig. 3.45 Introducing variety in a set of identical shapes

Harmony through colour Harmony can be brought out by colour. One way is to use related colours, i.e., colours that appear together on the colour wheel or colours having the same hue. For example, brown, gold, and yellow are related colours.

Harmony through texture Harmony can be created using texture by matching the textural quality, say, by carving and ribbing.

Harmony in tones Tones can be harmonious only in the sense of having the same tonal values. For example, harmony in tonal values can be brought out in a single building constructed with stone of one colour, but two different textures, one with a smooth finish and another with a rough finish. When two materials such as brick work and concrete are used together, there exists variety in colour, texture, and tone.

Harmony through proportion Proportion has been discussed elaborately in the previous chapter. It can be well appreciated in the Ducal Palace, Urbino, Italy (Fig. 3.46). This is a classical ideal building in which each part is related to every other part and to the whole through related geometric proportion.

Fig. 3.46 Harmony through proportion: Ducal Palace, Urbino, Italy

Harmony in direction Harmony in direction literally means using the same directions. The direction of the total building shape must relate to the directions of the elements within the building. An equal effect of direction will impart visual balance (Fig. 3.47).

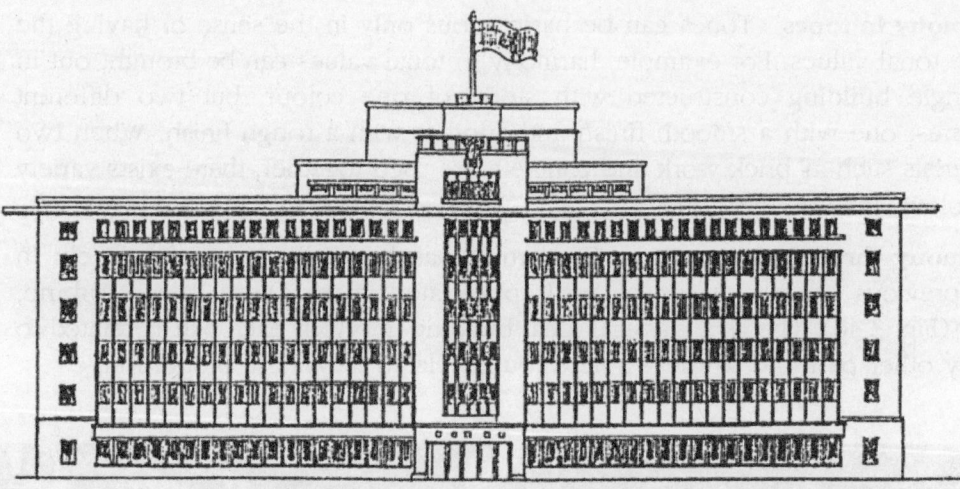

(a) Harmony in horizontality: Sachivalaya, Mumbai

| Empire State building, | Chicago Tribune building, |
| New York | Chicago |

(b) Harmony in vertically

Fig. 3.47 Harmony in direction

Principles of harmony

Variety and interest should be brought out in a composition to create harmony and unity. Asymmetrical arrangements can incorporate a variety of shapes, colours, and textures into their layouts (Fig. 3.48).

Fig. 3.48 Asymmetrical arrangement of a variety of shapes, colours, and textures

In any pattern of arrangement of various elements, dissimilar characteristics may be present. Dissimilar objects can be arranged in an asymmetrical way (Fig. 3.49).

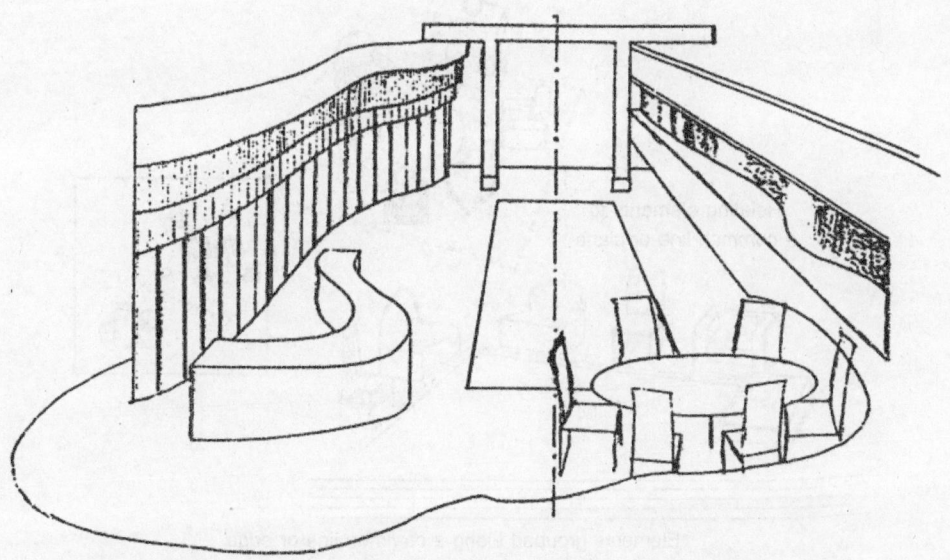

Fig. 3.49 Dissimilar objects can be arranged in an asymmetrical way

Harmony can be produced by elements that have a common characteristic. It can also be produced with similar elements having a variety of individual or unique characteristics. Figure 3.50 is an example of how variety can be introduced among a group of objects with similar dimensions by varying contours and other details.

Fig. 3.50 Harmony in dimensions but variety in contour and other details

Another method of arranging dissimilar elements is to simply group or arrange them in close proximity to one another or relate them to a common line, plane, or edge (Fig. 3.51). The plane could be an overhead plane or just a vertical plane, which could be a wall acting as backdrop. Other elements placed far away do not form elements of the group (Fig. 3.52).

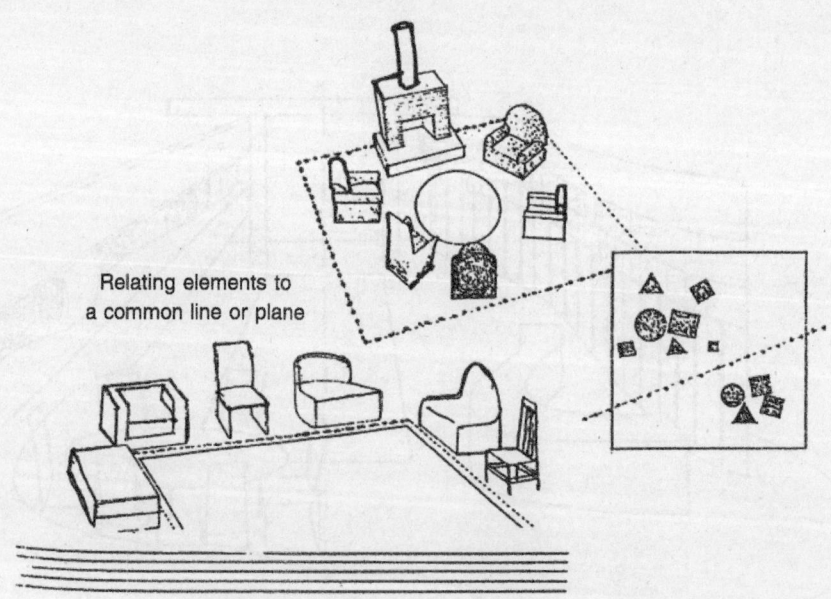

Relating elements to a common line or plane

Elements grouped along a common line or edge

Fig. 3.51 Dissimilar elements arranged by grouping

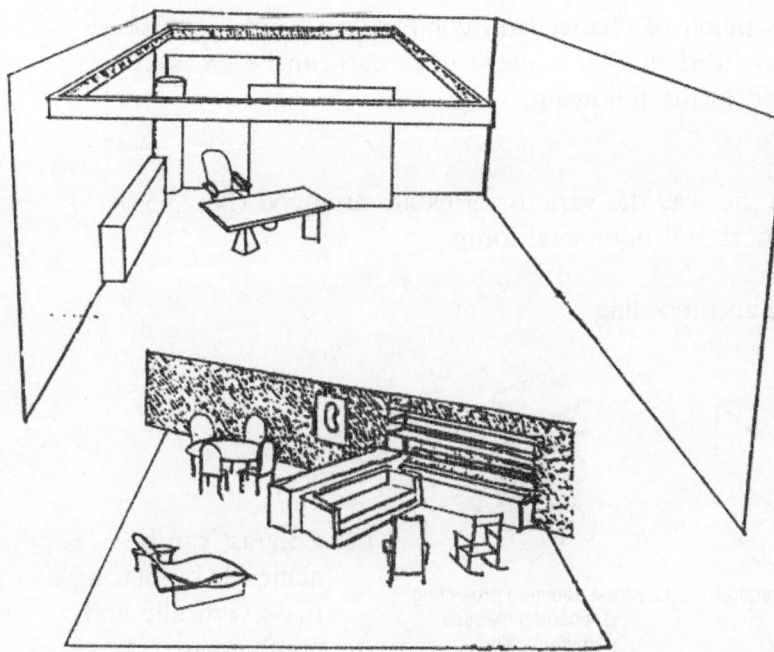

Fig. 3.52 Grouping related elements with respect to an overhead plane

One more method of creating harmony in the composition is by using a continuous line.

3.2.2 Contrast

The human eye functions in such a way that it can perceive or understand more than one difference between contrasting objects at the same time.

Visual monotony is the opposite of contrast. For example, if we consider a street elevation, there are buildings with different forms, heights, and purposes (Fig. 3.53). The eye is able to see and understand all these differences simultaneously.

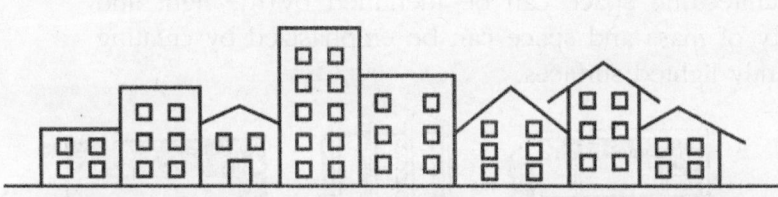

Fig. 3.53 Street elevation with different buildings of different forms

Our eyes are used to such differences and changes. Without these contrasts, visual tiredness and lack of concentration will result (Fig. 3.54). When human beings are placed in an environment that is static and lacks variety, they experience feelings of monotony and depression.

Fig. 3.54 Street elevation with static, visual monotony causes tiredness and lack of concentration

On the other hand, too much of change and contrast can cause confusion and an impression of lack of orderliness. Contrast in design can be created by various means, as explained in the following.

Contrast by mass

There can be a contrast in the way the various forms are arranged (Fig. 3.55):

- Contrast between vertical and horizontal forms
- Contrast of direction
- Contrast of projecting and receding

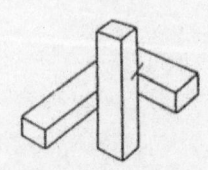

Contrast between vertical and horizontal forms

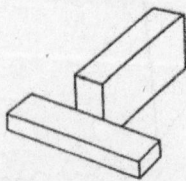

Contrast in direction

Contrast between projecting and receding masses rhythmically

Fig. 3.55 Contrast by mass

(a) Contrast can be achieved by placing a mass vertically and another one horizontally.

(b) Contrast in direction can be achieved between an upright slab and a low horizontal mass, both of which are placed at right angles to each other.

(c) Contrast can be achieved between projecting and receding masses which are rhythmically superimposed over each other.

Contrast by light and shade

Any object within an architectural space can be identified by the light and shadow effect. The quality of mass and space can be emphasized by creating deep shadows and brilliantly lighted surfaces.

Every opening, projection, and recess creates a shadow (Fig. 3.56). The presence of a shadow makes the adjacent area look brighter in contrast. By changing the shadows, different surface effects can be created. The effect of light and shade can occur only when there is excessive natural light falling on the particular area. This light and shade effect

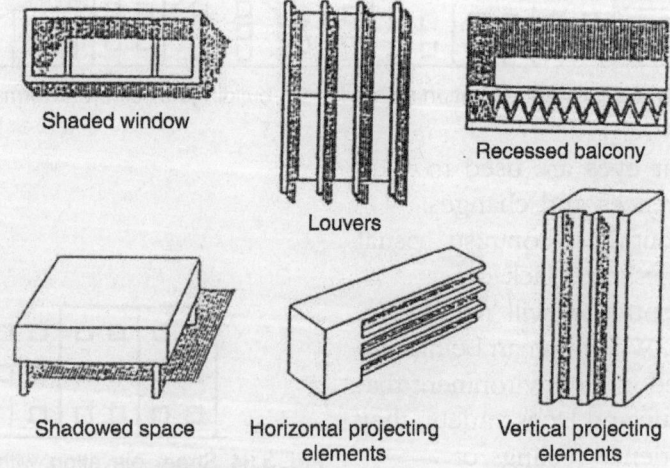

Shaded window

Louvers

Recessed balcony

Shadowed space

Horizontal projecting elements

Vertical projecting elements

Fig. 3.56 Contrast in light and shadow

depends on the climate, as, e.g., during the rainy season, natural light will be dim and dull.

Effect of light and shade on interiors   In interior areas, strong and contrasting light affects concentration. On the other hand, a small amount of contrast keeps one alert. Dull and diffused light causes monotony. Bright light, like bright colours, creates a festive atmosphere. Various moods can be created with light and the architect must use this effect of light appropriately and to his advantage. The following text provides examples of buildings in which various aspects of contrast are depicted.

Contrast in direction, texture, and mass   These aspects can be well appreciated in Kauffmann House, *Falling Water*, Pennsylvania, USA. The building was designed by Frank Lloyd Wright. This design is famous because of its imaginative use of materials and the great sense of vitality it conveys. The bold, floating, horizontal cantilevers in the balconies contrast with the main vertical mass both

in direction and texture (Fig. 3.57). The whole fits into and yet contrasts with the beautiful natural setting. It is probably the delightful use of the stone, trees, and water which gives this structure such a dramatic quality. There are also a number of minor rhythms and contrasts in the fenestration (arrangement of windows), steps, and structural forms, which adds to the composition and makes it interesting.

Fig. 3.57 Contrast in direction, texture, and structural forms: Kauffmann house, 'falling water' built by Frank Lloyd Wright

Contrast in shape, tone, height   The best example of this kind of contrast is the Russian Church, in which contrast is seen in shapes, between arches, and in pinnacles (Fig. 3.58). Contrast in tone, height, and detail characteristics is seen in the middle pinnacle, which is taller than the rest. The multiple repetition of the lower arch forms the building up to a crescendo in the five domed pinnacles. The middle one, being slightly higher, richer, and crested, is an example of repetition of shapes contrasting with another group of shapes, using distinctive and unifying forms with a strong contrast of tone.

Fig. 3.58 Contrast in shape, tone, and height: Russian Church

The importance of the composition is enhanced by the rich gilding of the domes and the decoration and piercing of the visual elements.

Contrast in form, tone, and colour   The best example of this is the Chateau de Chambord in France (Fig. 3.59). The fenestration of the three main floors gives

Fig. 3.59 Contrast in form, tone, and colour: Chateau de Chambord, France

a strong horizontal emphasis reflecting the overall form of the building. Above these floors, the roof provides contrasts in form, tone, and colour. The repeating, domed roof pinnacles, chimney groups, and dormer-type windows give an impression of a busy cluster of visual elements arising from relatively simple, functional, or structural elements.

3.2.3 Dominance

Dominance may be provided by one colour, tone, or texture being visually stronger than the remaining. That is, one effect should be more dominant than the other. Dominance is an aspect of unity. It could be established with respect to

(a) direction
(b) relationship between solid and void

Dominance in direction

If the horizontal mass is collectively much stronger than the vertical, then it means the dominance of horizontality. It can also be the other way around, where the vertical is much stronger than the horizontal. This means the dominance of verticality.

Dominance of horizontality This concept is well explained in the structure of City Theatre, Helsinki, Finland (Fig. 3.60). Strong dominance of horizontality is reflected by the roof line and overhang, with contrast from the columns. The directional emphasis is reinforced by the use of strongly ribbed cladding tile, which can be seen to run horizontally on the eaves soffit, on the beams linking the column heads, vertically on the columns, and on the solid parts of the wall behind it.

Fig. 3.60 Dominance of horizontality: The City Theatre, Helsinki, Finland

Fig. 3.61 Dominance of verticality: flats of Bremen, Germany

Dominance of verticality This concept is well understood from the structure of the flats of Bremen, designed by Alto (Fig. 3.61). The building as such has a dominant vertical composition. Contrast is provided by the horizontal emphasis of the fenestration.

Dominance in the relationship between solid and void

When solid and void are used in a building, there should be a dominance of solid over void or void over solid to give a composition which has unity. Unity cannot exist with visually equal elements.

The relationship between solid and void can be seen in the structure of the Cemetery Chapel in Turku, Finland (Fig. 3.62). The dominance of solid material contributes to the sense of unity. The porch and the opening above give some limited contrast.

The difficulty is that, in practice, when we try to overcome one visual weakness, we end up producing another weakness. Dominance is only one aspect of unity, our visual objective will be fulfilled only when we give due regard to other principles.

Fig. 3.62 Dominance of solid over void: Cemetery Chapel, Finland

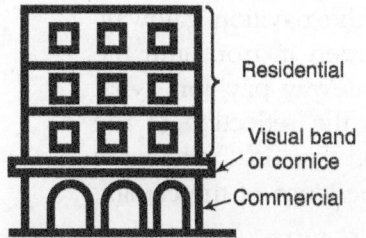

Fig. 3.63 Louis Sullivan building: punctuation created by a band or cornice

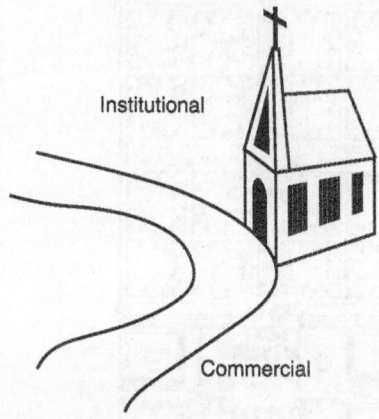

Fig. 3.64 Punctuation in land use pattern created by a visually dominant element

3.2.4 Punctuation

In a sentence, we use punctuation marks to indicate pauses or changes in tone. Punctuation in architecture is a pause in architectural visual continuity and refers to changes in function and pattern, acknowledged by some physical signal. For example, in the Louis Sullivan building (Fig. 3.63), the ground level of commercial units is separated from the upper multi-storey apartments by a band or a cornice. Similarly, a change in the land use pattern can also be indicated by using an element that is visually dominant and prominently separates, for instance, a commercial building and a religious institution. In Fig. 3.64, a change in the width of the street brings out the change of land use pattern. A church is used as a 'punctuating' element to bring into prominence, the pause—and change—in the land use pattern.

3.2.5 Climax

Climax is a visual phenomenon, in which interest is slowly built up, and as one slowly moves towards the point of interest, the actual form or the total structure is seen or perceived. For example, it could be the portion of the church spire seen from far, or a portion of the Gopuram of a temple or any structure that creates interest.

From a European market square one gets a view of only a portion of the church. As one moves towards the church (form or building), there is a feeling of surprise that is created. *This surprise that is viewed finally is the climax* (Fig. 3.65).

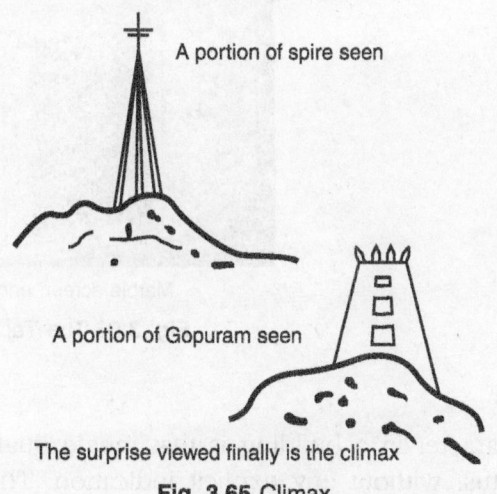

A portion of spire seen

A portion of Gopuram seen

The surprise viewed finally is the climax

Fig. 3.65 Climax

Another example is Taj Mahal at Agra. As one enters the pavilion gateway, after a flight of steps, one gets a framed view of the garden in front and the feeling of earth or ground. As one ascends the second gateway pavilion, what one sees is the water body and the landscape and then the reflection of the grand structure on the water. Then, as one looks up, the final climax is viewed—the marble tomb of the Taj Mahal. This is the experience of the climax.

General view

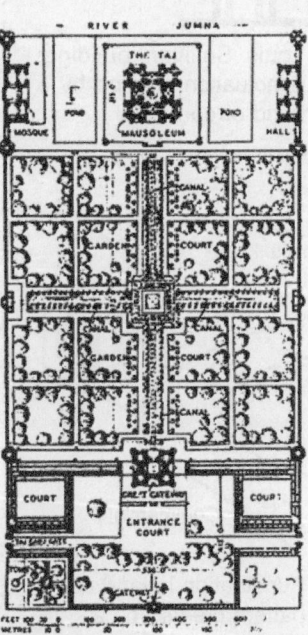

Plan showing entrance court gateway, garden court, and mausolem

Marble screen enclosing tombs

Fig. 3.66 The Taj Mahal, Agra

Summary

Character in a building is that innate quality which expresses its function and status, without any explicit indication. The nature of character in the field of architecture may be examined under the categories of functional character, associated character, and personal character.

Style is a quality that provides distinctiveness to the artistic expression in a building as, for instance, in the Gothic style of arches, the Gupta style, and the Buddhist style.

The history of traditional and modern architecture yields a wealth of information to help us understand how style and character are expressed in buildings. For ease of study, traditional architecture can be categorized as the architectural styles prevalent in the Greek period, Roman period, and medieval period. The medieval era can be further classified into early Christian, Byzantine, Romanesque, Gothic, and Renaissance architecture. Similarly, modern architecture can be studied under the styles of three periods—the industrial era, twentieth century architecture, and the Art Nouveau movement.

Harmony, contrast, dominance, punctuation, and climax are aesthetic elements that contribute to the style and character of a building. Harmony is achieved through the use of colour, texture, tones, etc. Contrast is depicted through texture, mass, tone, height, form, colour, etc. Dominance is expressed by one colour, tone, or texture being visually stronger than the rest. Punctuation is a physical signal, a pause in the visual continuity of a landscape, for instance, in order to express a change in pattern or function. Finally, climax is a visual phenomenon in which a viewer's interest is gradually built up to the point where the actual form of the building is seen. The climax is the feeling of surprise or delight experienced at finally seeing the building.

A basic knowledge of these principles of design and aesthetics is essential for us to understand the various factors to be considered when designing a building.

REVIEW QUESTIONS

Part A (2 marks each)

1. What do you understand by character and style in buildings?
2. What do you understand by traditional architecture?
3. What do you understand by modern architecture?
4. What were the contributions of Greeks towards optical correction?
5. The present-day theatres (auditorium) owe their structures to the Greeks. Explain. (Write about the fan shaped theatre.)
6. Explain the construction of vaults by Romans.
7. Compare the theatres of Greek and Rome.
8. Define Harmony.
9. What do you understand by contrast?
10. What is dominance?
11. Define punctuation and climax.

Part B (16 marks each)

1. Explain in detail the meaning of character and style in buildings and their realization in traditional and modern architecture.
2. Write notes on harmony, dominance, contrast, punctuation, and climax.

Factors Influencing Architectural Design

4.1 Meaning of Architectural Design

Architectural design means the design of any space; e.g., a kitchen, a hospital, a workshop, etc. It is the process by which an object which is to be physically constructed later is first visualized as an idea. This idea is also known as a concept and can exist in two different forms.

- Mental idea or concept
- Representational idea

The mental idea helps to

- identify the purpose of the object (building)
- analyse the aesthetic and functional aspects
- think of a good solution (design)

The representational idea is the reproduction of the concept in the form of sketches or diagrams or miniature models (Fig. 4.1).

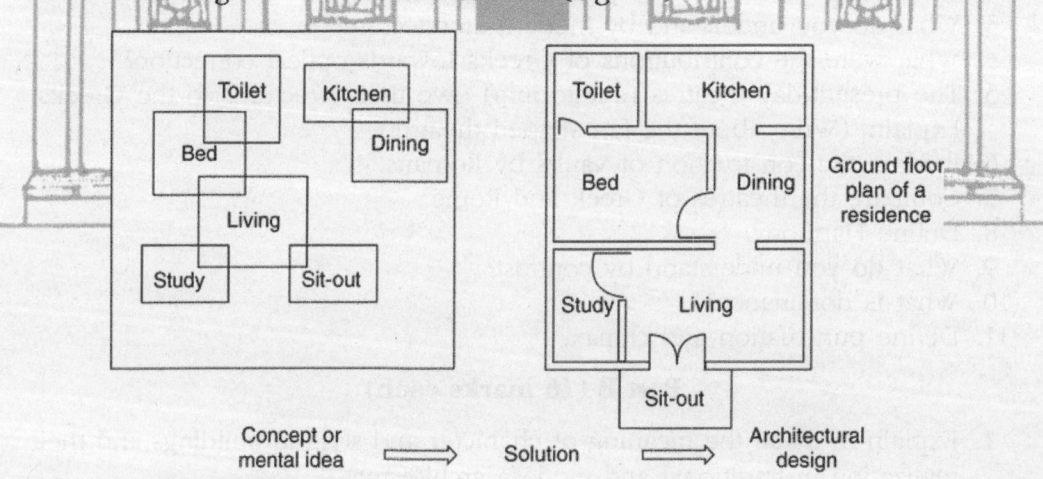

Concept or mental idea $\implies$ Solution $\implies$ Architectural design

Concept, bubble diagram

Fig. 4.1 Representational idea

The representational idea is not exactly the same as the mental idea or concept, as it may not be practically possible to reproduce all the mental ideas.

The object or the actual building to be created in architectural design is architectural space defined or enclosed by a structure that should satisfy both aesthetics and function. That is, it is the actual building to be designed. Thus, a *building* is defined as an object which integrates space and structure aesthetically and functionally.

Object in architectural design = architectural space + structure

Building = architectural space + structure

4.1.1 Relationship between aesthetics and function

While considering an architectural task (job or design) certain aspects of the job are influenced more by technical considerations and others by aesthetic factors. The proportion of technical (functional) considerations and aesthetic factors that influence a class of building varies or changes. This is clear from Fig. 4.2.

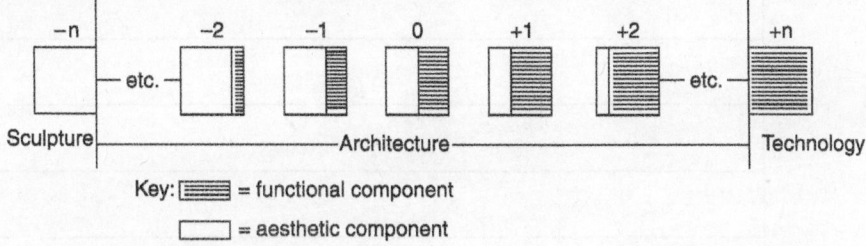

Fig. 4.2 Buildings classified according to the ratio of aesthetics to function

In the figure, each square represents a class of buildings (hospital, school, office, dwellings, etc.) showing the aesthetic to functional component ratio. The square in the centre represents the class of buildings (for example, schools and office buildings) in which the aesthetics and functional aspects are almost equal. As we move towards the right, each class of buildings becomes more technical and less aesthetic, whereas towards the left each class of buildings is progressively more aesthetic and less technical.

From this we may conclude that in certain classes of buildings the aesthetic factor predominates and in others the functional aspect is given more importance.

The aesthetic factor is highly subjective, as it concerns human emotions. For example, the entrance portico of a building can be designed in different styles (Fig. 4.3) according to the imagination and taste of the designer (architect). Both the elevations may not be liked by a few.

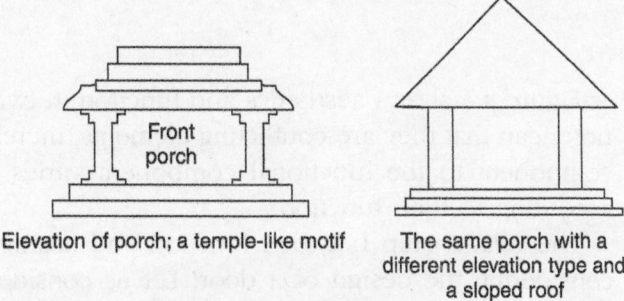

Elevation of porch; a temple-like motif

The same porch with a different elevation type and a sloped roof

Fig. 4.3 Different portico styles

Therefore, buildings can be classified according to function (Fig. 4.4) across a broad spectrum—hospitals, schools, factories, etc., in which the aesthetic requirement varies.

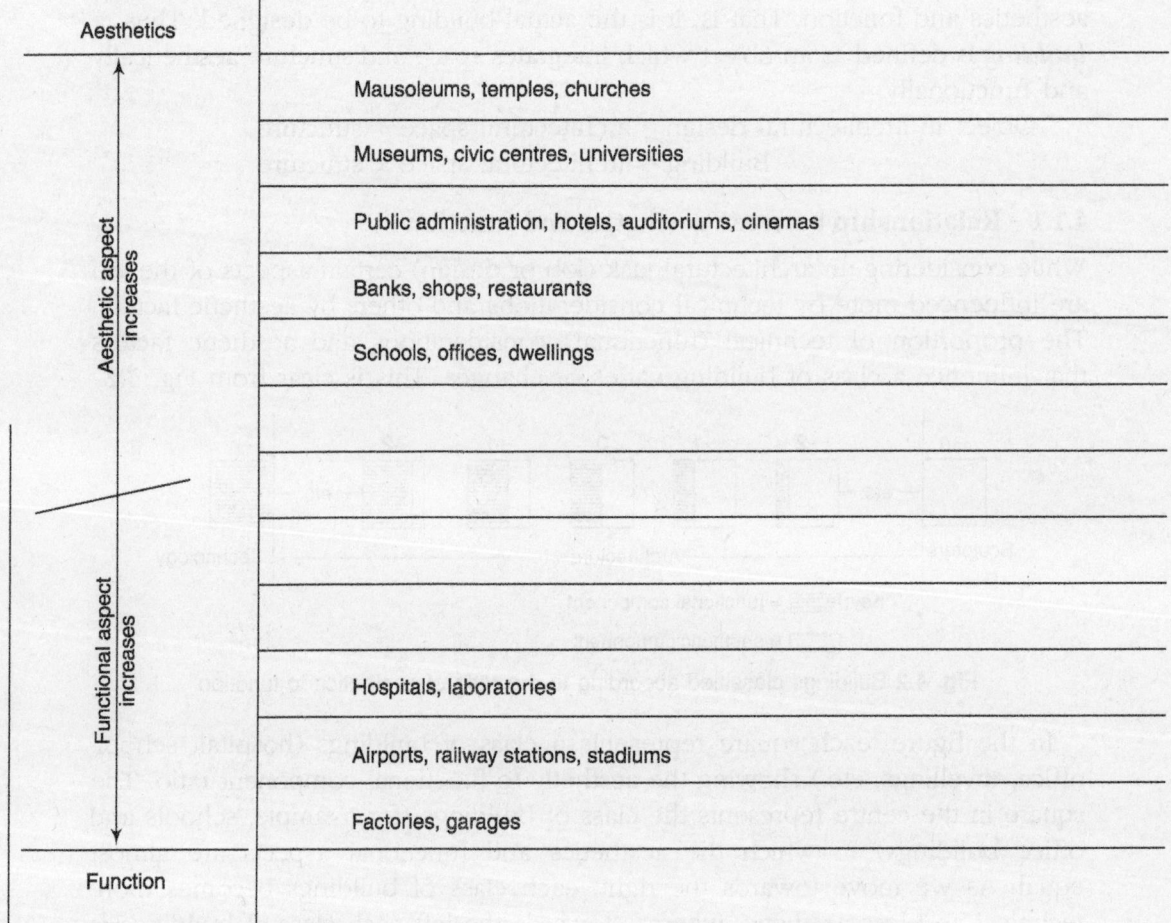

Fig. 4.4 Classification of buildings according to their aesthetics to function ratio

Figure 4.4 shows aesthetics and function in two opposite positions. This does not mean that they are conflicting elements, merely that the ratio of the aesthetic component to the functional component varies in each of the building types according to their function.

The relationship between function and aesthetics can be well explained by considering the design of a door. Let us consider the design of three kinds of doors—those of a hotel, a residence, and a factory.

Hotel	Residence	Factory
		A rolling shutter
Entrance door of a hotel	Entrance door of a residence	Entrance door of a factory
A hotel should have an attractive and inviting exterior to convey an image of warmth and hospitality. It also has to advertise itself in order to fend off competition.	The dwelling door may represent the personality or occupation of the owner.	The workshop door has only a utilitarian function to perform.
The hotel door design must fulfil all the aesthetic elements of design—proportion, contrast, pattern, and decoration	The dwelling door may be proportional and contrast with the surrounding wall.	This door might only have proportion and little or no aesthetic elements.

All the three doors must be
- functionally safe
- solid
- properly dimensioned (with proper width and height)

If these three doors are classified according to Fig. 4.4, the factory door would be at the bottom of the chart, giving importance to function, the hotel door at the top, signifying the predominance of aesthetics, and the dwelling door in the middle, satisfying both aesthetics and function.

All the three doors have the same function to perform (allowing entrance into the building), but have different designs. The difference is only due to the aesthetic component, which in turn is due to the difference in the purpose of the building.

4.2 Factors to be Considered in Architectural Design

When a design problem is given to an architect, there are many factors to be considered and analysed in stages before the final design is produced. These factors are
- requirements
- circulation
- anthropometrics
- space standards

- site planning
- landscape design
- climate
- integration of basic services
- safety regulations
- building rules and regulations
- layout regulations

In this section, we will examine the requirements, circulation, anthropometrics, and space standards.

4.2.1 Requirements

A building provides spaces for human activities. Designing or planning a building in architecture means planning the spaces provided by the building. The kinds of spaces to be designed are decided based on the kind of activities to be performed in the building. Each building must provide for a particular set of activities or, in other words, for a particular set of spaces. The set of spaces that is required for a particular building forms the design requirements for that particular building.

For example, in considering the design of a residence, the set of activities generally performed within a residence must be identified. The corresponding spaces required to perform these activities are the following:

Activities	Space required
1. Sitting/entertaining guests	Living space
2. Eating	Dining space
3. Cooking	Kitchen space
4. Sleeping	Bed space
5. Reading/learning	Study space
6. Family get together/entertainment	Family space
7. Bathing	Bath room
8. Sitting outside	Verandah/sit-out

Buildings are divided into classes according to the function to be performed within them. Buildings within which the same kind of activities are to be performed are classified into one segment. For example, the kind of activities performed in any particular hospital will bear a great similarity to the kind of activities performed in another hospital. Due to this similarity, they can be put under one class. All office buildings fall into one class because of the similarity between their activities.

In practice, architectural design does not deal with a specific class of buildings but with a particular building within a class. So it is not the design of a hospital, but that of a particular kind of hospital such as a maternity hospital, an orthopaedic hospital, or a general hospital. Each of these hospitals should perform a particular set of activities, for which a particular set of spaces is required. For example, the activities and requirements for a general hospital are the following:

Activities	Requirements
Outpatient department:	
Medical outpatient	Medical outpatient room with waiting room
Surgical outpatient	Surgical outpatient room with waiting room
Eye specialist	Ophthalmology examination room with waiting room
ENT specialist	1 room + waiting room
Dentist	1 room + waiting room
Gynaecology	1 room + waiting room
Pediatrics	1 room + waiting room
Facilities:	
Radiology	1 x-ray room + waiting room
Scan	1 Scan room + waiting room
Medicines	Dispensary

Consider the design of the institutional complex of an engineering college, which requires many spaces for many acitivites. Analysing the activities, we arrive at the following requirements:

Activities	Requirements
1. To receive the students, teachers, staff members, the heads of the institution, etc.	General reception and lobby space
2. Administration	Office space
3. To accommodate heads of institution	1 room for the Principal 1 space for the Director 1 space for the Vice-Principal
4. Learning	Classrooms, lecture halls, laboratories
5. Computer usage	Computer labs
6. Book reference, reading	Library
7. Eating	Canteen
8. Relaxation for teachers	Staffrooms
9. Relaxation for students	Restrooms
10. Public conveniences	Toilets
11. Vehicle parking	Vehicle stand

These are the space requirements for the institution building. Now let us consider the space inside a classroom.

General requirements for all classroom designs

There are several internal requirements for a classroom space. Sufficient space is required near the front of the room for setting up audio-visual equipment such as projection screens, OHP, and LCD projectors. The ceiling should be high enough, say, 3 m. Lighting and ventilation should be provided adequately by a sufficient number of windows. No teacher should face the windows when addressing the class from the normal teaching position.

The ceiling and walls should be finished with proper materials that will create an impression of brightness within the classroom, i.e., the colours used for the walls should be bright in order to create an atmosphere conducive to learning. The material used for the ceiling or walls should also be acoustically sound to prevent echoes. The classrooms should be located in quiet areas.

Conclusion

The starting point for any architectural design is the identification of the activities which are to be satisfied by the design. The nature of these activities will determine the minimum space required.

4.2.2 Circulation

Circulation is basically movement through space (Fig. 4.5). A circulation path is visualized as a thread that links the spaces of a building or a series of interior and exterior spaces.

Fig. 4.5 Circulation—movement through space

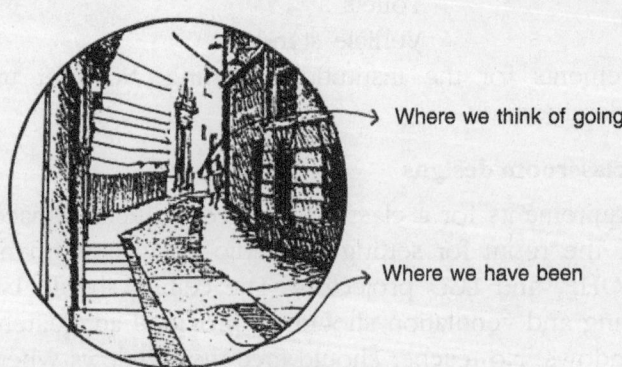

→ Where we think of going

→ Where we have been

Fig. 4.6 Circulation—experience of a sequence of space

As one moves along the circulation path, one experiences a sequence of spaces, from where one has been to where one is going (Fig. 4.6).

This section presents the principal components of a building's circulation system that affect our concept of a building's form and space. The circulation elements can be listed as (Fig. 4.7)

(a) The building approach: the distant view
(b) The building entrance: from outside to inside
(c) Configuration of the path: the sequence of spaces
(d) Path–space relationships: edges, nodes, and terminations of the path
(e) Form of the circulation space: corridors, balconies, galleries, stairs, and rooms

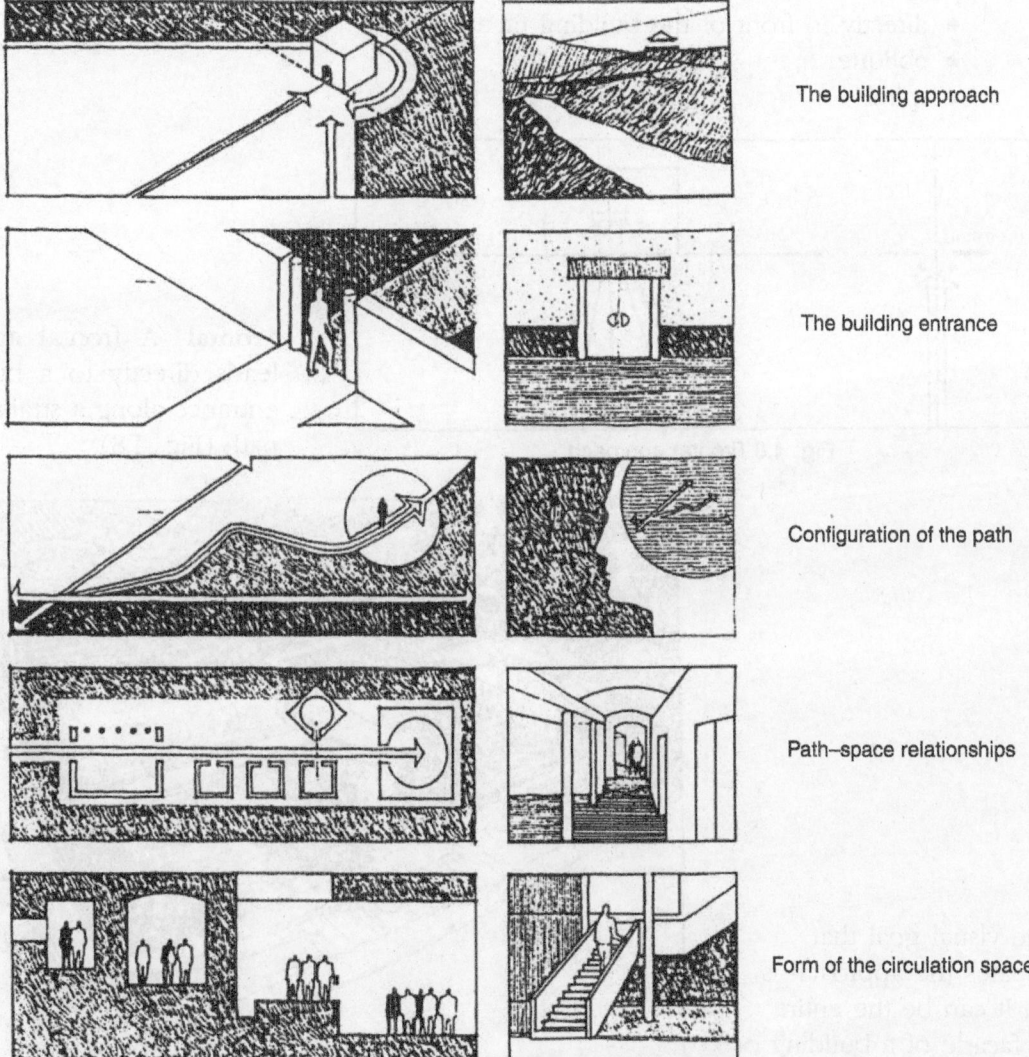

The building approach

The building entrance

Configuration of the path

Path–space relationships

Form of the circulation space

Fig. 4.7 Circulation types

Building approach

Before actually entering a building's interior, we approach the entrance of the building along a path. This is the first phase of the circulation system, in which we are prepared to

- see
- experience
- use the building's space

The approach to a building and its entrance may be separated by a few metres of distance or by a lengthy and circuitous route. The nature of the approach depends on whether an element has been used to terminate the approach (the end of the approach) or the approach continues into the building interior. The approach could be

- directly in front of the building face (frontal)
- oblique
- spiral

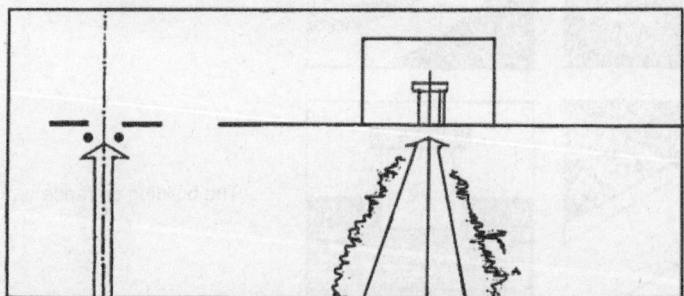

Fig. 4.8 Frontal approach

Frontal A frontal approach leads directly to a building's entrance along a straight axial path (Fig. 4.8).

The visual goal that terminates the approach is clear. It can be the entire front façade of a building or an entrance which is very large (see Figs 4.9–4.12).

Fig. 4.9 Frontal approach: Villa Barbaro, Italy

Fig. 4.10 Frontal approach: link from the imperial city to the outer city in China

Fig. 4.11 Frontal approach: Catholic Church, Mexico

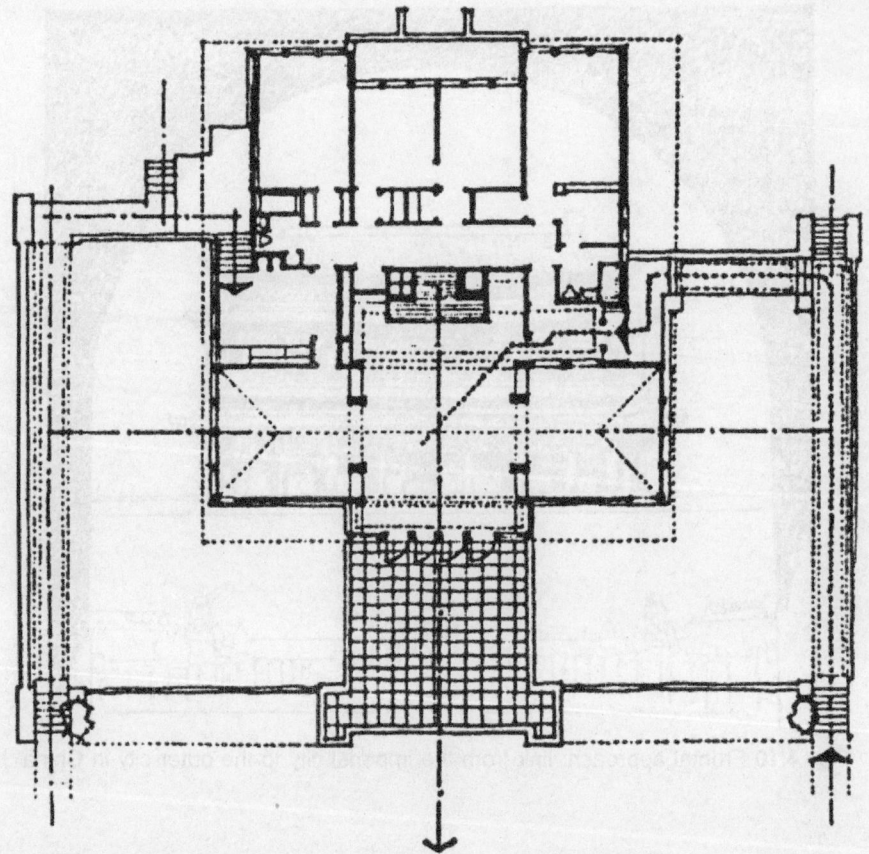

Fig. 4.12 Frontal approach: Edwin House, Illinois (built by F.L. Wright)

Oblique An oblique approach provides the effect of perspective of the building's front façade and form (Figs 4.13–4.15).

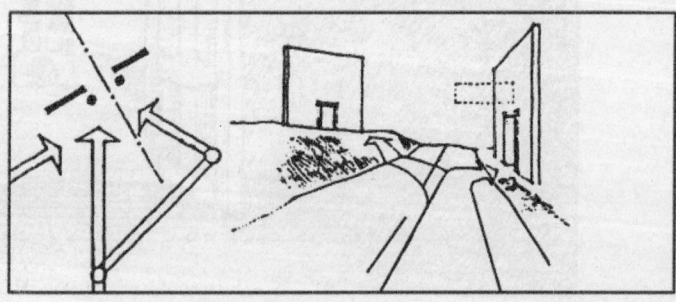

Fig. 4.13 Oblique approach

Fig. 4.14 Oblique approach to Notre Dame, Ronchamp (Le Corbusier)

The path can be redirected one or more times to delay and prolong (extend) the sequence of approach.

If the building is placed at a real slant angle, then the building entrance can be projected beyond its facade.

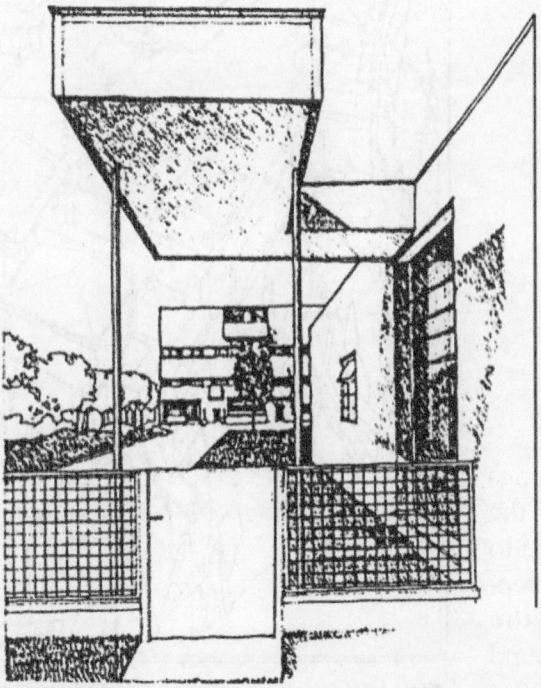

Fig. 4.15 Oblique approach, Villa at Garches (Le Corbusier)

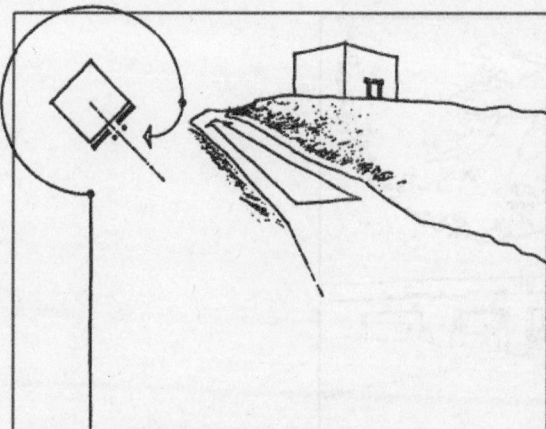

Fig. 4.16 Spiral approach

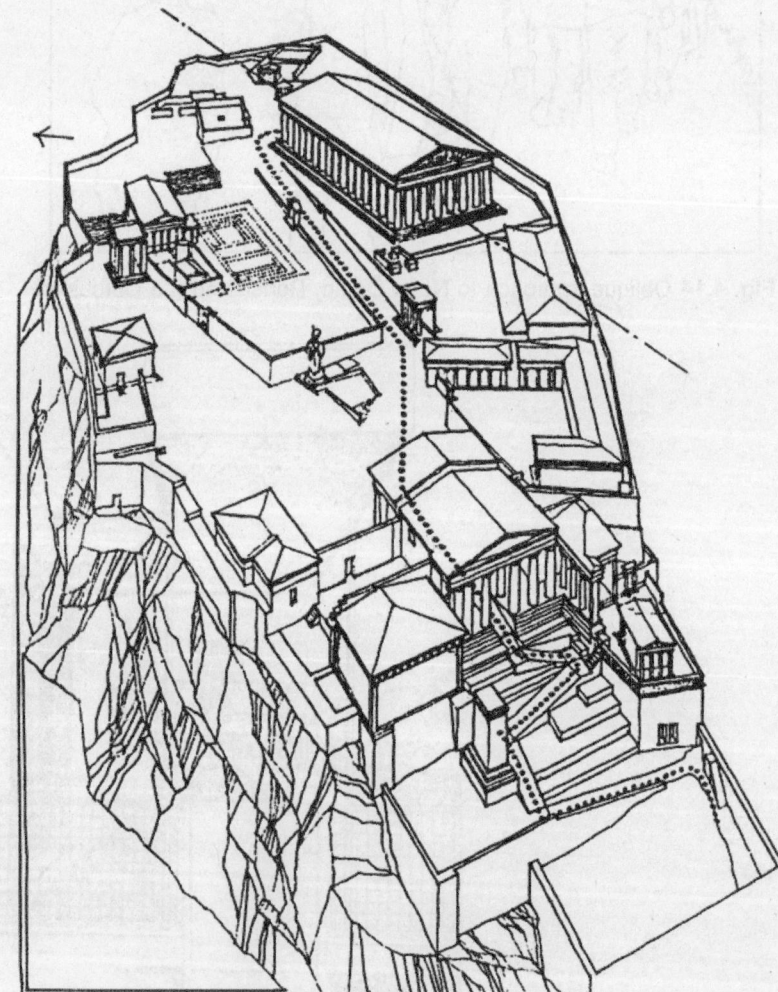

Spiral A spiral path prolongs (extends) the sequence of the approach, and can be used to give importance to the three-dimensional form of the building, as one moves around the building (Figs 4.16–4.18).

Fig. 4.17 Spiral approach: aerial view of the Acropolis, Athens, Greece

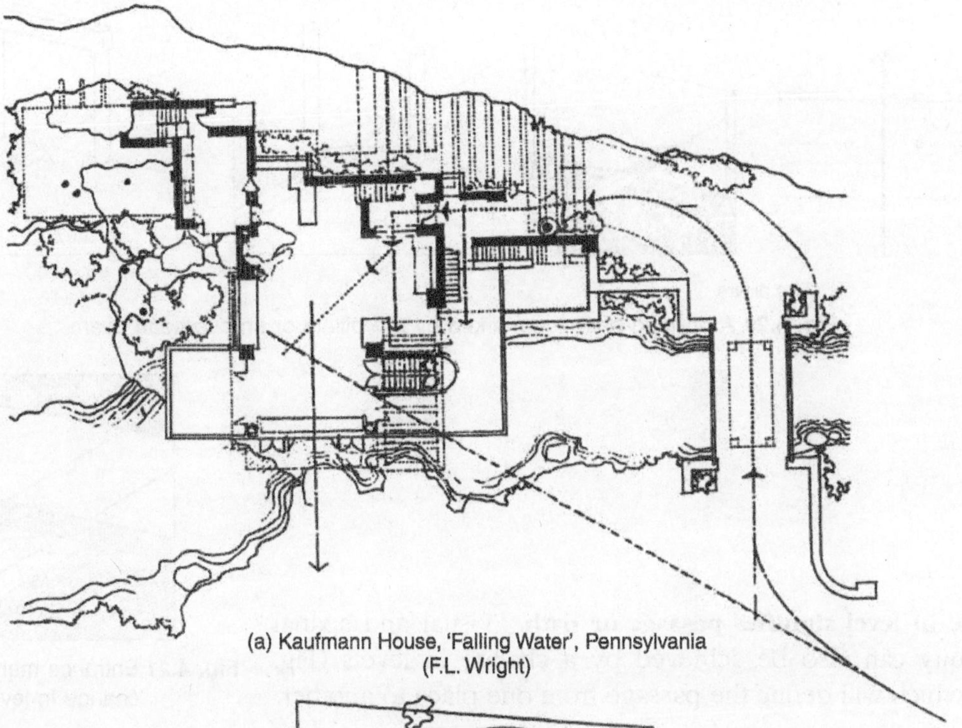

(a) Kaufmann House, 'Falling Water', Pennsylvania
(F.L. Wright)

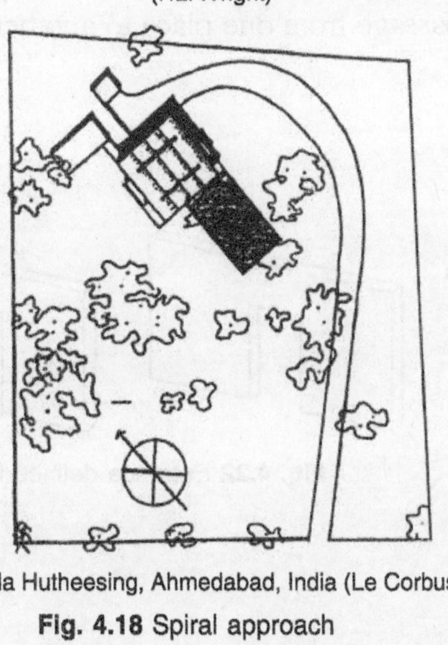

(b) Villa Hutheesing, Ahmedabad, India (Le Corbusier)

Fig. 4.18 Spiral approach

Building entrance

Building entrances can be of varied types. Entering is the act of penetrating a vertical plane (Fig. 4.19) which separates one space from another, 'here' from 'there'.

Entrance defined by two pillars or an overhead beam   An entrance can be defined by two pillars or an overhead beam (Fig. 4.20), which will define the passage.

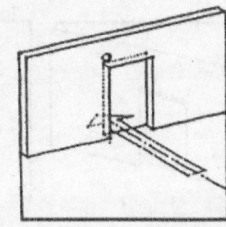

Fig. 4.19 Penetrating a vertical plane

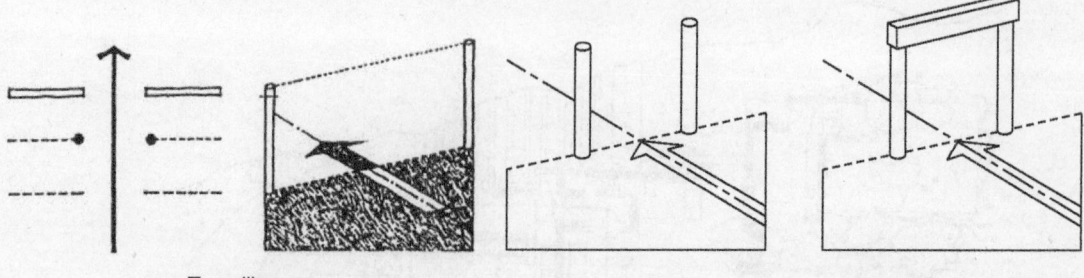

Two pillars

Fig. 4.20 A building entrance marked by two pillars or an overhead beam

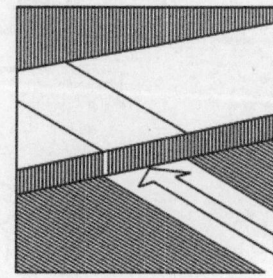

Change in level signifies passage or path Visual and spatial continuity can also be achieved by a change in levels (Fig. 4.21), which will define the passage from one place to another.

Fig. 4.21 Entrance marked by change in level

Entrance marked by an opening or an articulated gateway In a normal situation, where a wall is used to enclose a space or a series of spaces, the entrance can be defined by an opening in the wall or an articulated gateway (Fig. 4.22).

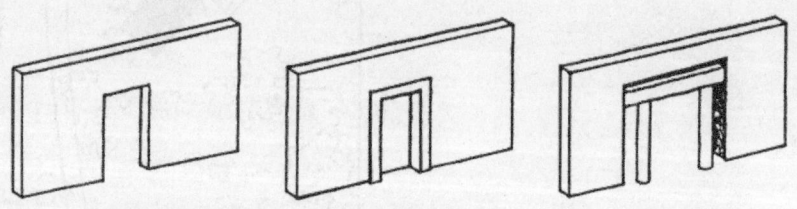

Fig. 4.22 Entrance defined by an opening/articulated gateway

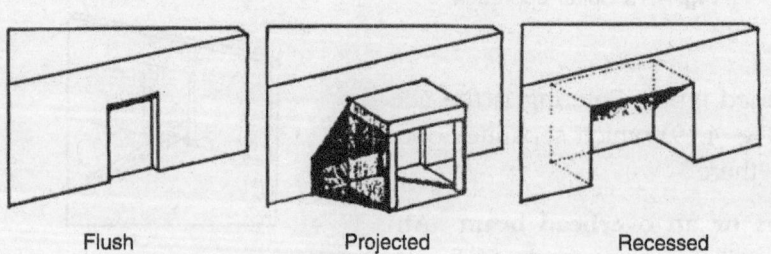

Flush   Projected   Recessed

Fig. 4.23 Flush, projected, and recessed entrances

Entrances can be flush, projecting, or recessed Figure 4.23 shows a flush, a projected, and a recessed entrance. A flush entrance maintains the continuity of a wall surface. A projected entrance gives importance to the function of the approach and provides a shelter overhead. A recessed entry also provides shelter and takes the exterior space inside the building.

Figures 4.24–4.38 illustrate the concept of building entrance with examples of historic buildings.

Fig. 4.24 Building entrance framed by two columns: the entrance to the Piazza Venice, is marked by two granite columns

Fig. 4.25 Building entrance marked by an opening on the wall: Legislative Assembly Building, Chandigarh Capital Complex, India (Le Corbusier)

Fig. 4.26 A building entrance in Japan marked by a gateway

Fig. 4.27 The entrance to Pylons, Egypt, marked by a plane gateway

(a) Entrance to the S.C. Johnson and Son, Administration Building, Wisconsin (F.L. Wright)

(b) Mill Owner's Association Building, Ahmedabad, India (Le Corbusier)

Fig. 4.28 Building entrances marked by change in level

Fig. 4.29 A flush entrance to a house in Pennsylvania

(a) The High Court, Chandigarh Capital Complex, India (Le Corbusier)

(b) Articulated entrances

Fig. 4.30 Articulated gateway entrances

Fig. 4.31 Building entrance articulated by defining the arch: Row House, Illinois

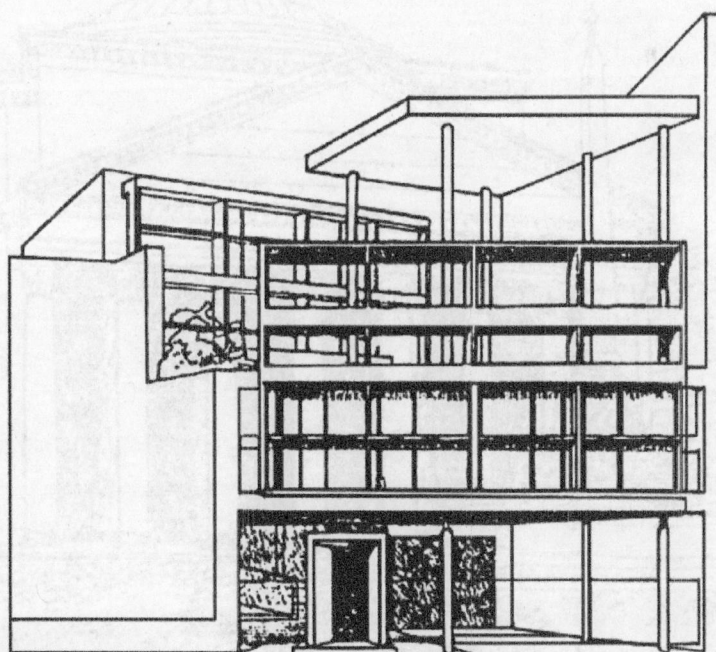

Fig. 4.32 Articulated entrance to a house in Argentina (Le Corbusier)

Fig. 4.33 Articulated entrance: Giorgio, Venice

Fig. 4.34 Projected building entrance: entrance portico, The Pantheon, Rome

Fig. 4.35 Recessed building entrance: The Chapel, Florence, Italy

Fig. 4.36 Recessed building entrance: National Gallery of Art, Washington

Fig. 4.37 Recessed building entrance: S. Andrea, Italy

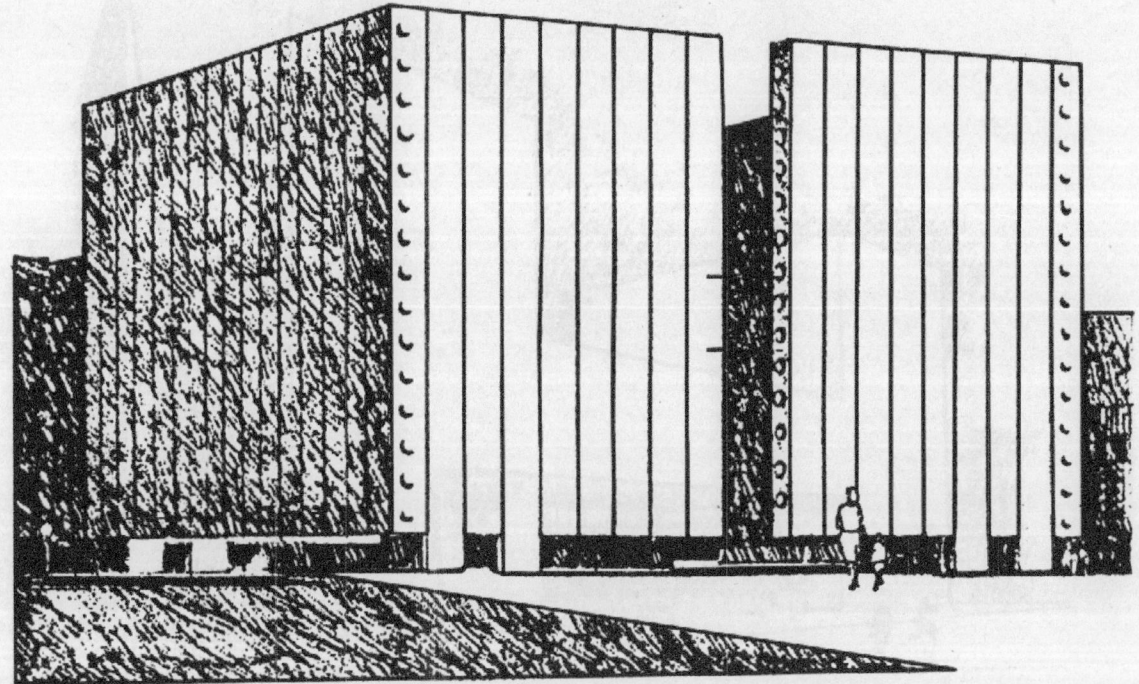

Fig. 4.38 Recessed building entrance: John Kennedy Memorial, Dallas (Philip Johnson)

Location of entrance   While considering the location of an entrance, the architect could decide to centre it on the front façade of the building or place it off-centre (Figs 4.39–41).

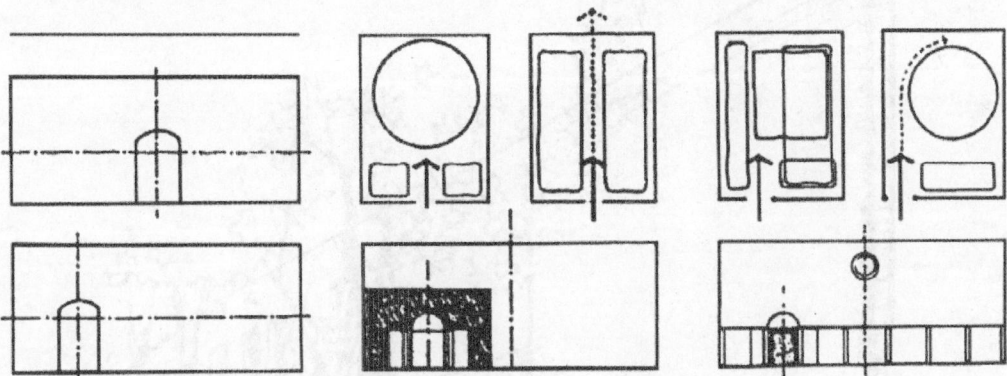

Fig. 4.39 An entrance could be central or off-centre

Fig. 4.40 Centralized entrance: Merchants National Bank, Iowa (Louis Sullivan)

Fig. 4.41 Off-centre entrance: Morris Gift Shop, San Francisco, California

Path configurations

All paths of movement, whether they are for people, vehicles, goods, or services are linear in *nature*. All paths have a starting point, from which one is taken through a sequence of spaces to the destination. The path could be in various forms (Fig. 4.42) in architectural design as discussed below.

(a) Linear—A straight path, which is a linear path, could be the primary organizing element and take up many shapes such as curvilinear, segmented, or intersecting; branched or looped.

(b) Radial—A radial configuration has paths extending from or terminating at a central common point.

(c) Spiral—A spiral path is a single, continuous path that starts from a central point and revolves around it, becoming progressively distant from it.

(d) Grid—A grid configuration consists of two sets of parallel paths that intersect at regular intervals and create square or rectangular fields of space.

(e) Network—A network configuration consists of random paths that connect points in space.

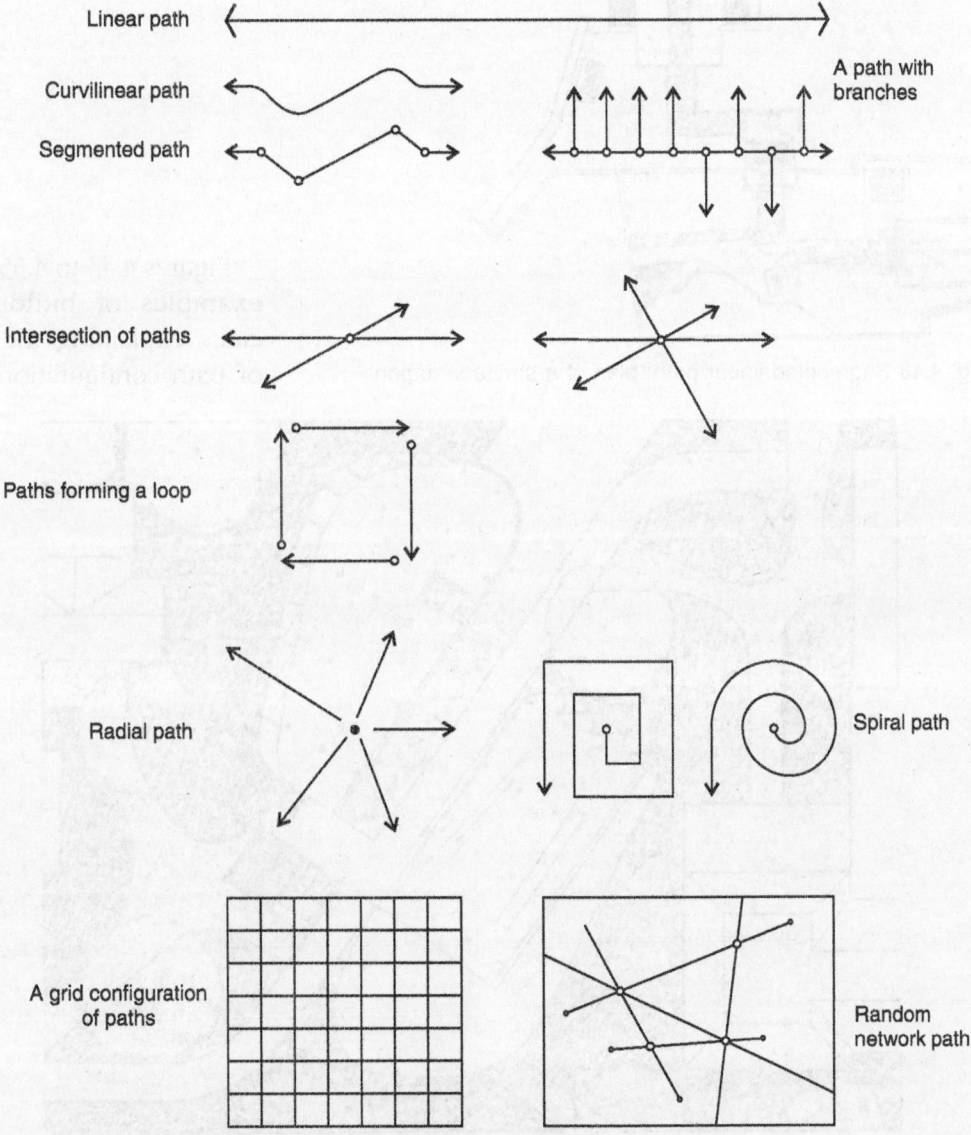

Fig. 4.42 Path configuration

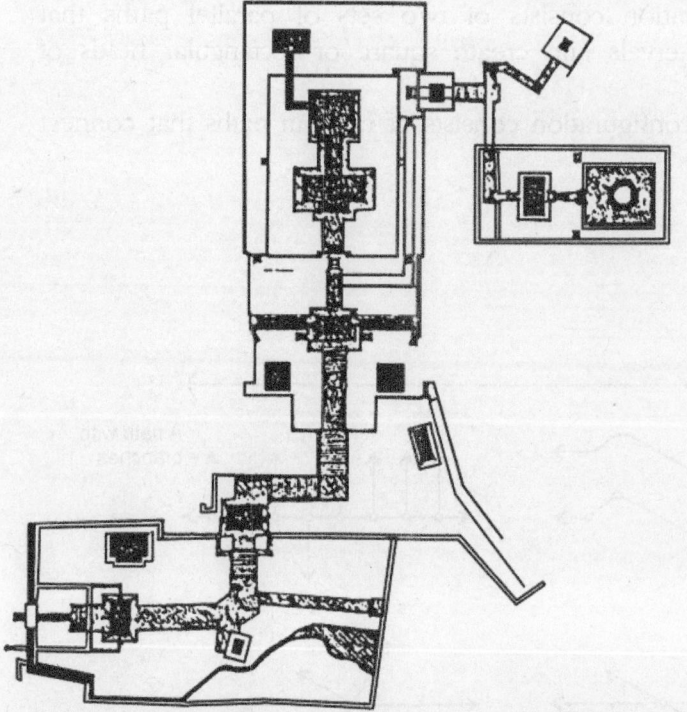

Fig. 4.43 Segmented linear path: plan of a shrine in Japan

Figures 4.43 to 4.55 illustrate examples of buildings and cities explaining the concept of path configuration.

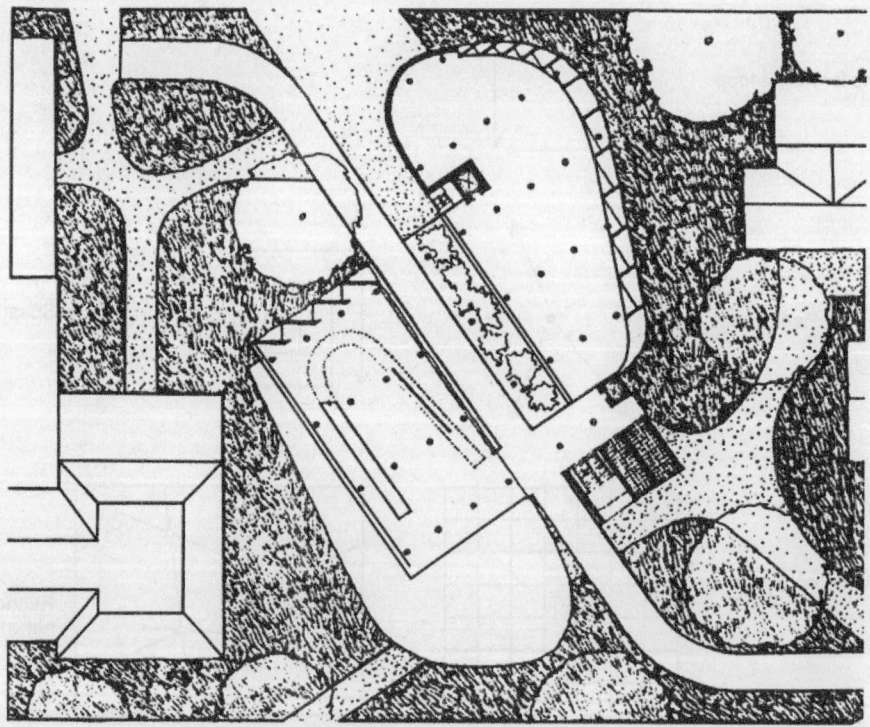

Fig. 4.44 Curvilinear path: Carpenter Center, Harvard University, Cambridge

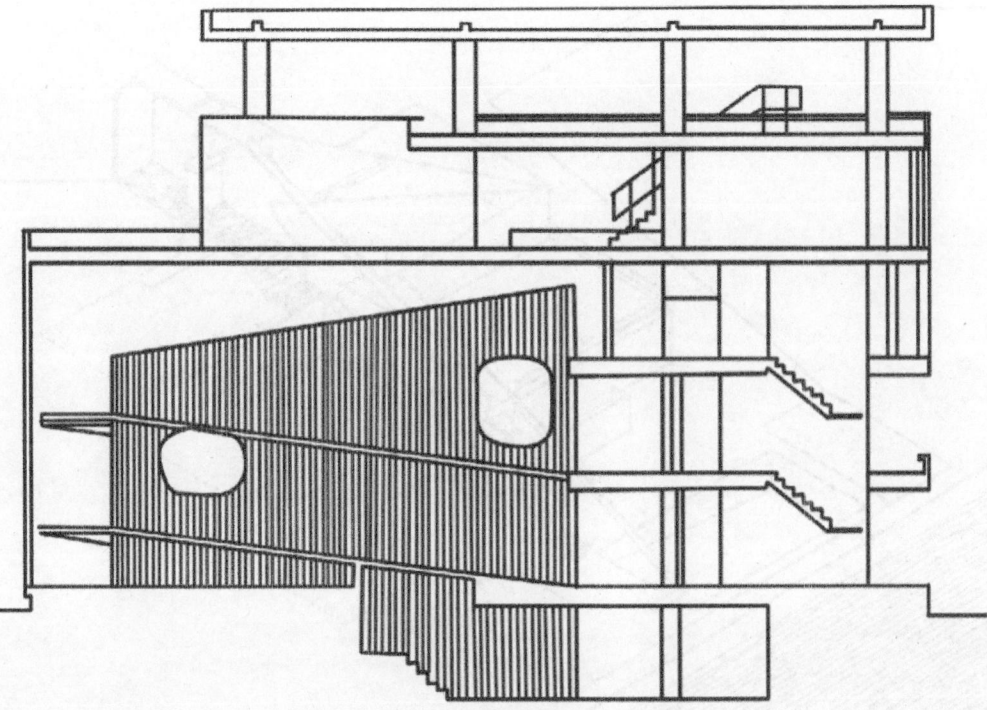

Fig. 4.45 A loop section formed using ramps and stairs: Shodhan House Ahmedabad, India (Le Corbusier)

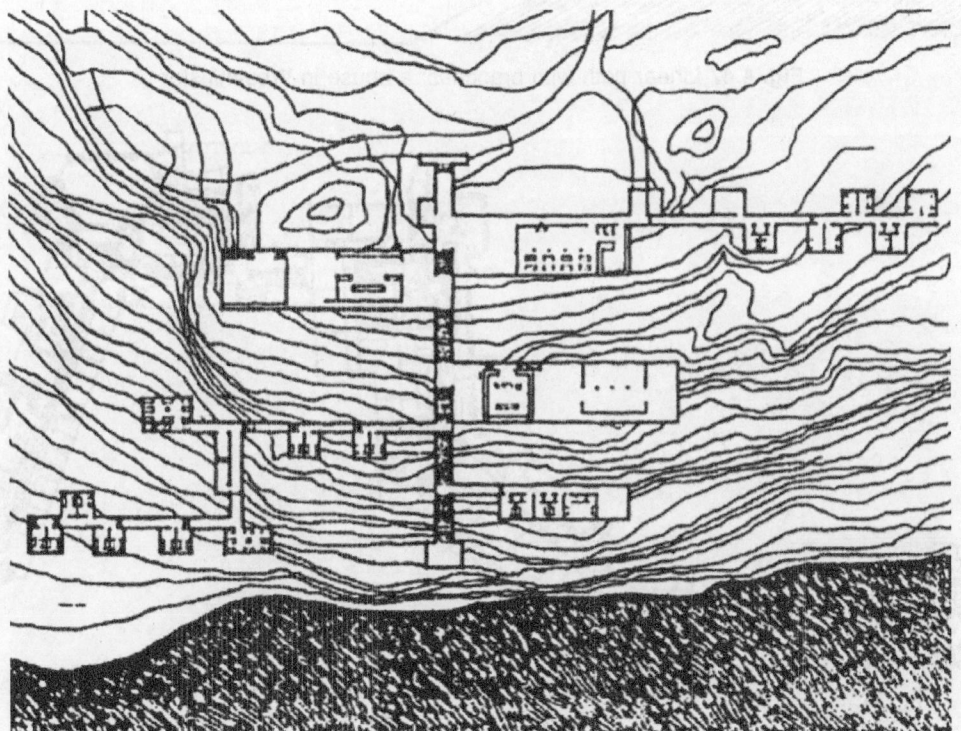

Fig. 4.46 Linear path with branches: Mountain School of Arts and Crafts, Maine

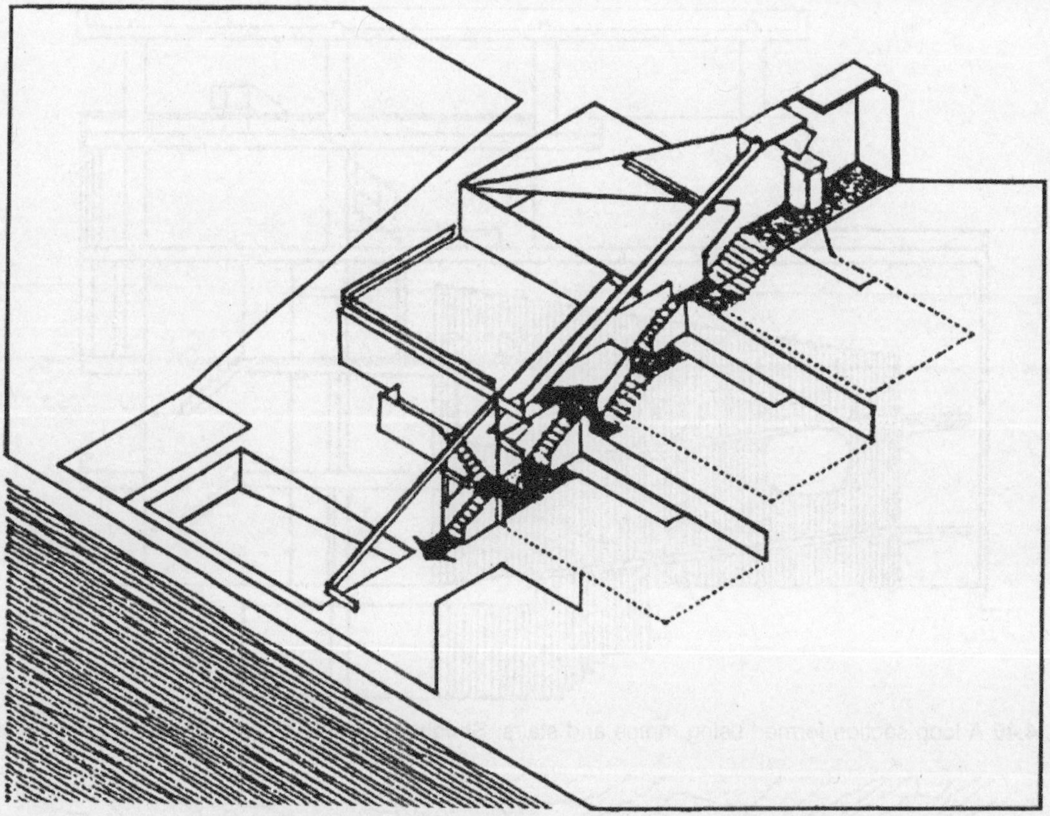

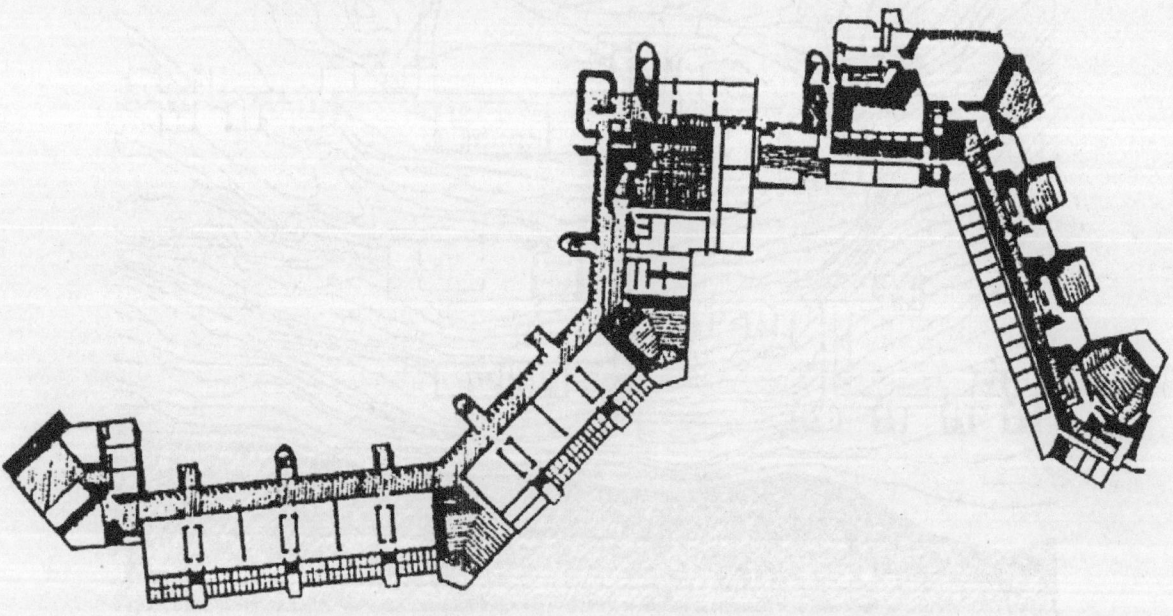

Fig. 4.47 Linear path with branches: a house in Westminster

Fig. 4.48 Segmented linear path: a college in West Hill, Ontario

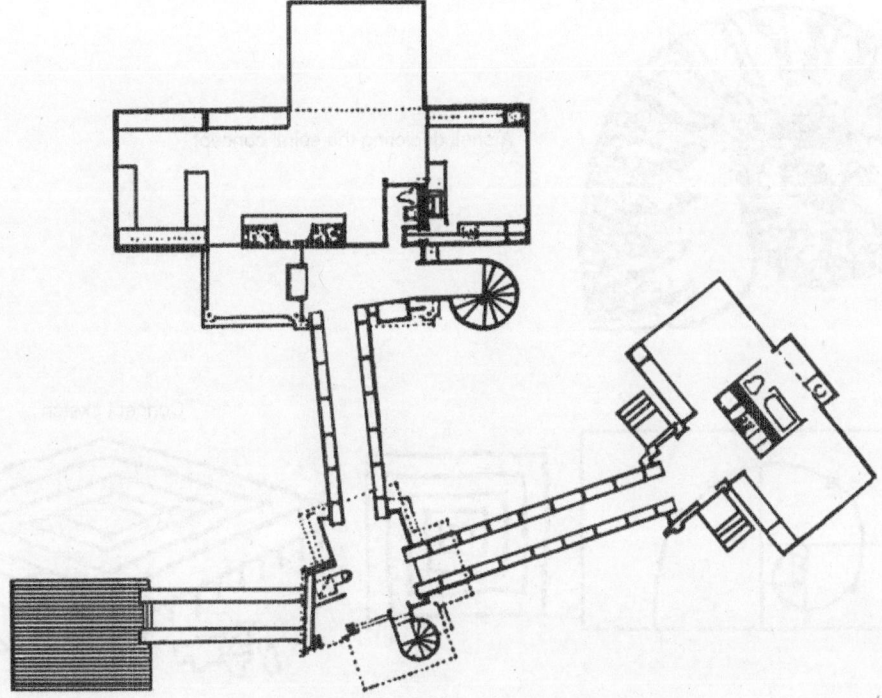

Fig. 4.49 Radial path: Pope House, Connecticut

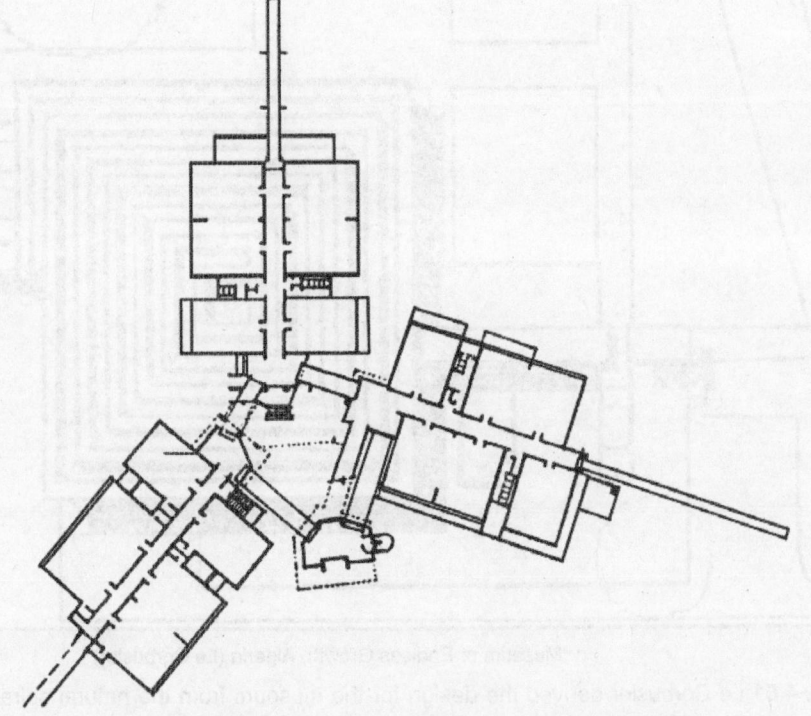

Fig. 4.50 Radial path: Smith Elementary School, Indiana

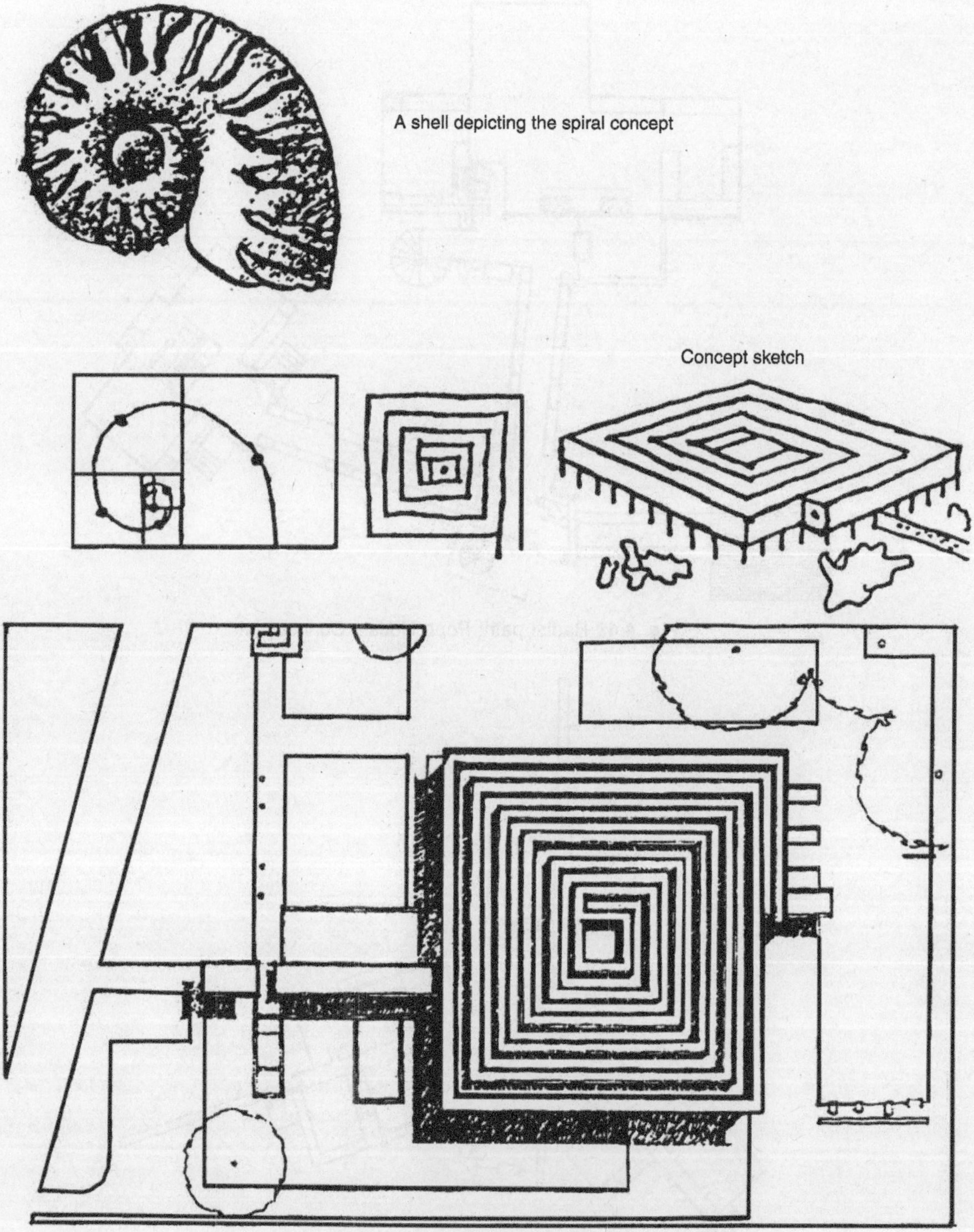

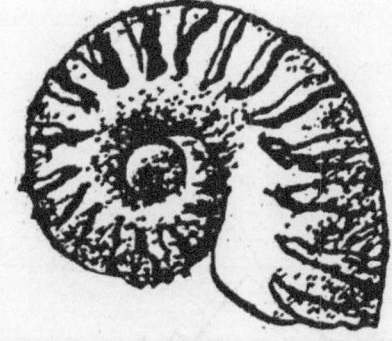

A shell depicting the spiral concept

Concept sketch

Museum of Endless Growth, Algeria (Le Corbusier)

Fig. 4.51 Le Corbusier derived the design for the museum from the natural spiral shell

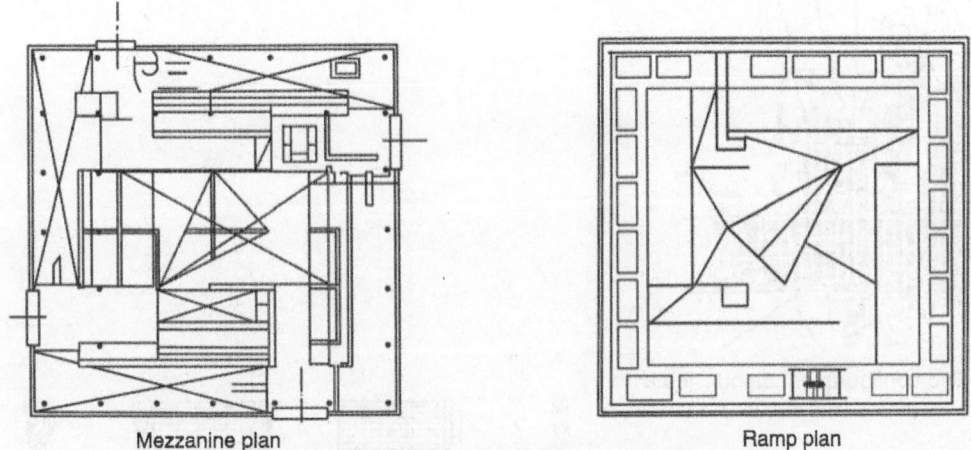

Mezzanine plan   Ramp plan

Fig. 4.52 Spiral path: Museum of Western Art, Tokyo (Le Corbusier)

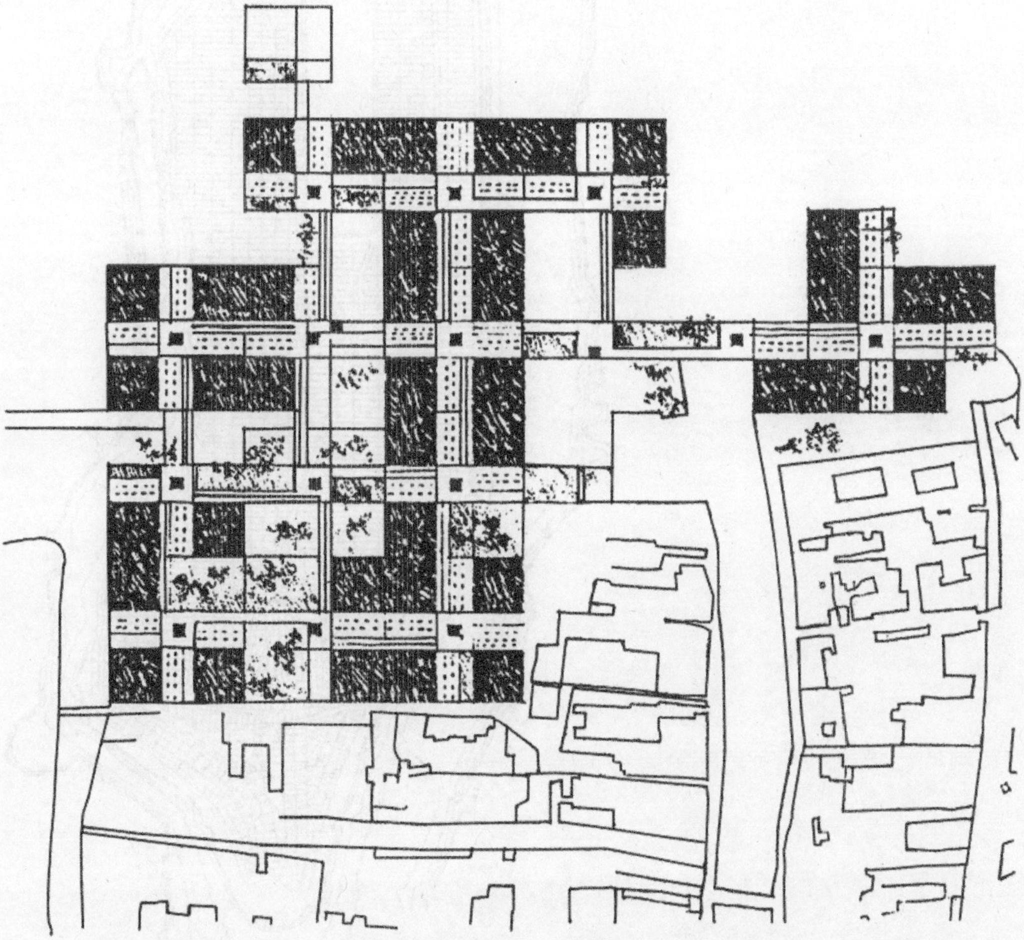

Fig. 4.53 Grid configuration of the path: Hospital project, Venice (Le Corbusier)

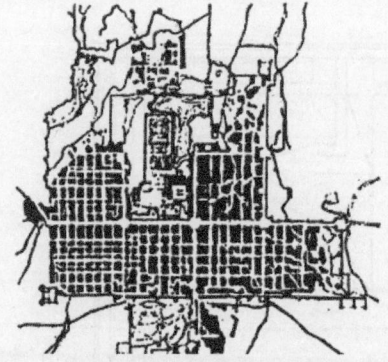

Fig. 4.54 Grid configuration: Jaipur, India

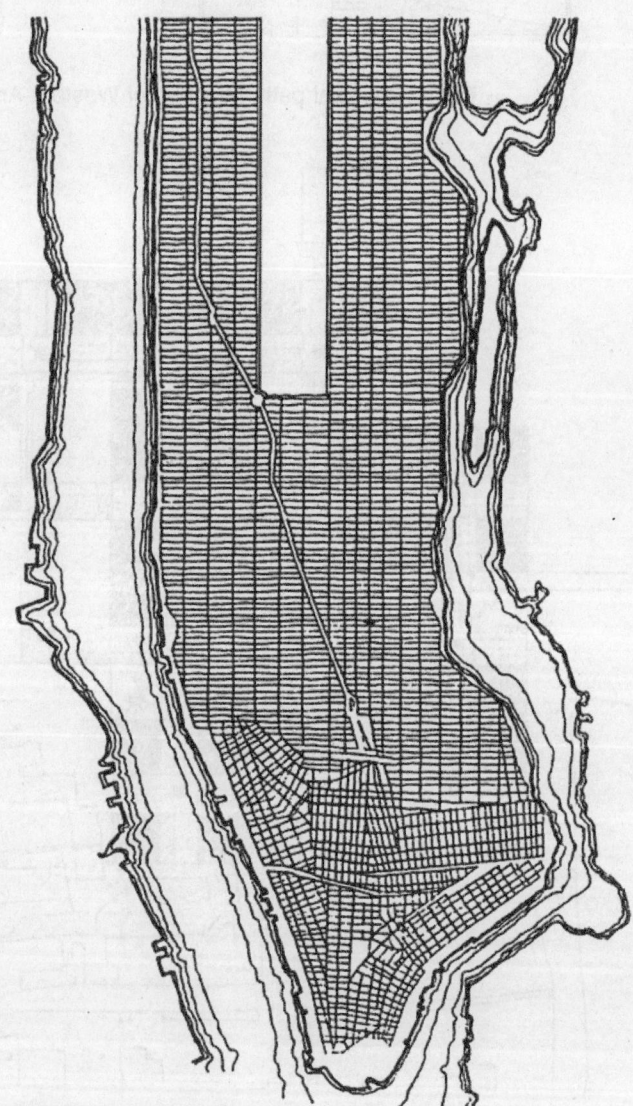

Fig. 4.55 Grid configuration: New York City, Manhattan

Path–space relationship

Paths may be linked to spaces in the following ways.

Pass by spaces The path may 'pass by' the spaces. This allows a passer-by to look into the spaces. The configuration of the path is flexible. The spaces in between the buildings or structures can be used to link the path with the spaces.

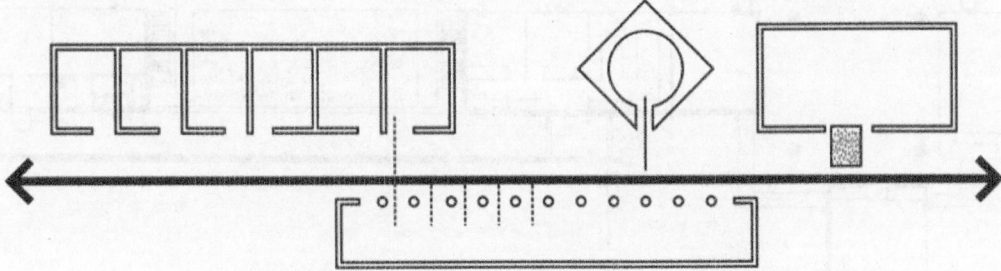

Fig. 4.56 Path–space relationship: pass by spaces

Pass through spaces The path may pass through a space axially, obliquely, or along the edge. In cutting through a space, the path creates patterns of rest and movement within it.

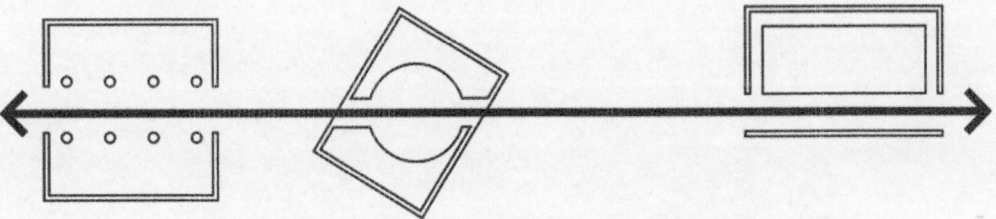

Fig. 4.57 Path–space relationship: pass through spaces

Terminate in a space The location of the space decides the path. This path–space relationship is used to approach and enter a functionally and symbolically important space.

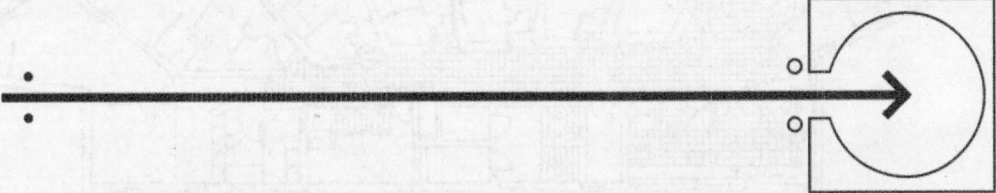

Fig. 4.58 Path–space relationship: terminate in a space

Figures 4.59–4.61 illustrate examples of buildings explaining the concept of the path–space relationship.

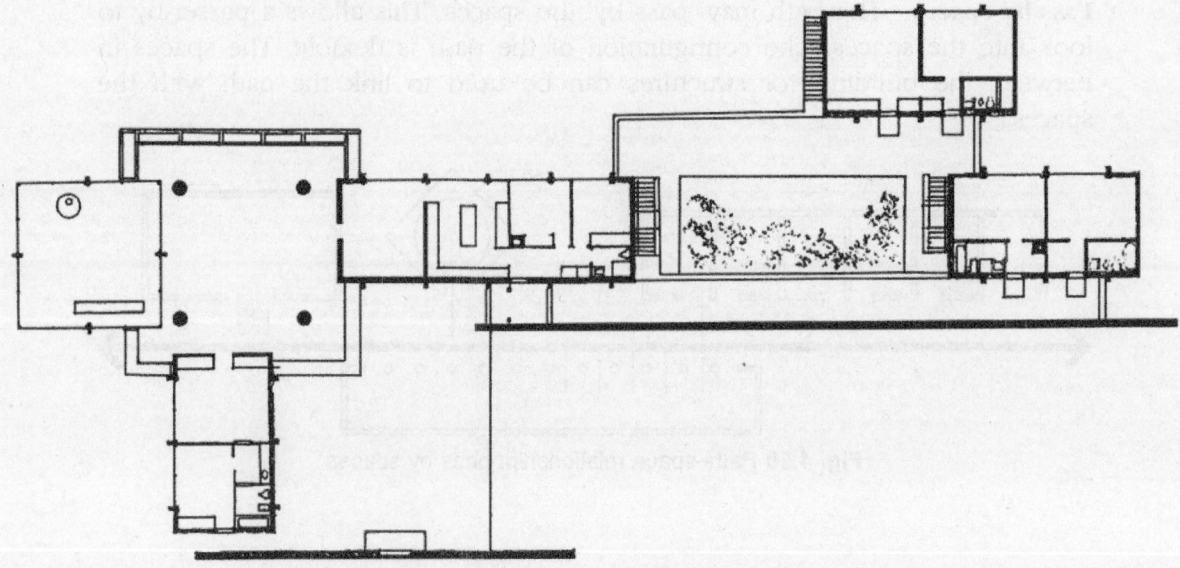

Fig. 4.59 Pass by spaces: a house in France (Philip Johnson)

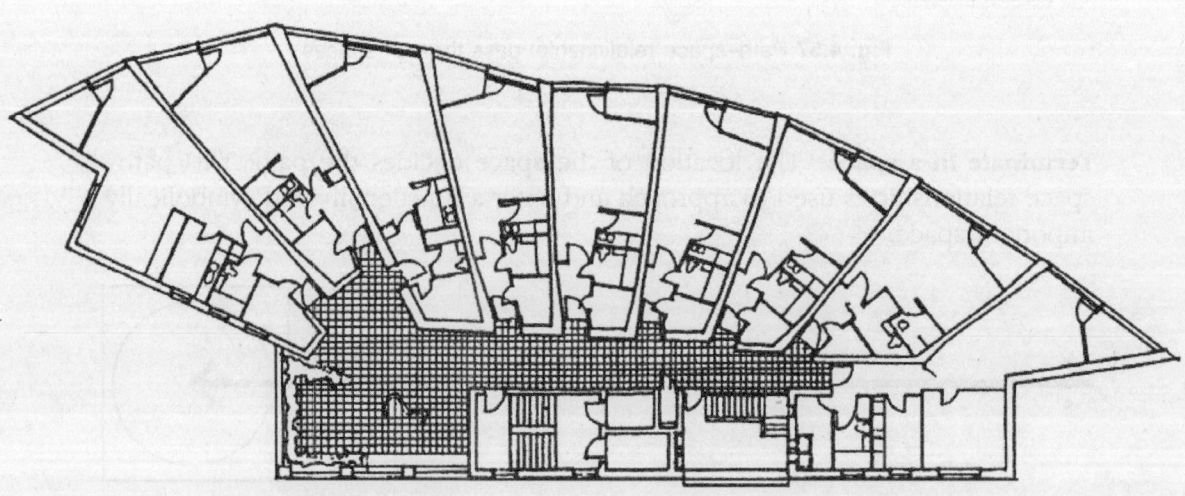

Fig. 4.60 Pass by spaces: an apartment building in Germany (Alvar Aalto)

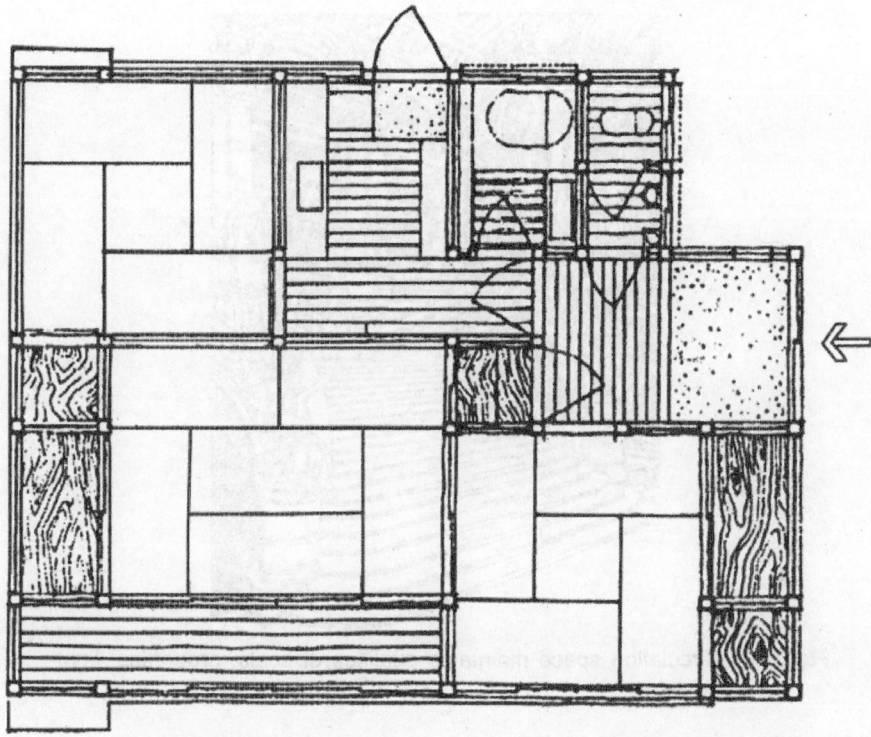

Fig. 4.61 Pass through spaces: a typical Japanese house

Form of circulation space

Circulation spaces are the main part of any building and occupy some amount of space within the building. A circulation space, e.g., a corridor connects the various functional areas of the building. The form and scale of a circulation space give it a certain significance, and provide spaces for rest as well as a good view along its path. The form of a circulation space can change according to

- the boundaries of the space
- the form of the space it links (or connects)
- the qualities of scale, proportion, light and view maintained (Fig. 4.62)
- an entrance opening onto the circulation space
- changes in levels with stairs and ramps (Figs 4.63 and 4.64)

A circulation space may be the following (Fig. 4.65).

- *Enclosed* Here space is enclosed and forms a corridor which is linked to the rest of the space through entrances in the wall plane.
- *Open on one side* In this, there is a visual and spatial continuity with the spaces it links.
- *Open on both sides* The circulation path passes through the main space.

The width and height of a circulation space should be proportionate to the type and amount of traffic it must handle. If the path is *narrow*, it will suggest only movement.

Fig. 4.62 Circulation space maintains qualities of scale, proportion, light, and view

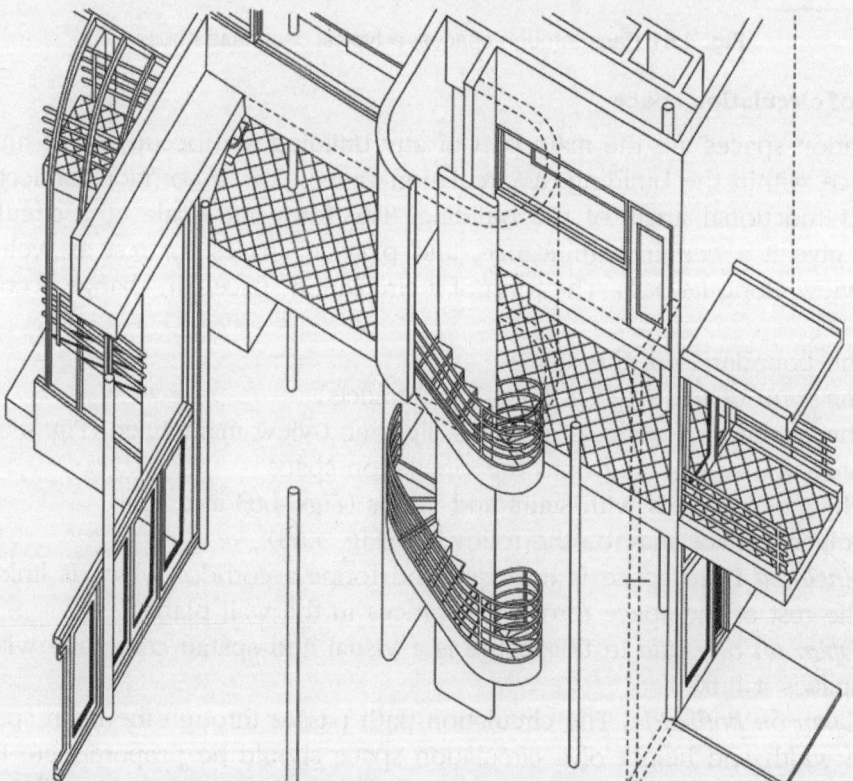

Fig. 4.63 Form of circulation space: changes in level can be achieved with stairs and ramps

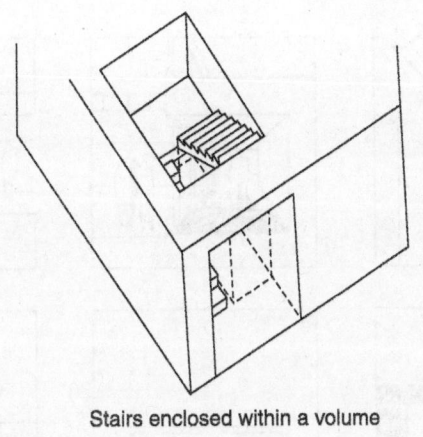

Stairs enclosed within a volume

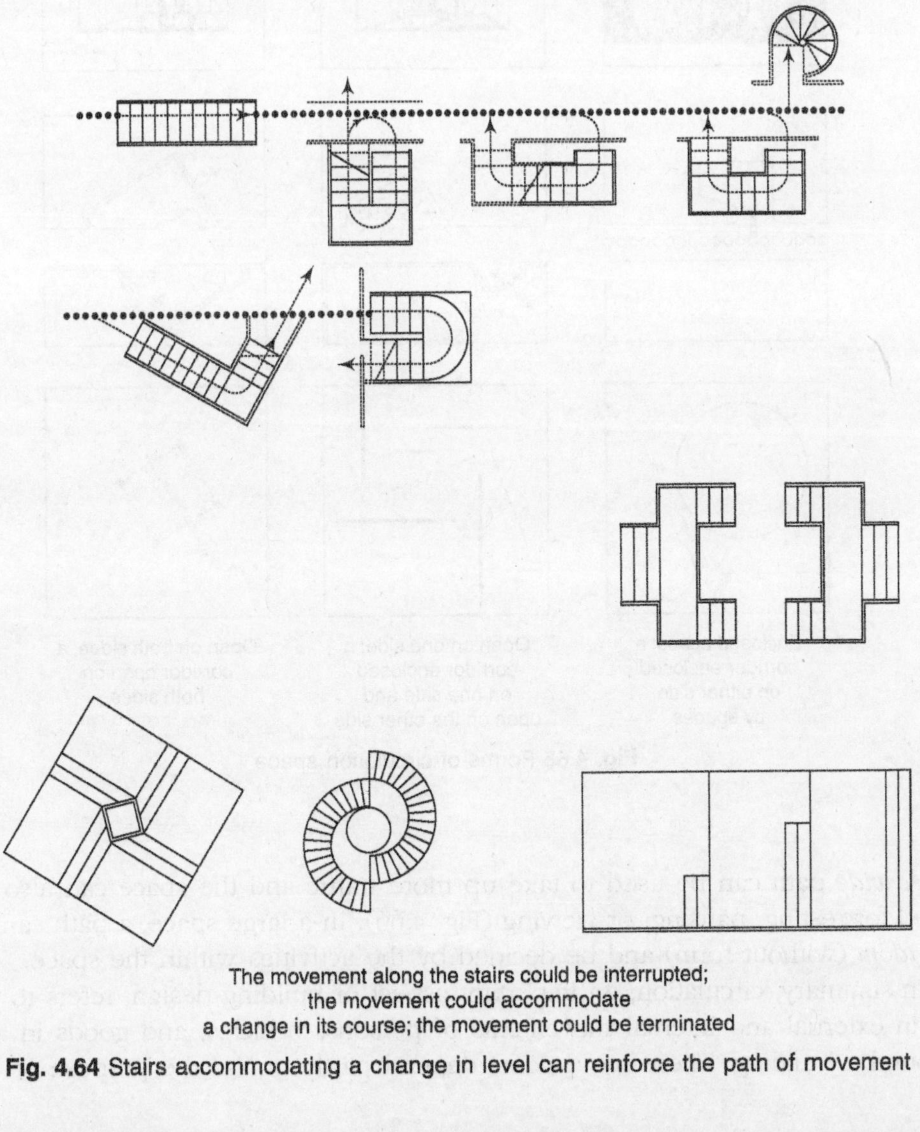

The movement along the stairs could be interrupted;
the movement could accommodate
a change in its course; the movement could be terminated

Fig. 4.64 Stairs accommodating a change in level can reinforce the path of movement

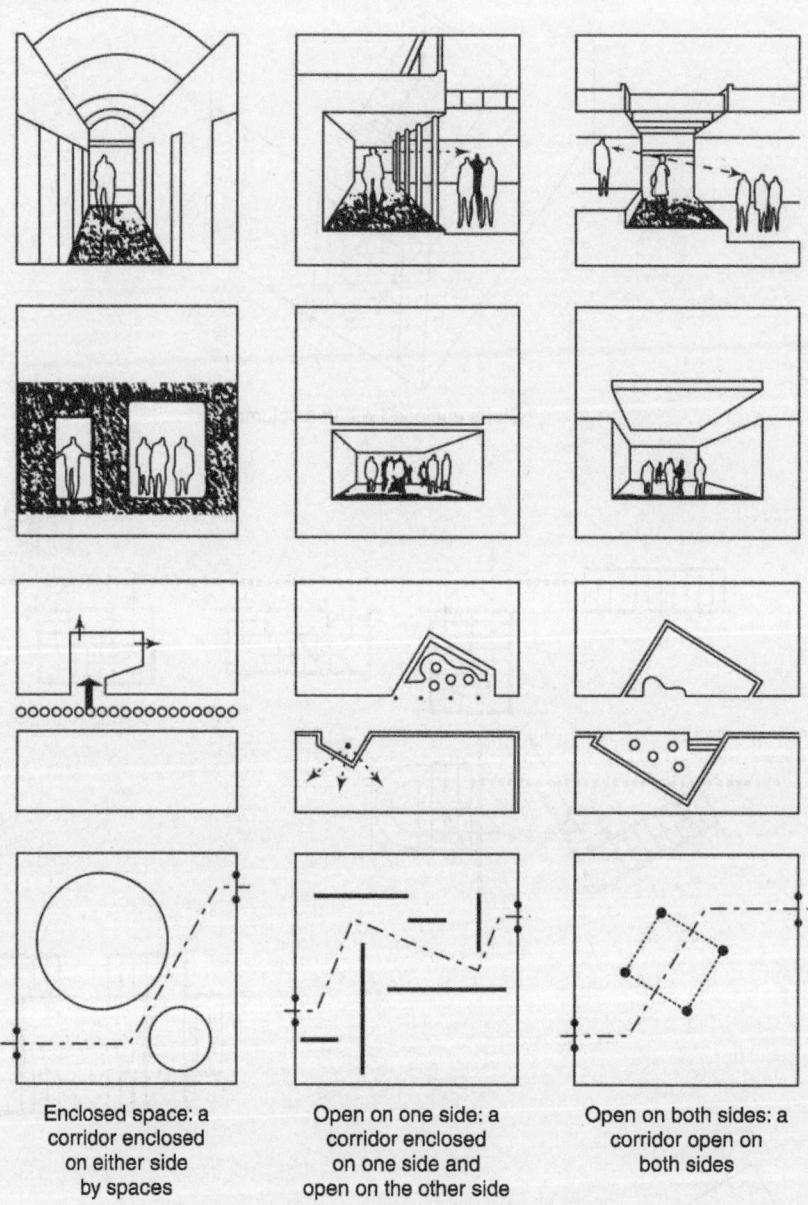

Enclosed space: a
corridor enclosed
on either side
by spaces

Open on one side: a
corridor enclosed
on one side and
open on the other side

Open on both sides: a
corridor open on
both sides

Fig. 4.65 Forms of circulation space

A *wide* path can be used to take up more traffic and the space can also be used for resting, pausing, or viewing (Fig. 4.66). In a large space, a path can be *random* (without form) and be decided by the activities within the space.

In summary, circulation, an important aspect of building design, refers to the main external and internal movements of persons, vehicles, and goods in and around a building. These movements depend partly on (a) the purpose of the

Fig. 4.66 Circulation space provides spaces for resting and viewing, and depicts the importance of the place

building and (b) the way the individual masses and spaces (mass means the building: space means the area around the building) are arranged. The purpose of the building decides the various categories of movement:

- Only vehicular, e.g., petrol bunk, multilevel car parking, automobile garage, etc.
- Vehicular + pedestrian, e.g., shopping complex, hospital, residential building, industrial building, school, institutional building, etc.

The diagramatic representation

The movement of vehicles, people, or goods can be represented in a simple, diagrammatic form. Figure 4.67 shows the circulation diagram for a residential

building. If the circulation diagram is drawn to scale, it becomes the plan of the building. The circulation diagram actually represents the movement between spaces and masses. It shows the external movement (movement outside the building) and the internal movement (movement within the building).

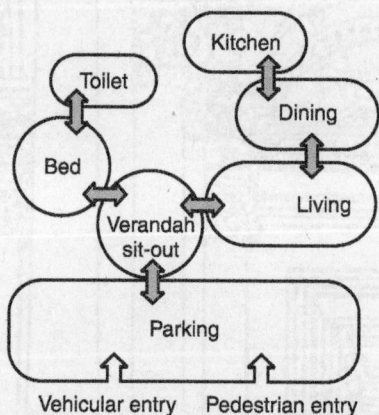

Fig. 4.67 Circulation diagram for a residential building

4.2.3 Anthropometrics

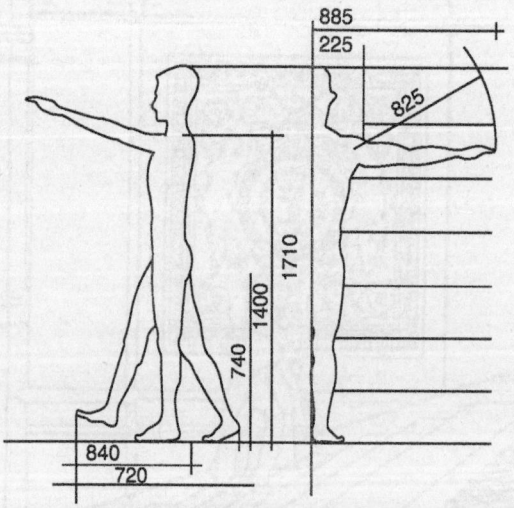

Anthropometrics are based on the dimensions and proportions of the human body (Fig. 4.68).

There is an assumption in architecture that the form (building) and space are decided by the dimensions of the human body.

Fig. 4.68 Dimensions and proportions of the human body decide the dimensions of objects and the spaces within and around them

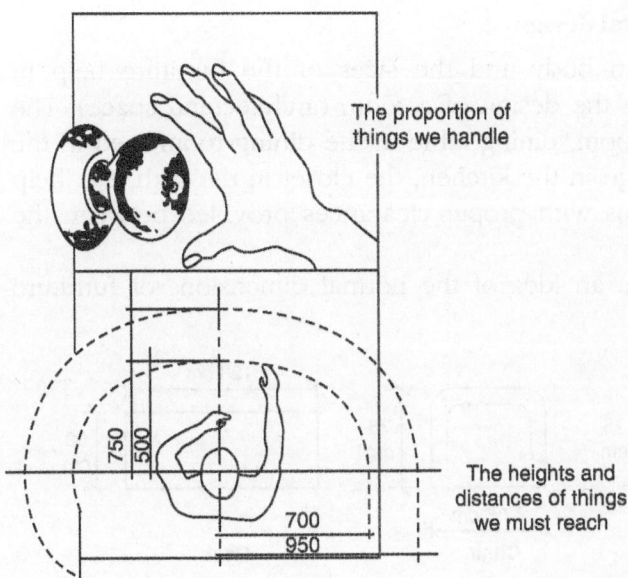

The proportion of
things we handle

The heights and
distances of things
we must reach

The dimensions and proportions of the human body affect the following (Fig. 4.69):

- The proportion of things we handle
- The heights and distances of things we must reach
- The dimensions of the furniture we use for sitting, working, eating, and sleeping (Figs 4.70, 4.72, 4.75, and 4.77)
- The volume of space required for movement, activity, and rest (for further details, refer to Section 2.3.4)

The volume of space required for movement, activity, and rest

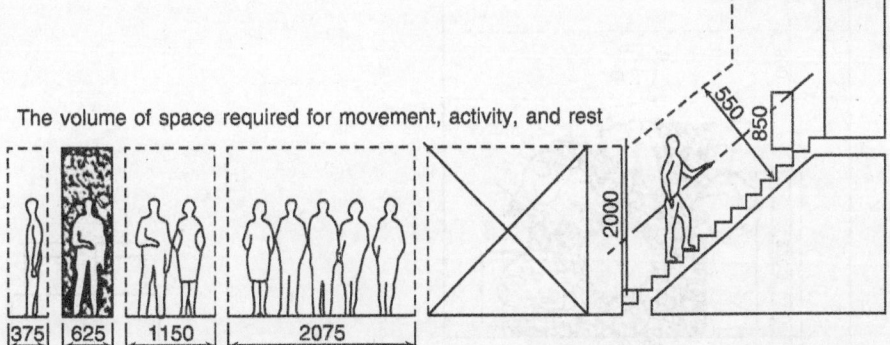

Fig. 4.69 Elements affected by anthropometrics

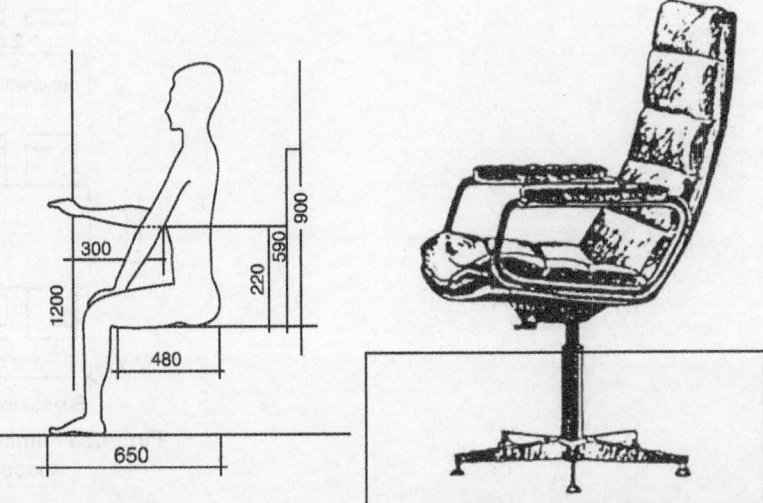

Fig. 4.70 The dimensions and form of a chair are decided by the dimensions of the human body

Anthropometrics in architectural design

The dimensions of the human body and the sizes of the furniture help in deciding the space required in the design of a room (architectural space). The size of the sofa in the living room, dining table in the dining room, bed in the bedroom, stove, sink, and fridge in the kitchen, the closet in the bath, etc. help to decide the size of the rooms with proper clearances provided between the furniture for easy movement.

Figures 4.71 to 4.80 provide an idea of the normal dimensions of furniture units in a residence.

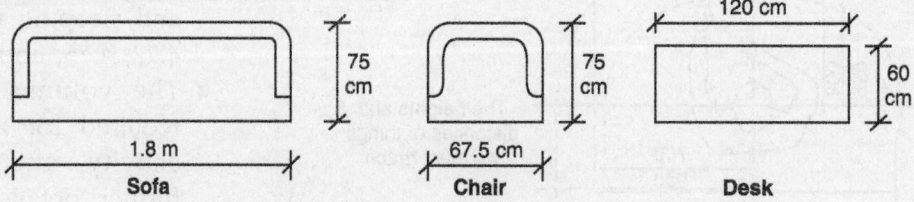

Fig. 4.71 Furniture units in a living space: sofa, chair, desk

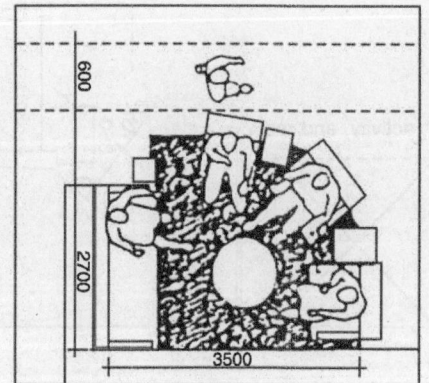

Fig. 4.72 Seating in a living room (all dimensions in mm)

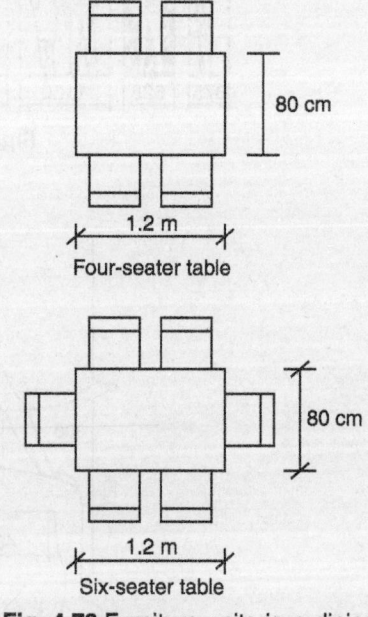

Fig. 4.73 Furniture units in a dining space

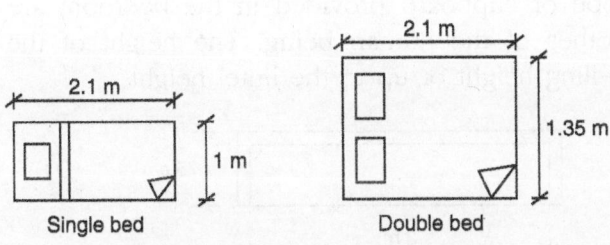

Fig. 4.74 Bed units in a bedroom

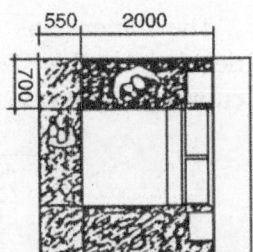

Fig. 4.75 Movement space in a bedroom (all dimensions in mm)

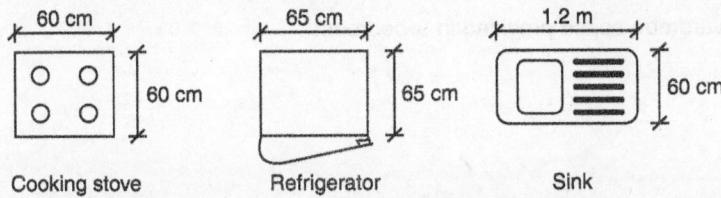

Cooking stove   Refrigerator   Sink

Fig. 4.76 Dimensions of typical appliances found in a kitchen

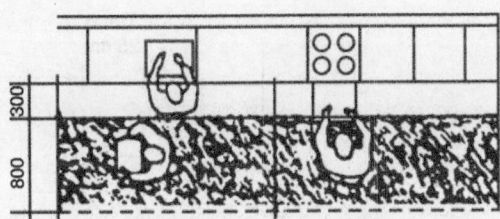

Fig. 4.77 Working in the kitchen

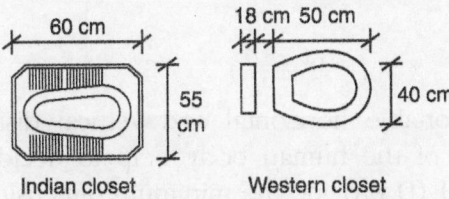

Indian closet   Western closet

Fig. 4.78 The closet in the bathroom

186 Principles of Architecture

The dimensions of a wardrobe or cupboard provided in the bedroom are decided by the size of the clothes of the human being. The height of the cupboard could be up to the ceiling height or up to the lintel height.

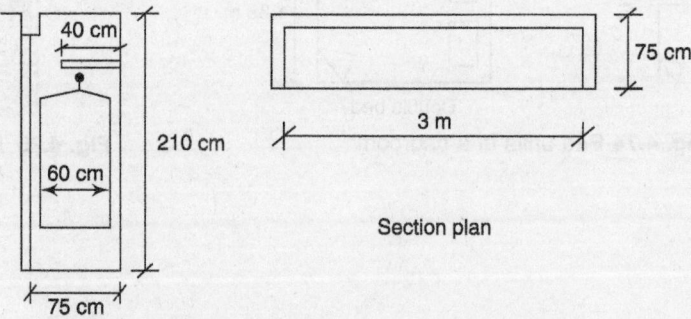

Fig. 4.79 Wardrobe space provided in a bedroom

Staircase dimensions are also based on human dimensions. The step width and height are based on the size of the human foot, making it comfortable for a human to ascend the stairs (Fig. 4.80).

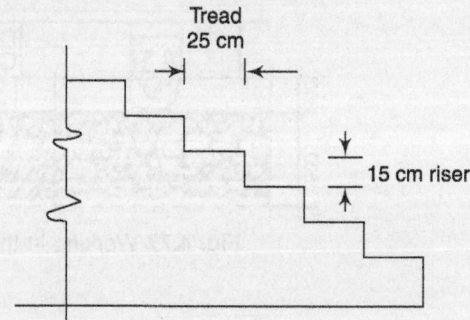

Fig. 4.80 Staircase section showing the treads and risers

Figure 4.81 shows how the comfortable horizontal and vertical distances from equipment and the dimensions of the human body help to decide the optimal sizes of rooms. Figures 4.82(a)–(f) provide the minimum dimensions of spaces required for basic human activities.

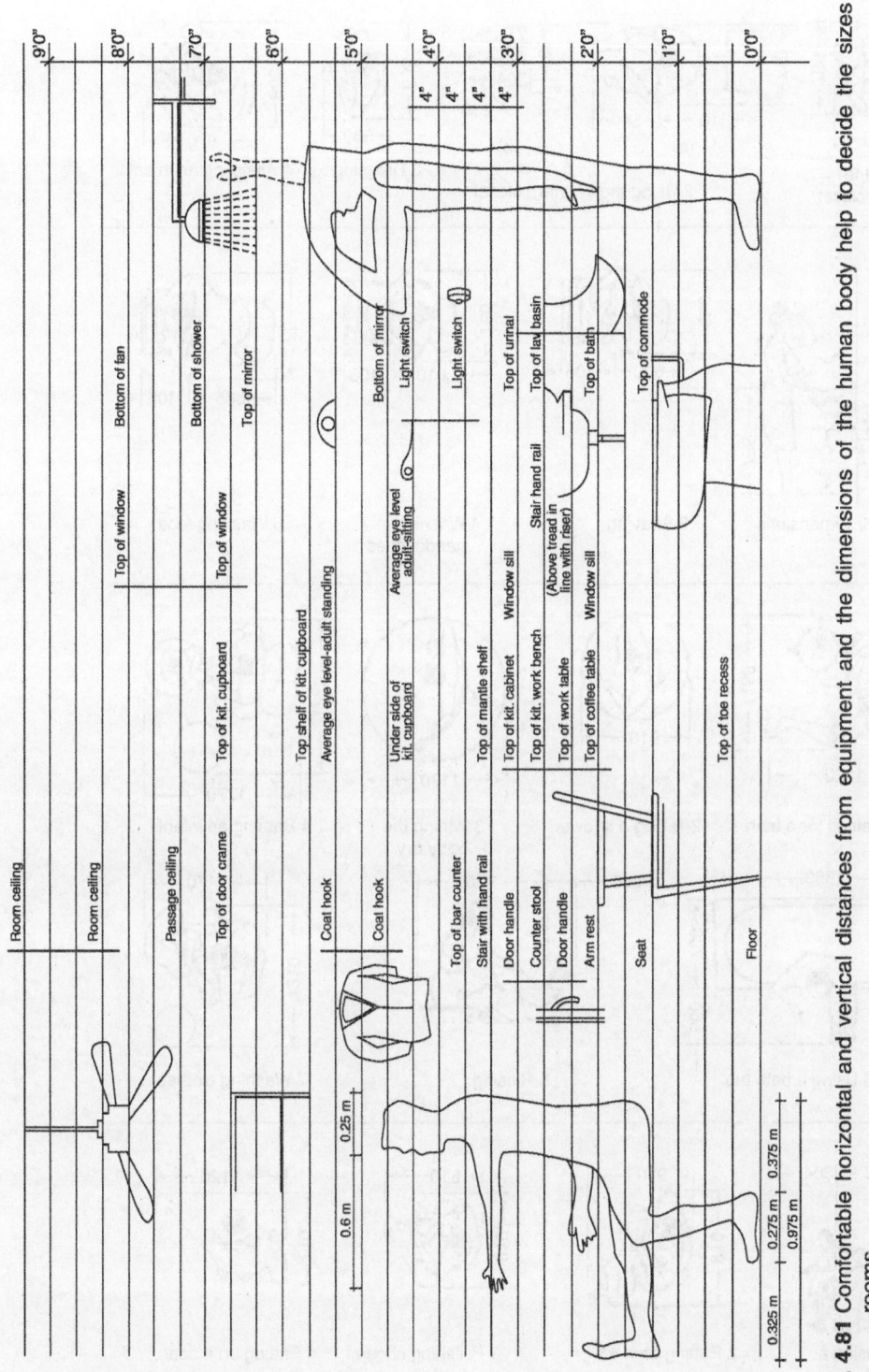

Fig. 4.81 Comfortable horizontal and vertical distances from equipment and the dimensions of the human body help to decide the sizes of rooms

All dimensions in mm

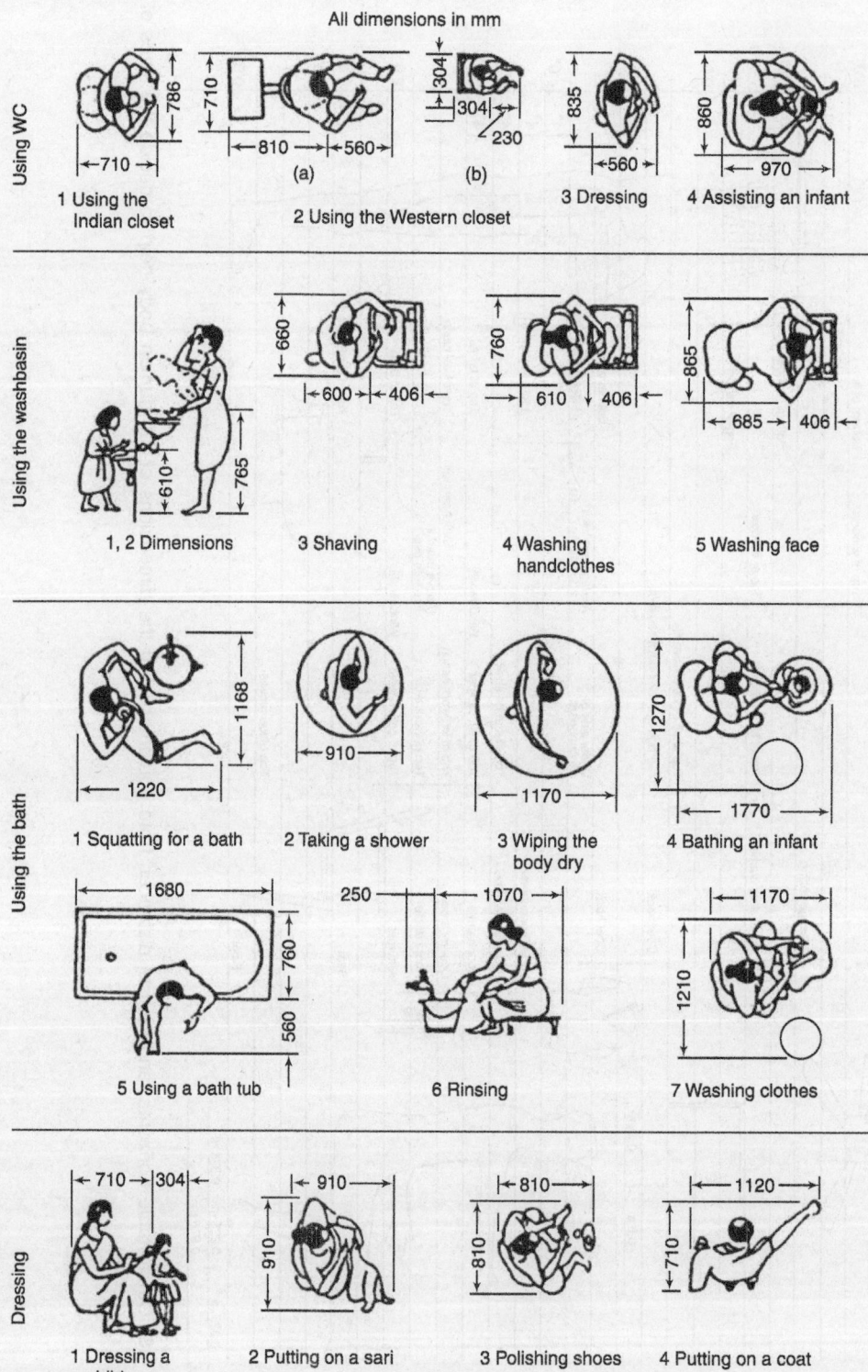

Using WC

1 Using the Indian closet

2 Using the Western closet

3 Dressing

4 Assisting an infant

Using the washbasin

1, 2 Dimensions

3 Shaving

4 Washing handclothes

5 Washing face

Using the bath

1 Squatting for a bath

2 Taking a shower

3 Wiping the body dry

4 Bathing an infant

5 Using a bath tub

6 Rinsing

7 Washing clothes

Dressing

1 Dressing a child

2 Putting on a sari

3 Polishing shoes

4 Putting on a coat

Fig. 4.82(a)

All dimensions in mm

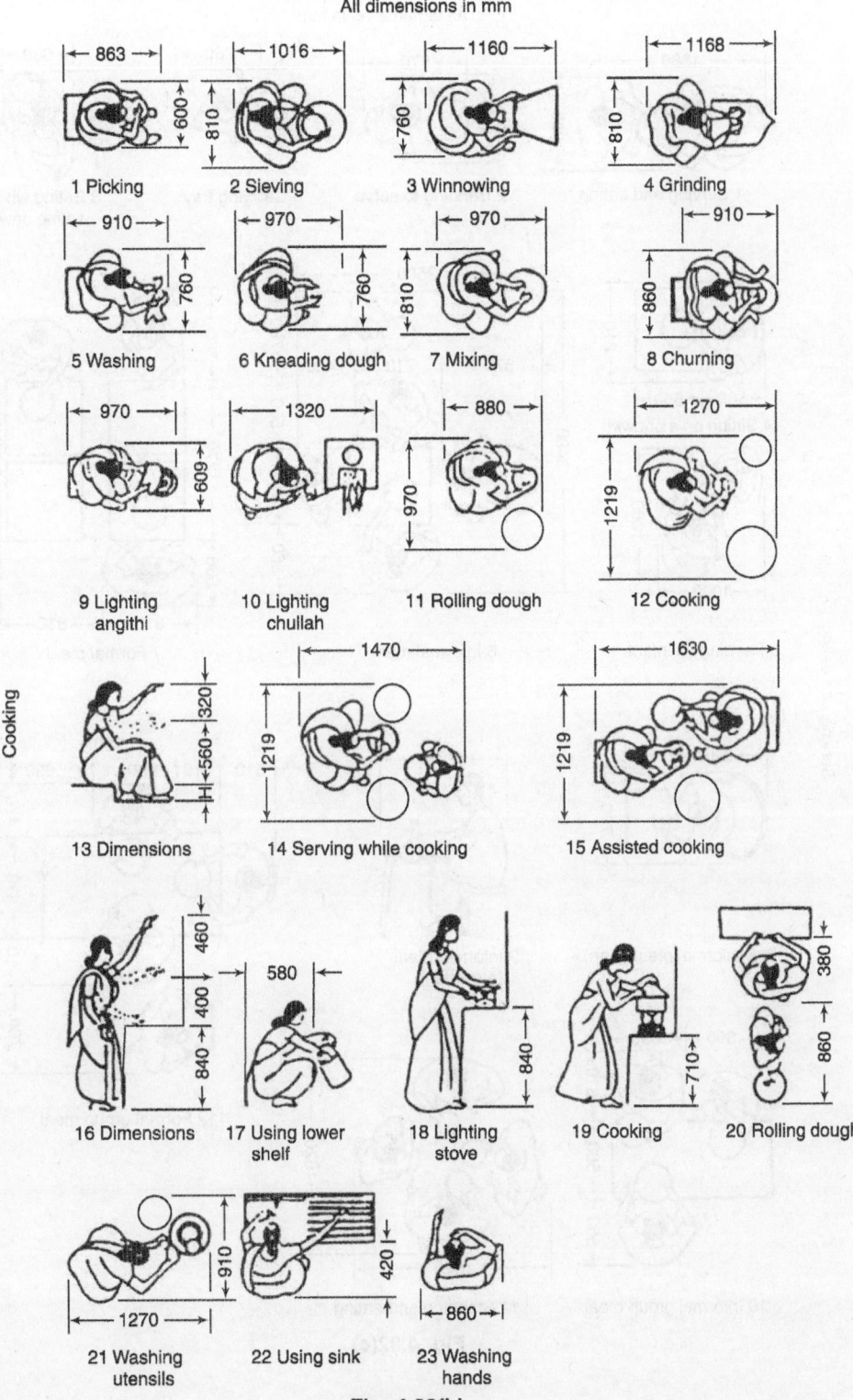

1 Picking

2 Sieving

3 Winnowing

4 Grinding

5 Washing

6 Kneading dough

7 Mixing

8 Churning

9 Lighting angithi

10 Lighting chullah

11 Rolling dough

12 Cooking

Cooking

13 Dimensions

14 Serving while cooking

15 Assisted cooking

16 Dimensions

17 Using lower shelf

18 Lighting stove

19 Cooking

20 Rolling dough

21 Washing utensils

22 Using sink

23 Washing hands

Fig. 4.82(b)

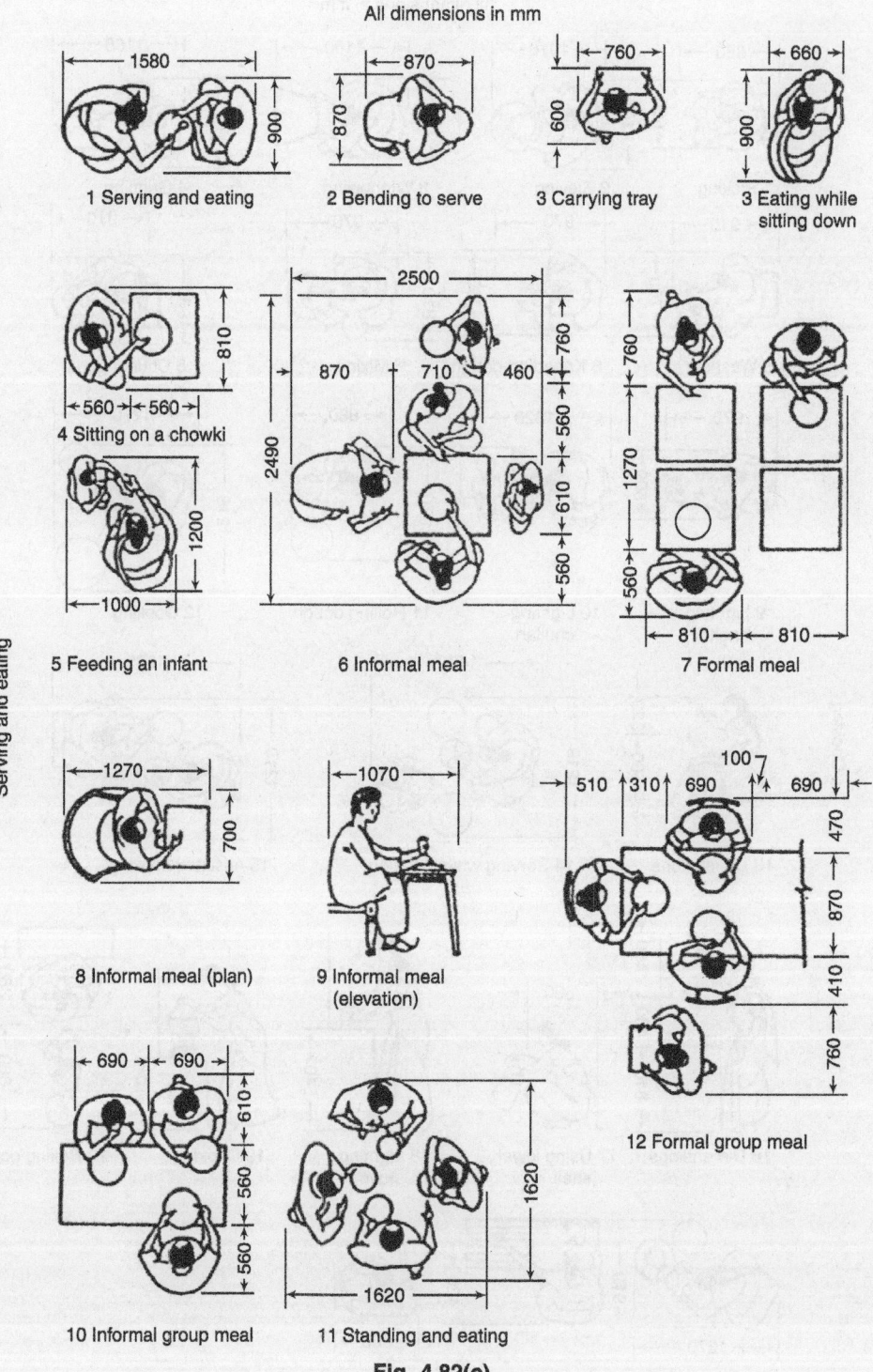

All dimensions in mm

1 Serving and eating

2 Bending to serve

3 Carrying tray

3 Eating while sitting down

4 Sitting on a chowki

5 Feeding an infant

6 Informal meal

7 Formal meal

8 Informal meal (plan)

9 Informal meal (elevation)

12 Formal group meal

10 Informal group meal

11 Standing and eating

Serving and eating

Fig. 4.82(c)

All dimensions in mm

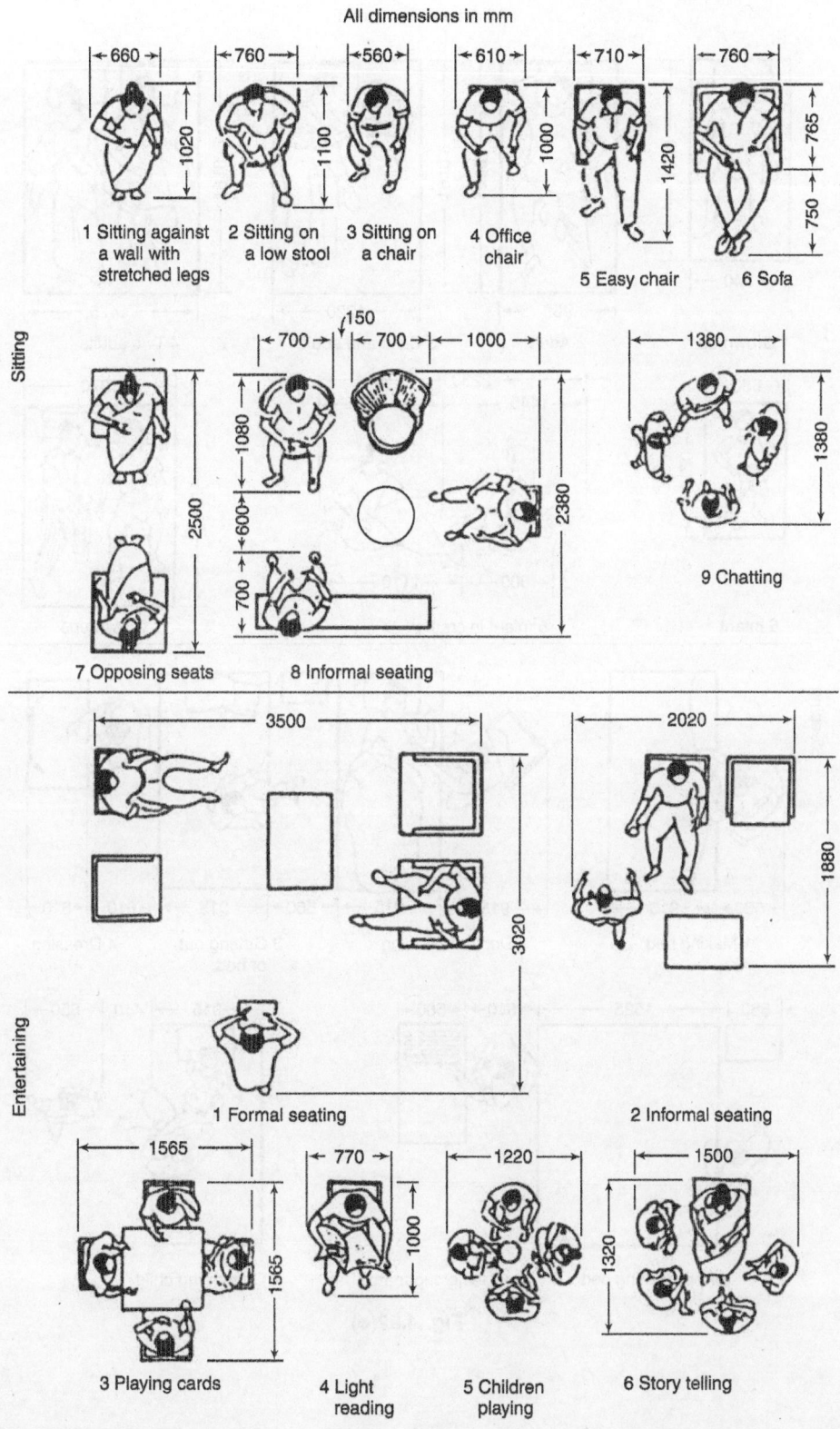

Sitting

1 Sitting against a wall with stretched legs
2 Sitting on a low stool
3 Sitting on a chair
4 Office chair
5 Easy chair
6 Sofa

7 Opposing seats
8 Informal seating
9 Chatting

Entertaining

1 Formal seating
2 Informal seating

3 Playing cards
4 Light reading
5 Children playing
6 Story telling

Fig. 4.82(d)

All dimensions in mm

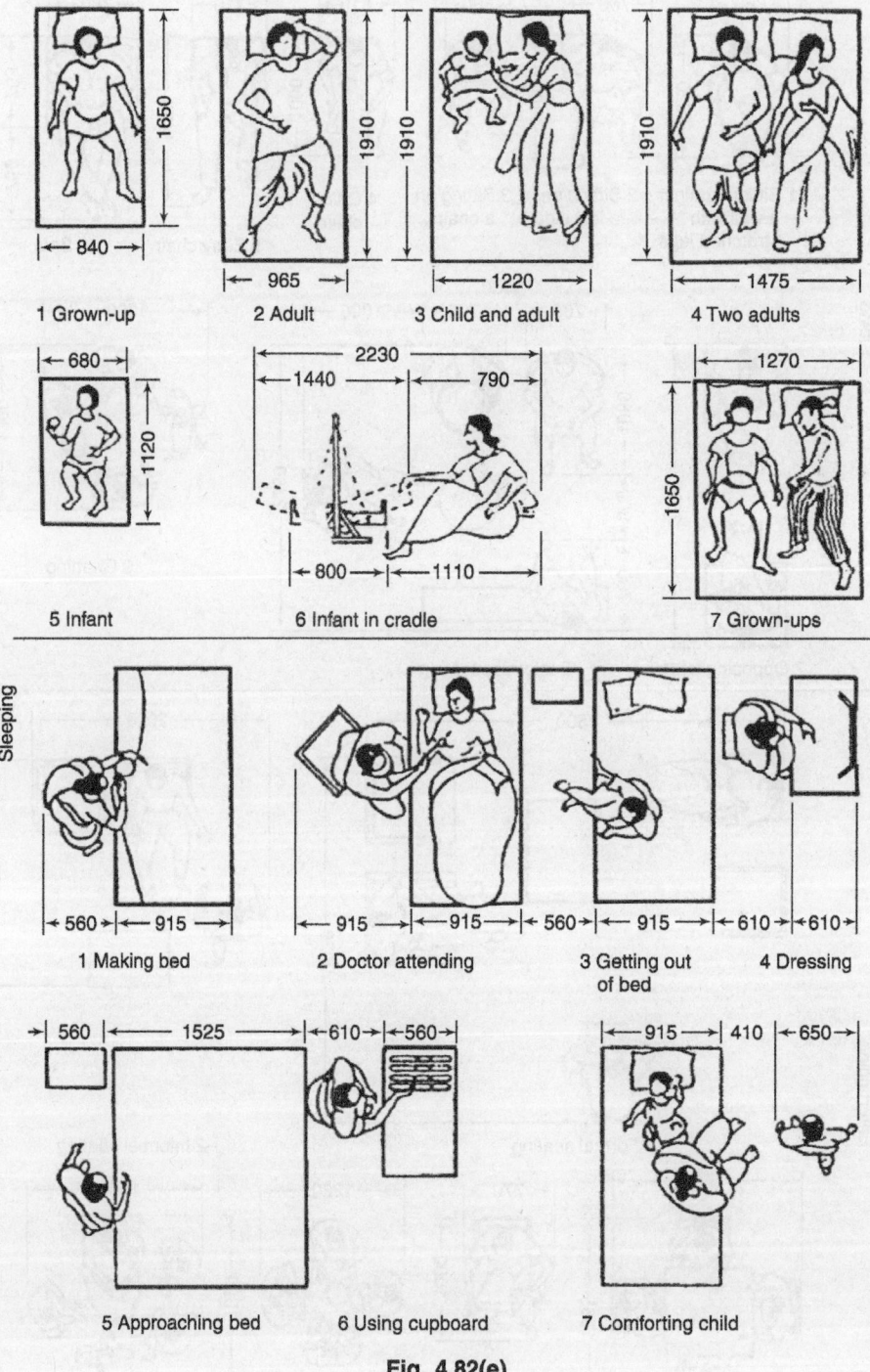

1 Grown-up   2 Adult   3 Child and adult   4 Two adults

5 Infant   6 Infant in cradle   7 Grown-ups

Sleeping

1 Making bed   2 Doctor attending   3 Getting out of bed   4 Dressing

5 Approaching bed   6 Using cupboard   7 Comforting child

Fig. 4.82(e)

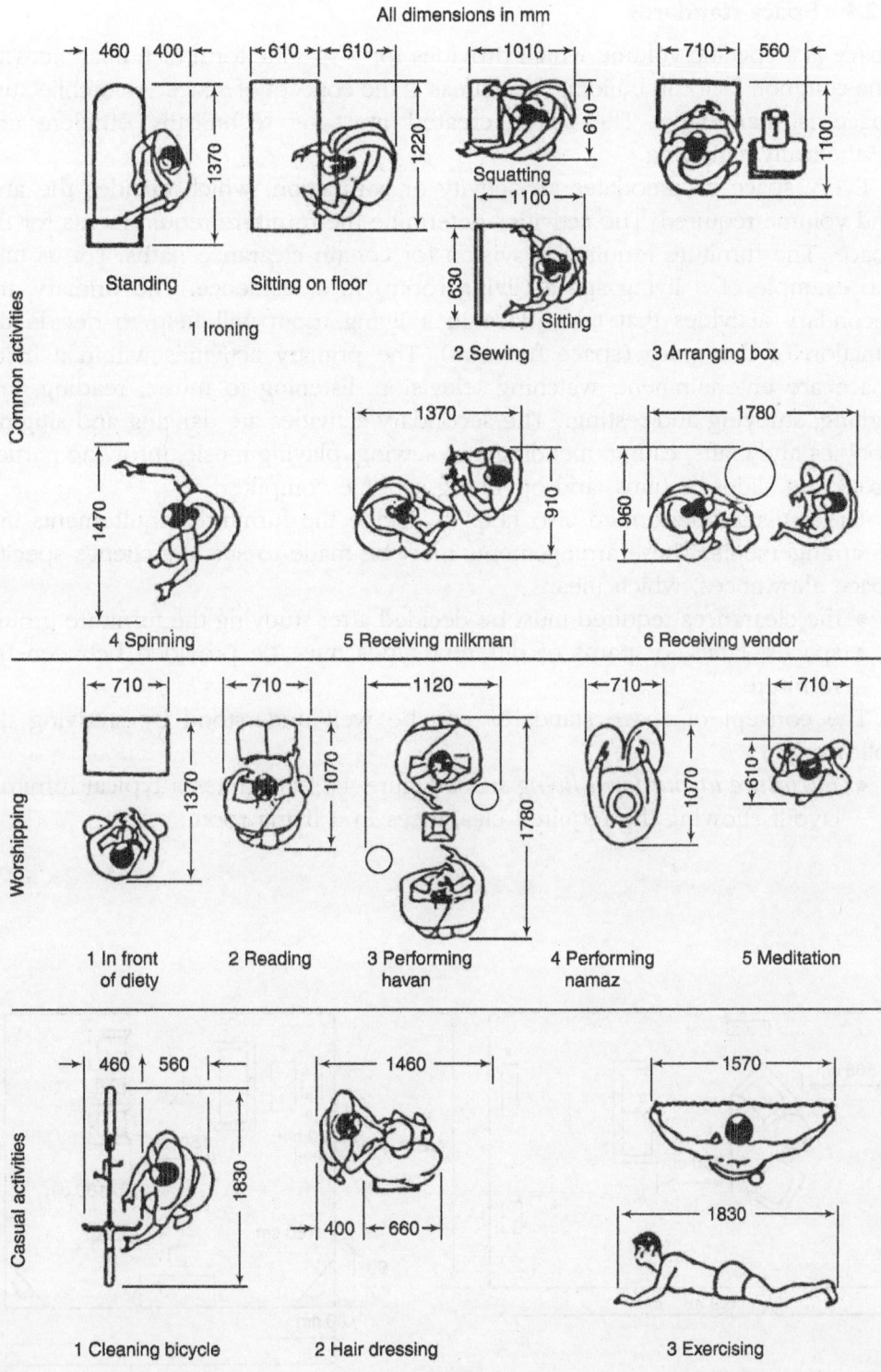

Fig. 4.82 Minimum dimensions of spaces for basic human activities

4.2.4 Space standards

Space is a specific volume which provides for a specific form of human activity. The common factor in buildings and areas is the concept of space. An architectural space is man-made. The space created must be technically efficient and aesthetically satisfying.

Every space accomodates an activity or a function, which decides the area and volume required. The activities determine the furniture requirements for the space. The furniture requires provision for certain clearance paths. Let us take the example of a living space (living room) in a residence. The primary and secondary activities that take place in a living room will help to decide the function of the space (space function). The primary activities within a living space are entertainment, watching television, listening to music, reading, and writing, studying and resting. The secondary activities are dancing and singing, hobbies and crafts, eating, mending and sewing, playing music, throwing parties, projecting slides or films, and operating a home computer.

The activities performed also help to decide the furniture requirements and its arrangements. These arrangements must be made to suit the client's specific space allowances, which means

- the clearances required must be decided after studying the furniture groups
- spaces, lanes, or paths of different types must be provided between the furniture

The concept of space standards can be well understood by studying the following.

- *Furniture layout for a living space* Figure 4.83 illustrates a typical furniture layout showing the required clearances in a living room.

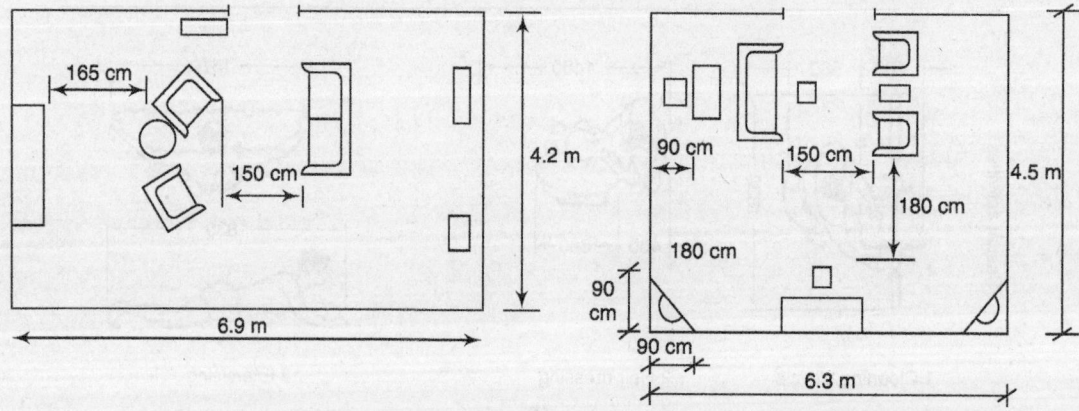

Fig. 4.83 Typical furniture layout showing the clearances in living room

- *Furniture layout for a dining space* The main criteria in dining space design is to provide adequate space for comfortable use of the dining area. Clearances should be provided in front of and sometimes around the furniture to allow activities to take place freely. See Figs 4.84(a) and (b).

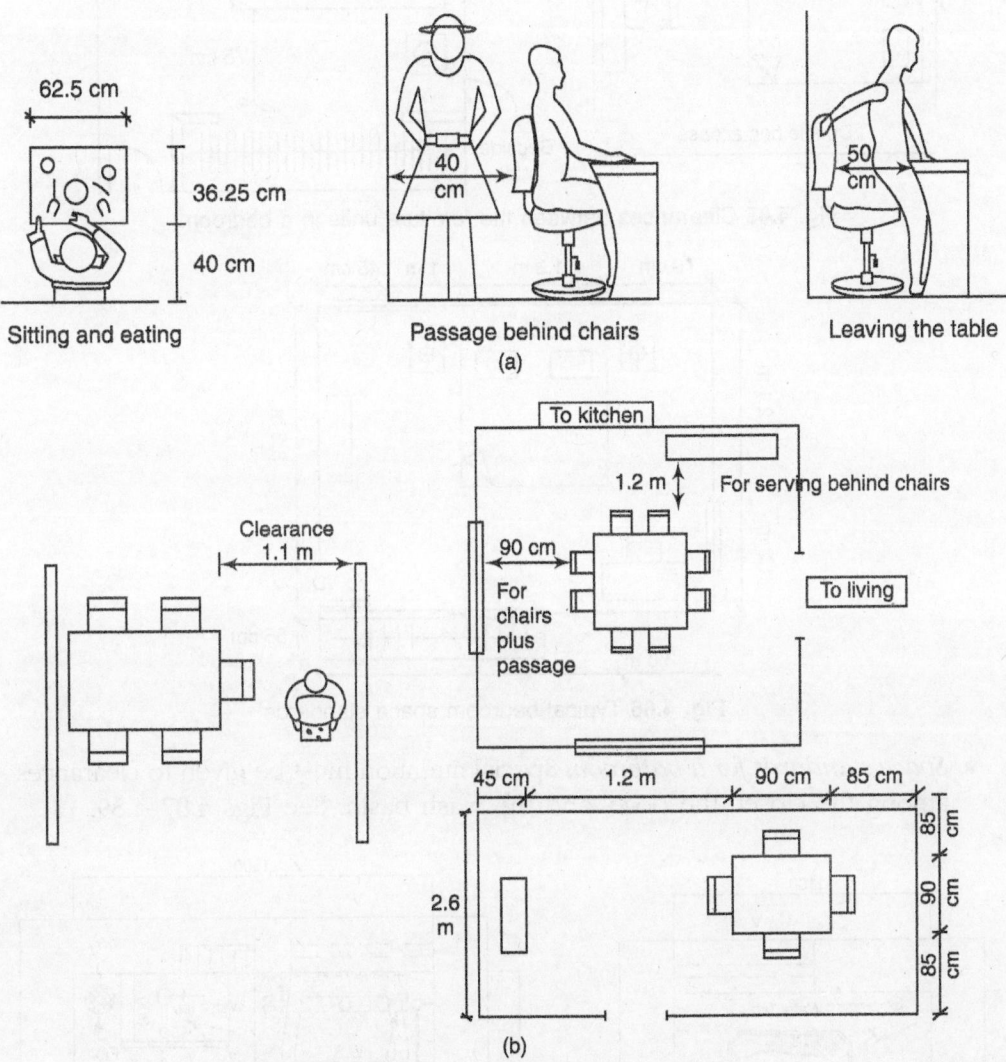

Fig. 4.84 Clearances required for a dining space

- *Space standards for a bedroom* Clearance must be provided in front of and around the furniture in a bedroom, so that the activities of sleeping and dressing can take place comfortably. The location of doors and windows should permit alternate furniture arrangements. See Figs 4.85 and 4.86.

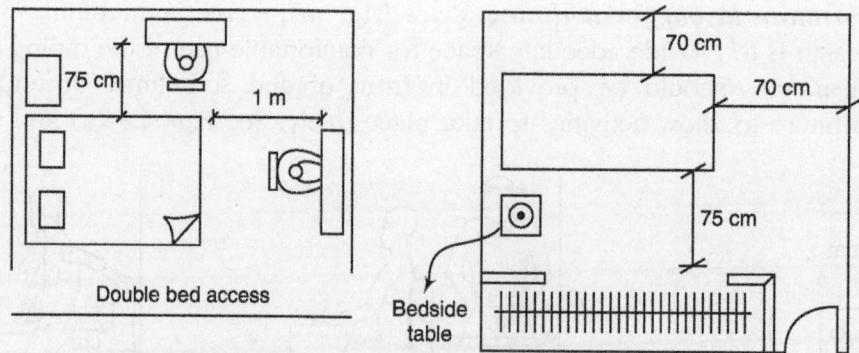

Fig. 4.85 Clearances between the furniture units in a bedroom

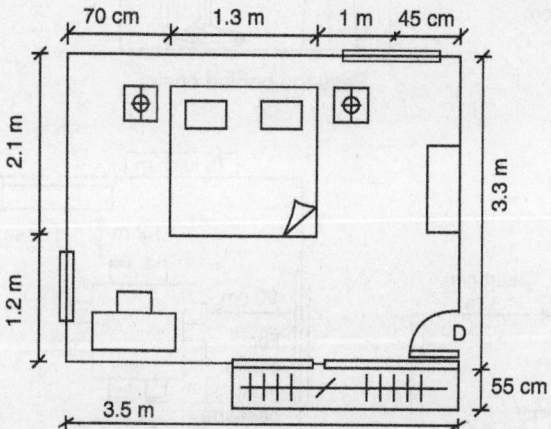

Fig. 4.86 Typical bedroom space standards

- *Space standards for a bathroom* Special attention must be given to clearances among the closet, the door, and the wash basin. See Figs 4.87–4.89.

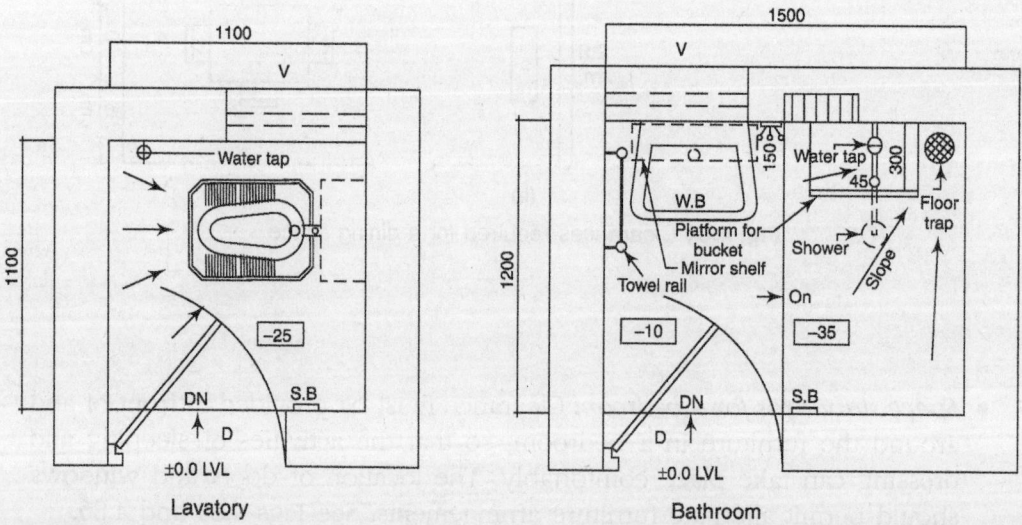

Fig. 4.87 Door clearance and space inside a lavatory: sink and bathing requirement (all dimensions in mm)

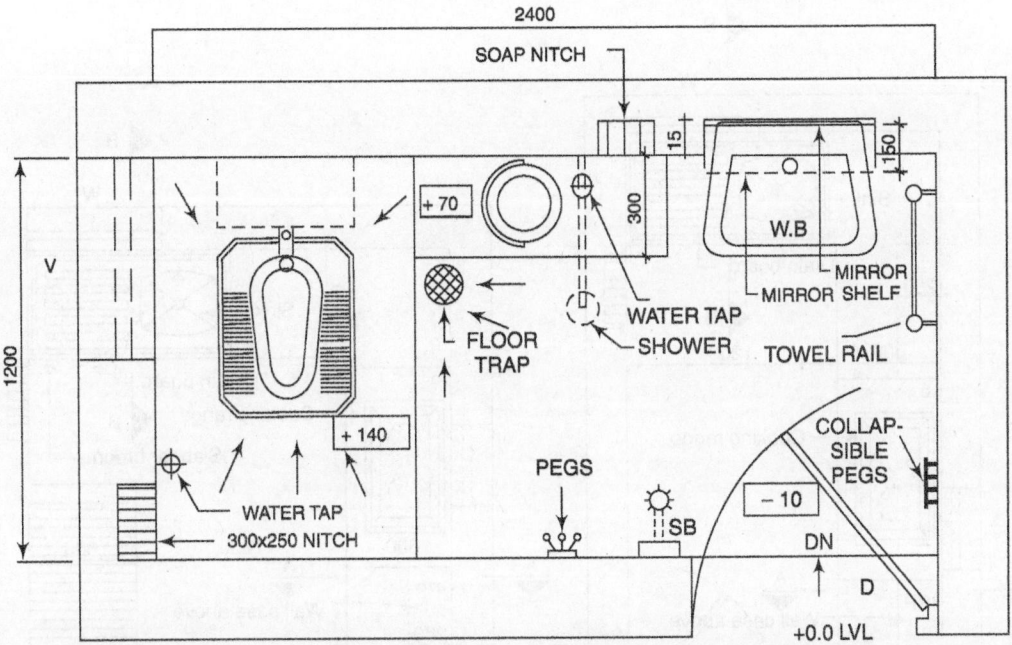

Fig. 4.88 Space standards for lavatory and bathing (all dimensions in mm)

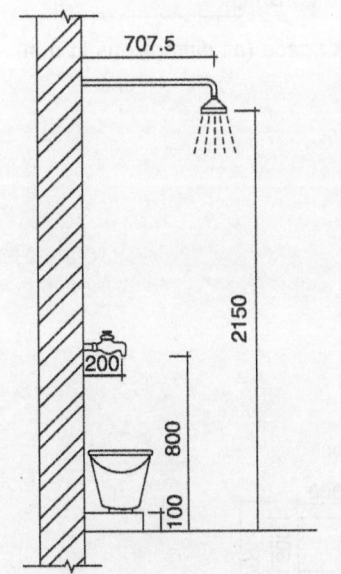

Fig. 4.89 Projection of water tap and shower (all dimensions in mm)

Door The minimum door width of 65 cm is enough. The door swing should not cross over the person using the closet. Good ventilation is necessary in bathrooms to reduce humidity and odour.

- *Space standards for a kitchen* A kitchen is a specialized work room, as it has many uses. It is used for food preparation, storage of food and utensils, and eating. In a kitchen, the work area (place to use the mixer, cut vegetables, etc.), the sink, and the cooking range (stove) should be so located that the movement inside the kitchen is easy. See Figs 4.90 and 4.91.

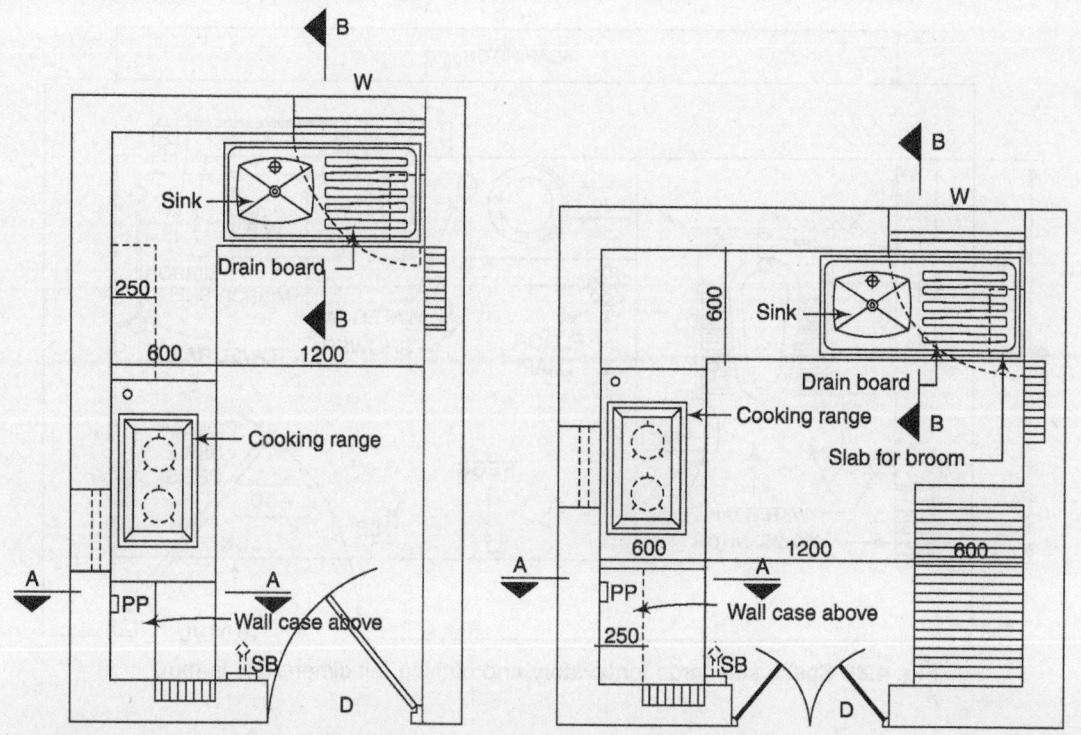

Fig. 4.90 Kitchen plan showing the position of cooking range, sink, and work space (all dimensions in mm)

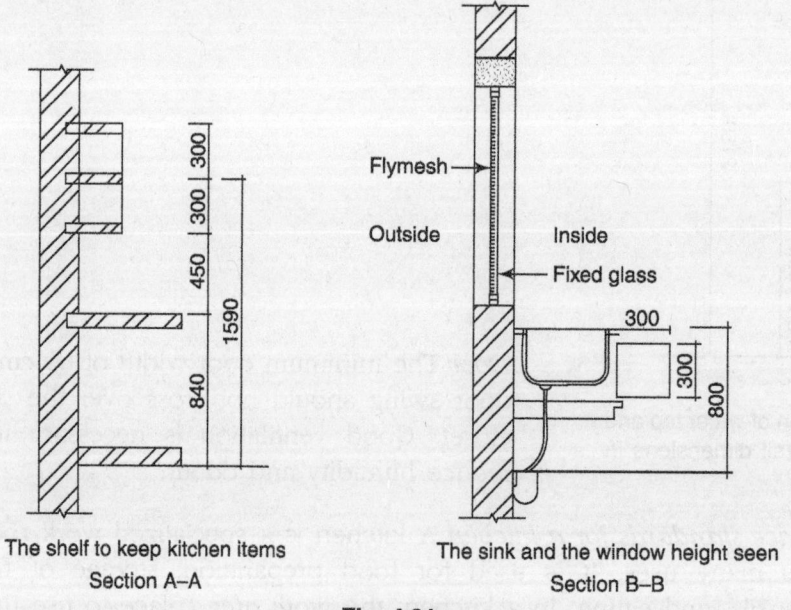

The shelf to keep kitchen items
Section A–A

The sink and the window height seen
Section B–B

Fig. 4.91
(all dimensions in mm)

- *Parking standards*The minimum space required for car parking is 2.5 m × 5 m. The minimum space for a bus parking is 4 m × 11 m. (Fig. 4.92)

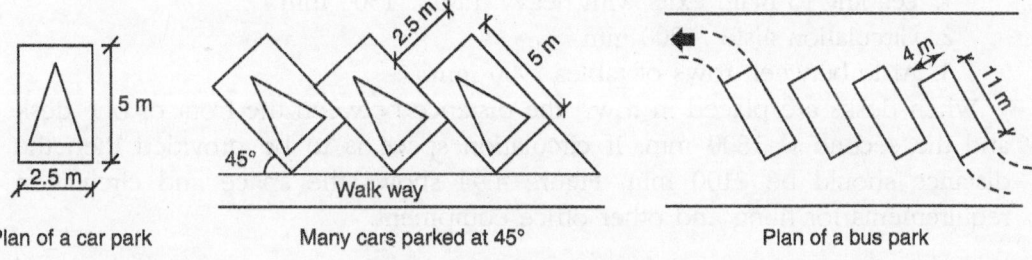

Plan of a car park   Many cars parked at 45°   Plan of a bus park

Fig. 4.92 Parking standards

Parking layout in a public building or commercial building (office building) The standard to be followed is five to six car parks for 95 sq. m of leasable area. The cars can be parked at an angle of 45 degrees or also at 90 degrees parallel to each other. Hedges and trees can be used to separate parking lots from the road (Fig. 4.93).

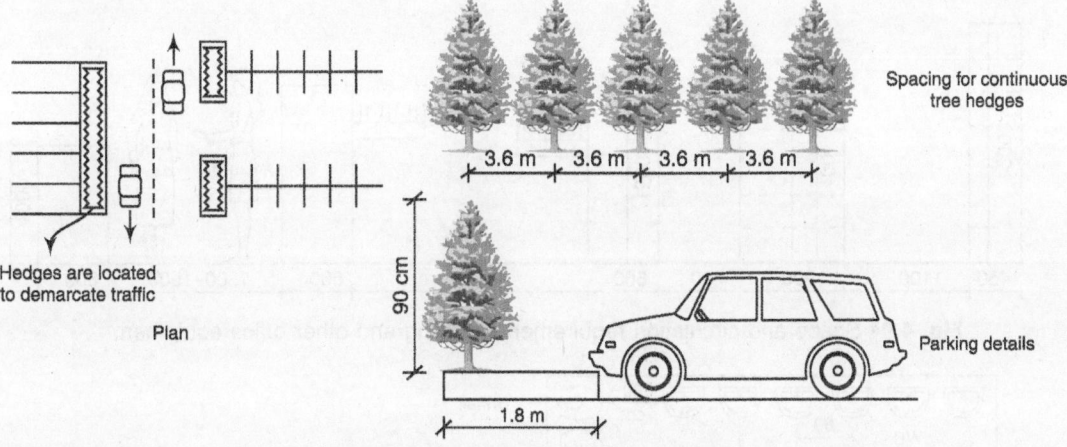

Hedges are located to demarcate traffic

Plan

Spacing for continuous tree hedges

Parking details

Fig. 4.93 Parking in a public building

Design of an office space

In an office, a lot of furniture and movement spaces for people are required. Office cubicles must be planned keeping anthropometrics in mind.

Space allowances (in sq. m)

Top executives	35.5
Junior executives	15–25
Supervisor	7.5–10
Operators @ 1500 mm desk	5.5
Clerk @ 1200 mm desk	5

(For seating side by side plus space for file cabinets.)

Reception: Waiting space per person is 1 sq. m (minimum 5 sq. m)

Aisle:

1. Leading to main exits with heavy traffic 1500 mm
2. Circulation aisle 1200 mm
3. Aisle between rows of tables 900 mm.

When desks are placed in rows, the distance between the front of one desk and the second is 1800 mm. If circulation space is to be provided then the distance should be 2100 mm. Figure 4.94 shows the space and circulation requirements for filing and other office equipment.

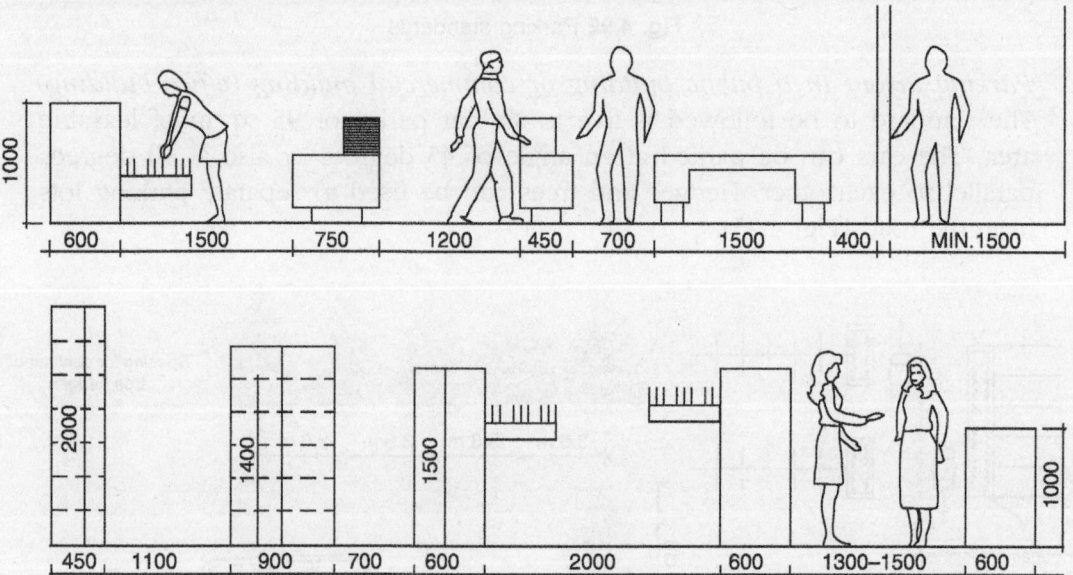

Fig. 4.94 Space and circulation requirements of filing and other office equipment

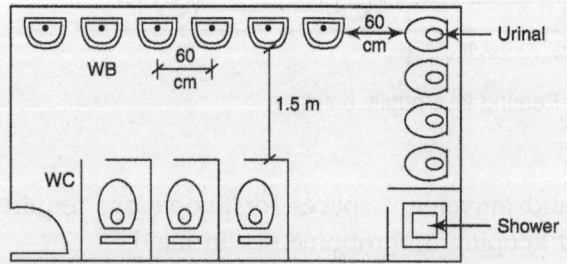

Fig. 4.95 Toilet clearances in a public building

The toilets provided in public and office buildings should also have the required clearances (Fig. 4.95).

Furnishings provided in an office

Office furniture includes filing cabinets, desks, chairs, tables, and storage units (Fig. 4.96). A desk is the universal item of an office workstation. It is a suitable workstation with drawers for stationery, office supplies, and tools. The major anthropometric consideration for the modern office is the computer workstation (Fig. 4.97). All stese components are designed to satisfy the physical and psychological needs of the user.

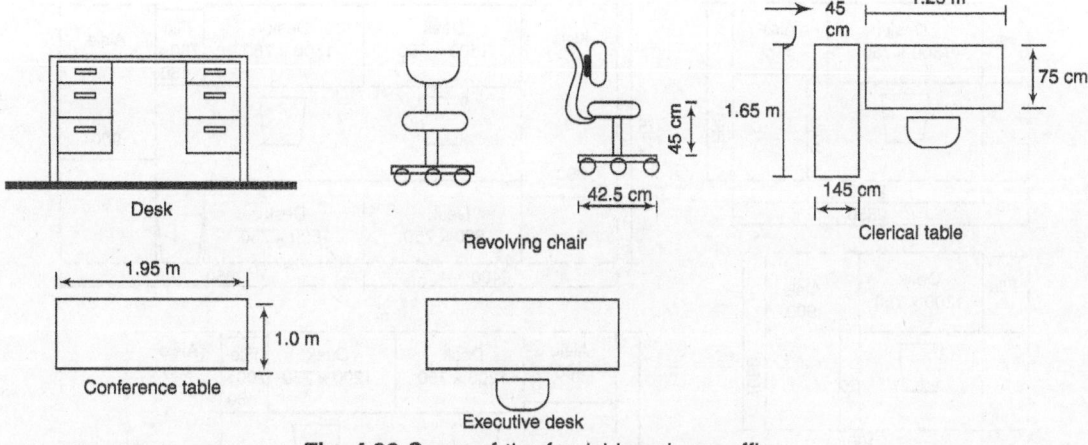

Fig. 4.96 Some of the furnishings in an office

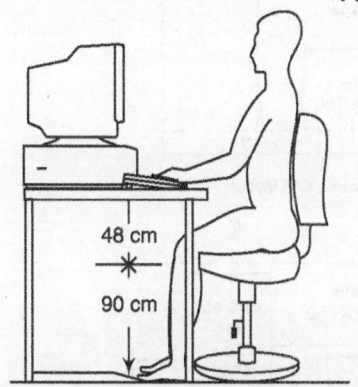

Fig. 4.97 The major anthropometric considerations for the use of a computer workstation

Office planning module or workstation An office layout may have a single workstation or clustered workstations, as shown in Fig. 4.98. These workstations are used to accommodate computers. Various arrangements could be worked out for the arrangement of the table unit, the desk, and the file cabinet. For example, the table unit could be placed beside the desk. The aisle space could also be worked out accordingly. Figure 4.99 shows various desk and file cabinet spacing layouts.

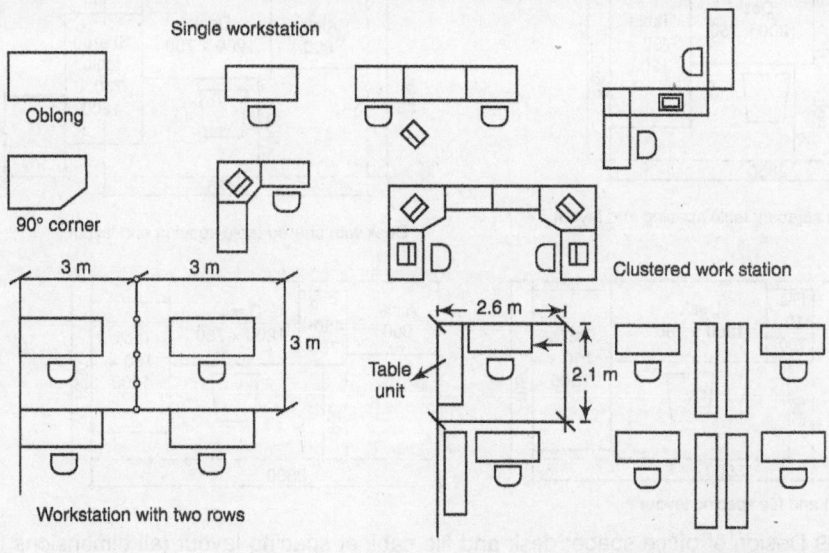

Fig. 4.98 Office planning module

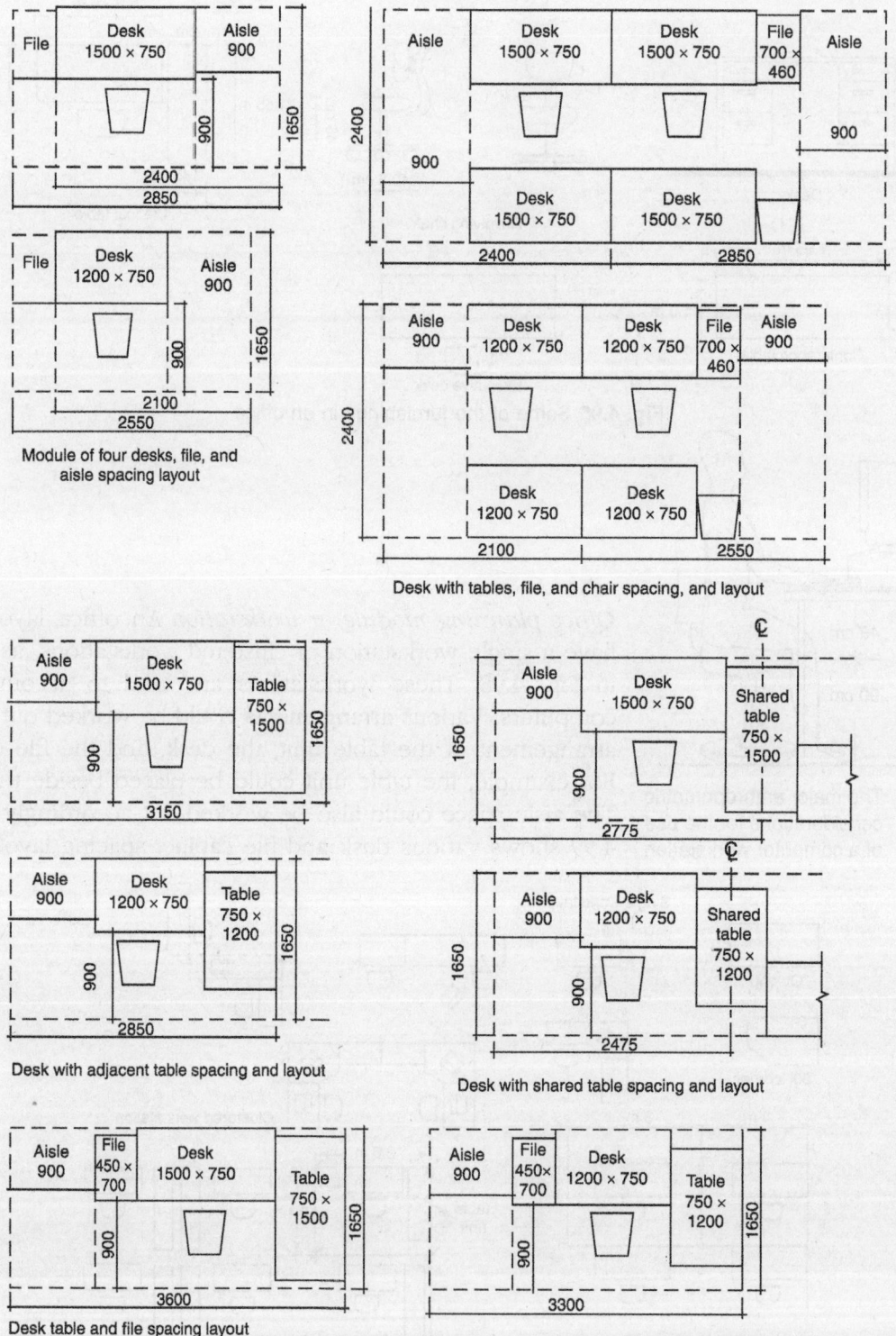

Module of four desks, file, and aisle spacing layout

Desk with tables, file, and chair spacing, and layout

Desk with adjacent table spacing and layout

Desk with shared table spacing and layout

Desk table and file spacing layout

Fig. 4.99 Design of office space: desk and file cabinet spacing layout (all dimensions in mm)

4.3 Environmental Factors

Environmental study for architectural design involves the collection of data, reconnaissance survey, creative ability and imagination, and the design of solutions to solve building problems.

When an architect is given a design assignment, there are many environmental factors that are to be considered. The site is the major factor that has to be considered. Site means the area or the land that is meant for the construction of the proposed project. Site planning is the art and science of arranging the various portions of a particular piece of land according to their uses. The site planner decides on the uses of the site in detail by selecting and analysing it for various characteristics of soil, slope, vegetation, etc.

The landscape involves the design of the outside space. This should be thought about carefully to make an architectural design complete. The climate at the location of the site is very important as it affects the building that is to be constructed. Services such as water supply, drainage, sanitation, electricity, fire protection, air conditioning system, etc. also have to be considered in order to make an architectural design complete.

4.3.1 Site planning

What is site planning?

Site planning involves arranging structures on the land and shaping spaces between them. It is an art linked to architecture, engineering, landscape architecture, and city planning. The site plan locates objects and activities in space and time. It may be concerned with a small cluster of houses, a single building and the surrounding space, or a small community built in a single operation.

Identification of site and its preparation

Each site has a unique nature of its own. The purpose for which it is to be used should be clearly understood. Every site when disturbed takes time to experience the mutual adjustment of its elements. For example, the flow of water creates a drainage pattern. Many factors are involved in the analysis of the site. These include the factors above the ground, below the ground, and on the ground, as discussed in the following.

Natural factors

- Geology
- Topography—slope analysis
- Hydrography—streams, lakes, swamps
- Soil—classification of types and uses
- Vegetation
- Wildlife

• Climatic factors—solar orientation, summer, and winter winds, humidity, precipitation

Geology The type of rock below the surface of the soil (if present as shown in Fig. 4.100), the depth, and the characteristic features of the rock should be identified. Such rocks could act as a foundation for many buildings. These are natural and could form visible land forms. The stability of such geological formations is also important.

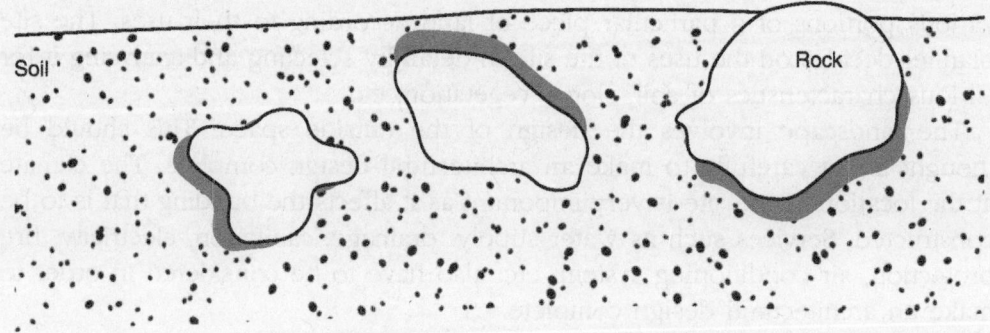

Fig. 4.100 Rock below the surface of the soil

Topography The form of land is called its topography. This is the most important factor to be analysed. Geology and the slow process of natural erosion (soil being worn away because of wind or water) are responsible for land forms and slopes. A topographic survey will reveal the badly drained areas and natural drainage channels. It will also reveal places that have good views and parts of the site that are visible or hidden from any selected point outside the site. The slopes will decide the roads and paths; a steep slope will increase building costs. See Fig. 4.101.

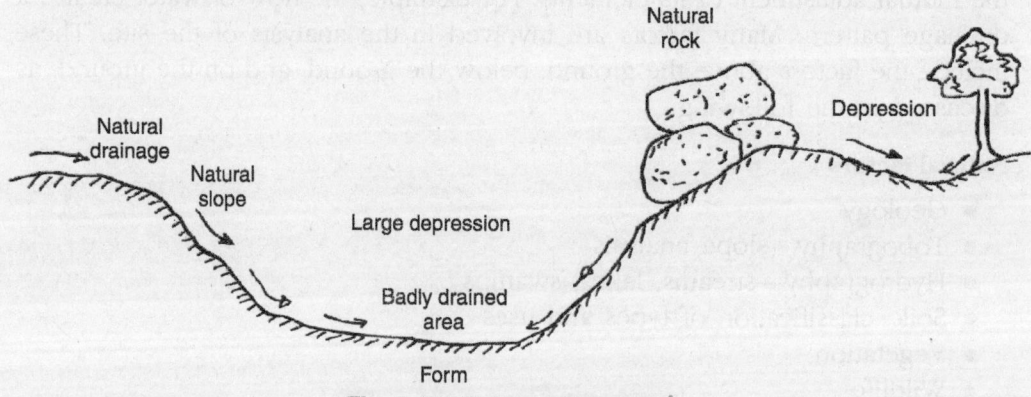

Fig. 4.101 Topographic survey

Hydrography   Hydrography provides information about all types of water bodies present in and around the site: lakes, streams, any marshy land (swamps), or natural wells. It also reveals the availability or otherwise of a ground water table and the depth at which it is available. See Fig. 4.102.

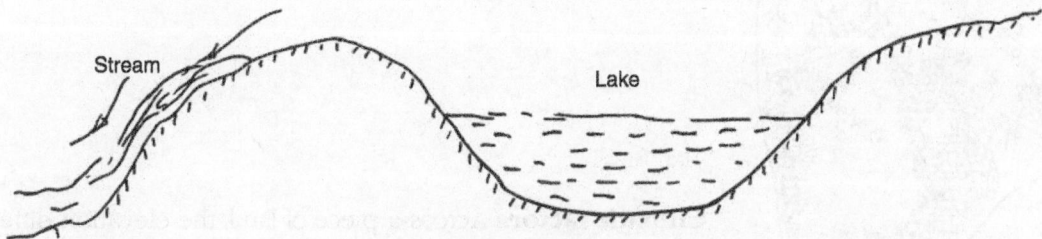

Fig. 4.102 Hydrographic survey

Soil   The different types of soil present are analysed. Soil decides the stability of land, foundation, suitability, excavation, erosion, drainage, and plant growth (as the top soil is essential for good plant growth). The bearing capacity of soil is an important factor to be considered while locating buildings.

Vegetation   A study of the vegetation helps in locating large existing trees, which can be retained. These can be used for providing seating (Fig. 4.103). The ecology of the area should also be examined to know what plants or shrubs would grow in that area.

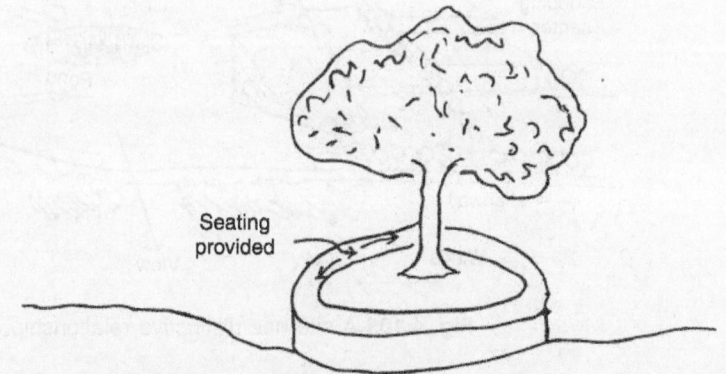

Fig. 4.103 Large existing trees on the site, which could be retained and seating provided around them

Wildlife   This is an important consideration when choosing sites for parks and recreation. Fishing and hunting are major recreational activities. The selection of land suitable for such activities depends on the natural wildlife present in the area. Wildlife also adds form, colour, and movement to the landscape.

One might also want to be informed about the wildlife present in an area to preserve it and not disturb the animals' natural habitats due to the construction.

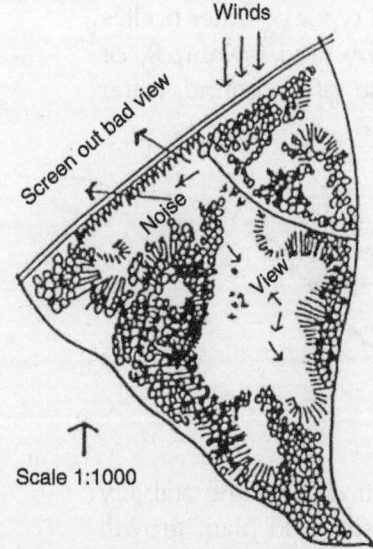

Scale 1:1000

Fig. 4.104 A site layout showing all external features

Climatic factors Across a piece of land, the elevation difference, character of topography, vegetation cover, and water bodies influence the climate of that area. On the other hand, precipitation and temperature are the major factors affecting vegetation. In cool and temperate climates, vegetation may be used to block winter winds. See Fig. 4.104. Figure 4.105 shows the layout of a site in relation to its surroundings.

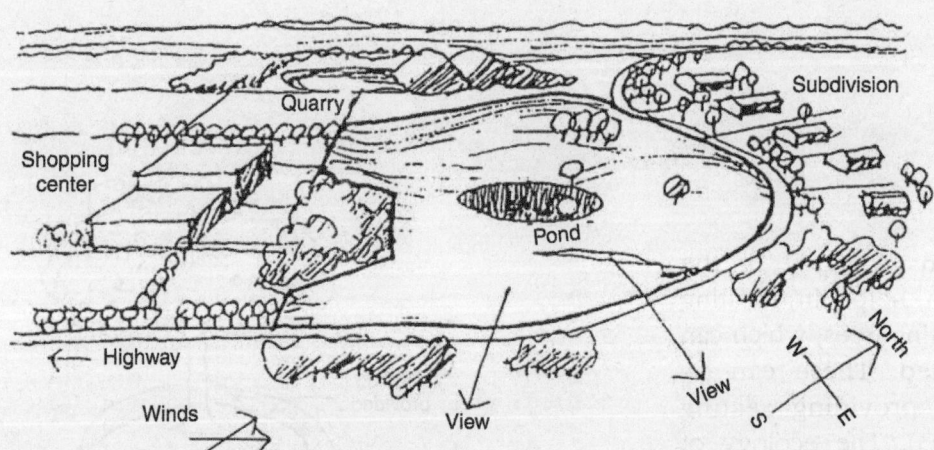

Fig. 4.105 A site has distinctive relationship characteristics

Cultural factors

- Existing land use—ownership of adjacent property and off-site nuisance
- Linkages
- Traffic and transit—vehicular and pedestrian circulation on or adjacent to site
- Density and floor area ratio
- Utilities—sanitation, water, gas, electricity, stormwater (rainwater) drainage
- Existing buildings
- Historic factors—historic buildings and landmarks

Existing land use This implies a survey of the present status of the land—whether it is residential, commercial, industrial, or recreational. The ownership of the adjacent site will also affect the land being surveyed.

Off-site nuisances Disturbances (nuisances) from outside and around the site (Fig. 4.106) have to be studied.

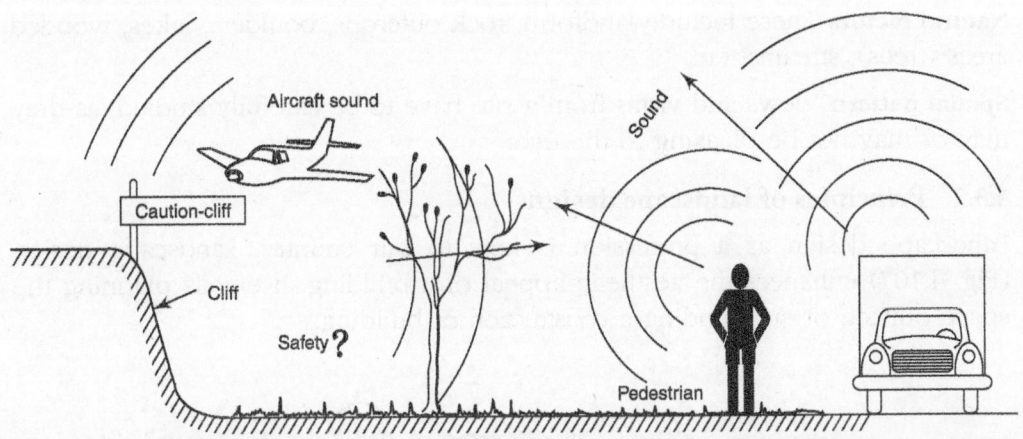

Fig. 4.106 Cultural factors affecting a site: off-site nuisances

Visual nuisance elements Power lines, water tower, certain industrial complexes, highways, advertisement boards, junkyards (waste dumps), etc., are some examples of eyesore elements that have to be taken into account.

Possible auditory (hearing) nuisance Noise produced by heavy automobiles, trains, air traffic, etc. and the surrounding population has to be studied.

Olfactory nuisance (nasal nuisance) Dumps, chemicals, and other wastes in and around the site have to be taken care of.

Safety hazards Severe or sudden changes in landform, such as a steep cliff at the edge of a site, have to be noted.

Linkages While studying the site, linkages to all surrounding social establishments and institutions should be taken into account to see whether, for instance, shopping centres, residential areas, churches, schools, parks, playgrounds, etc. are present in the close vicinity of the site.

Traffic and movement If the site is well connected with roads, the bus service, or the rail service for movement, this should be specified.

Density and zoning This is a legal element to be considered.

Utilities Water and electricity supplies are to be considered. Telephone connections should also be available at the site.

Existing buildings If a building already exists on the site, its plan has to be shown in the plan of the site.

History If the site is a big campus, then any building that has to be preserved has to also be considered.

Aesthetic factors

- Natural features
- Spatial pattern—views, spaces, and vistas

The natural features of the land and the pre-built forms (spatial pattern) will relate to the aesthetic factors.

Natural factors These include landform, rock outcrops, boulders, lakes, wooded areas (trees), streams, etc.

Spatial pattern Views and vistas from a site have to be carefully studied, as they may or may not be pleasing to the eye.

4.3.2 Principles of landscape design

Landscape design as a profession is new to our country. Landscape design (Fig. 4.107) enhances the aesthetic appeal of a building. It entails planning the space outside or surrounding a construction or building.

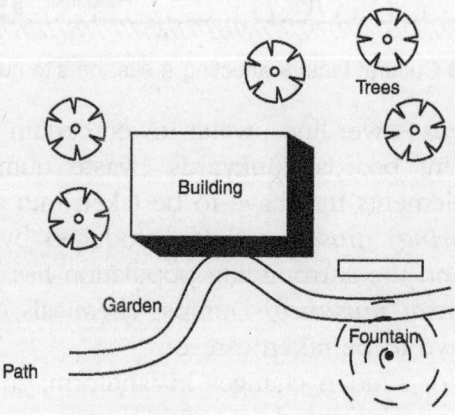

Fig. 4.107 Landscape development

A landscape designer manipulates and shapes the natural layout of a site (the outdoor scene) to suit his uses and produce aesthetic pleasure. The outdoor environment could be natural or artificial.

- **Natural environment** A landscape derived from the natural habitat of the region constitutes a natural environment and depends on the type of existing vegetation, such as a forest or a desert. Such an environment exists even in man's absence.
- **Artificial environment** Man alters the natural habitat by creating an artificial environment, such as farms or fields.
- **Man designs his own environment** Man designs the landscape outside the built environment, either by merging the building with the natural environment or by creating parks and gardens. In doing so, he makes the environment pleasant to live in.

Maintenance

A man-made landscape, once created, needs constant care and maintenance. Some landscape elements get worn out and destroyed in the course of time, and need to be replaced. Weeds should be constantly removed. Proper watering of plants should be done. Cutting and pruning must be carried out periodically. Fertilizers should be used. All this involves cost and skill.

Elements of landscape design

Landscape design can also be defined as the composition of masses and spaces. There are various elements of landscape. Some of these are explained in the following.

Path The path defines the passage in the landscape created. It could be paved or unpaved. Paving materials are used to eliminate hazards from mud and dust and to form a smooth surface for ease of circulation. The paving material could be natural or man-made. Man-made paving materials are available in a wide variety of textures and colours. Some of the materials used are explained in the following.

Stone Stone, one of the oldest paving materials, offers a good, durable, wearing surface with a minimum maintenance required. Rubble and ashlar masonry are the two forms of stone used for paving. Rubble masonry is rough stone, as it comes from the quarry, but may be trimmed somewhat where necessary. Ashlar masonry is hewed or cut stone from the quarry and is used much more often than rubble for the surfacing of walls. Figures 4.108–4.111 illustrate various stone pavements.

Fig. 4.108 Cobble stones used for paving children's play area, Scotland

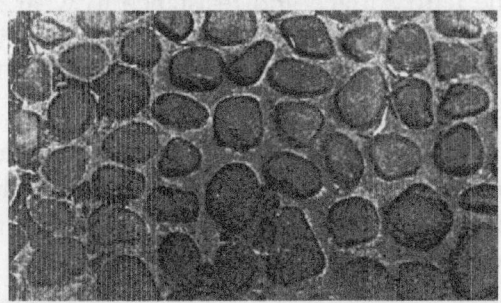

Fig. 4.109 Paving—pebbles laid flat in concrete to form an interesting texture

Fig. 4.110 A radial paving pattern used with a fountain, Lincoln Centre, New York City

Fig. 4.111 Tiles used in a fan pattern

Bricks Bricks are the oldest artificial building material in use today. They offer a great variety of textures and colours as well as flexibility in use. Composed of hard, burnt clay and shale, brick is available in many colours due to variation in the chemical content of clay. Bricks can be used to make tiles with varied shapes. Because of the hard surface and resistance to wear and cracking, these tiles are best suited for outdoor paving. These could be laid on stone bases or on concrete slabs. The most common patterns are running bond, herringbone, and basket weave. See Figs 4.112 and 4.113.

Concrete Concrete may be poured in situ, offers a variety in texture and colour, and forms a durable walking surface. It is extensively used as a paving material. Concrete lends itself to variations in finish, and may be smooth or rough, with aggregates exposed when desired.

Fig. 4.112 Brick pavers used in a herringbone pattern

Fig. 4.113 Hexagonal pavers are used for pedestrian walks or plaza areas

Asphalt Asphalt does not offer the variety of textures that concrete does, although it provides a softer walking surface. Asphalt is not as durable as concrete; it is less expensive and is used extensively for walkways on college campuses, in large parks and recreational areas, and in the construction of roads.

Rock gardens Arranged groups of rocks, big boulders, or large stones constitute a rock garden. The rock outcrop could be naturally present at the site, which could be converted into a rock garden. Alternatively, it could be created artificially by bonding large boulders with mud, cement, or concrete. This can be merged with plants growing in between the rocks. A water cascade can also be a part of the rock garden (Fig. 4.114).

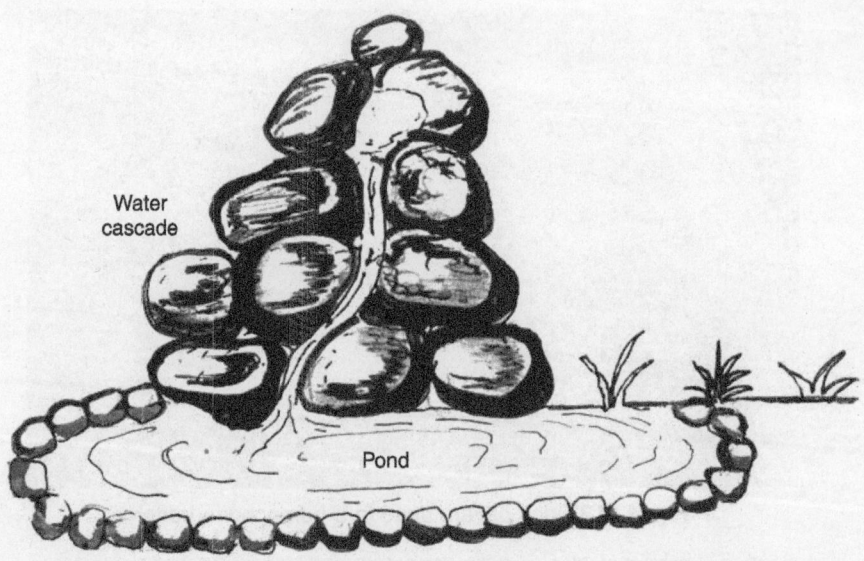

Fig. 4.114 Rock garden

Street furniture Any mode of seating provided in the external designed space (landscape space) is known as street furniture. These could be seats with backs or without backs. They are usually made of wood, concrete, stone, or metal casting. Concrete or stone seats may act as sculptural elements. They are easy to maintain and less prone to vandalism. Wooden benches with backrests are most comfortable. Concrete and metal cast seats can also have backs. Seating could also be combined with tree planters.

Street furniture provides for social and recreational gathering of people in outdoor spaces. Figures 4.115–4.120 illustrate various street furniture designs.

Fig. 4.115 Street furniture—wooden benches used in a sitting area at Constitution Plaza, Hartford

Fig. 4.116 Concrete slab bench with more detail for the stand

Fig. 4.117 Concrete slab bench

Fig. 4.118 Street furniture seating used in a small courtyard

Fig. 4.119 Benches made of cut stone

Fig. 4.120 Seating incorporated with planting as a design element, California

Plant containers These are tree planters or pots. Tree planters must be of appropriate sizes to enable trees to grow above structures such as parking garages. Trees and plants grow much better when they are planted directly in the ground. Pots are small plant containers which are versatile and may be moved or arranged for displays. Concrete is most commonly used for plant containers. These could be in the form of concrete rings, which are precast or made in situ (cast at site). Figures 4.121–124 show various types of plant containers.

Fig. 4.121 Plant container pots large enough to hold a variety of flowers or plants

Fig. 4.122 Precast concrete plant containers, Hartford

Fig. 4.123 Saucer-like flower pots, Pittsburgh

Fig. 4.124 Flower pots, Constitution Plaza, Hartford

Sculptures Sculptures sometimes act as focal points in courtyards or plaza areas. They may be created from natural or artificial materials and a great variety of forms, colours, and textures are possible. Stone and wood are some natural elements which may be employed to make sculptures.

Placement of a sculpture within a space depends upon sunlight and shadow patterns, which project the object in various ways during different times of the day. Night lights can also be adjusted to highlight the interesting aspects of the sculpture from various angles. The sculpture should be so located to take advantage of varying sight lines (Figs 4.125–4.127).

Fig. 4.125 Sculpture composed and oriented to be seen easily against a wooded backdrop, Connecticut

Fig. 4.126 Sculpture acting as a focal point, Institute of Technology, Cambridge

Fig. 4.127 Stone sculpture taking advantage of sunlight and shadow patterns

Water body This is the most interesting object in landscape design. Water, which is a natural element, can be a prominent feature in the landscape. It may be used in the form of fountains or pools for its reflective qualities (Fig. 4.128), differences in sound, or cooling effect.

Fig. 4.128 Water may be used in the form of a pool for its reflective qualities

Programming the flow of water in fountains is done by an electronic timing system, which may also control the night light sequence. The water flow and lighting must be coordinated to achieve maximum effect. Figures 4.129–4.132 shows various types of water fountains.

Fig. 4.129 Water flows out of a bowl in this fountain, University of Colorado

Fig. 4.130 Water body: water flows out of concrete elements, which appear to be abstract animals

Fig. 4.131 The plaza area steps down to the fountain, St. Louis

Fig. 4.132 A marvelous fountain at Los Angeles

Sculptural elements of granite or concrete can be integrated with a water cascade or fountain effect (Fig. 4.133).

Fig. 4.133 Water used in conjunction with a sculpture in this fountain, St. Louis

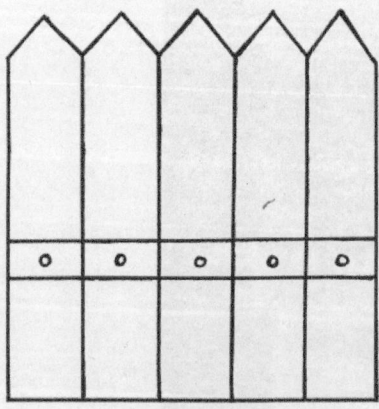

Fig. 4.134 Wooden fence used to separate walkways from landscape space

Fences or walls Fences or walls may be used to provide enclosures, articulate a space, or act as retaining elements. Brick, stone, and concrete are the materials most commonly used. Wood or metal grills can also be used as fences to separate the walkways from the planting areas and lawn spaces (Fig. 4.134). The height and type of wall vary with its use in the overall design concept of the project. They may be at seating height or up to 1.8 m or more in height to provide privacy.

Fig. 4.135 Walls—stone wall with coping, light used in conjunction with walls usually have metal covers so that they are not broken

Retaining walls may also have to be provided depending on the site conditions. Reinforced concrete is the most economical material for constructing such walls (Fig. 4.136).

Fig. 4.136 Dry stone retaining wall

Fig. 4.137 Precast concrete drain

Drainage Drainage mainly involves the provision of a stormwater drain. A utilitarian element, drainage can be integrated with the pavement in such a way that it becomes pleasing to the eye (Figs 4.137 and 4.138).

Fig. 4.138 Drain used in conjunction with paving pattern

Fig. 4.139 Precast concrete grates allow water to be absorbed by trees and add interest to the paving pattern

Precast concrete grates can be used around trees so that they look aesthetically pleasing and also allow water to be absorbed by trees (Fig. 4.139).

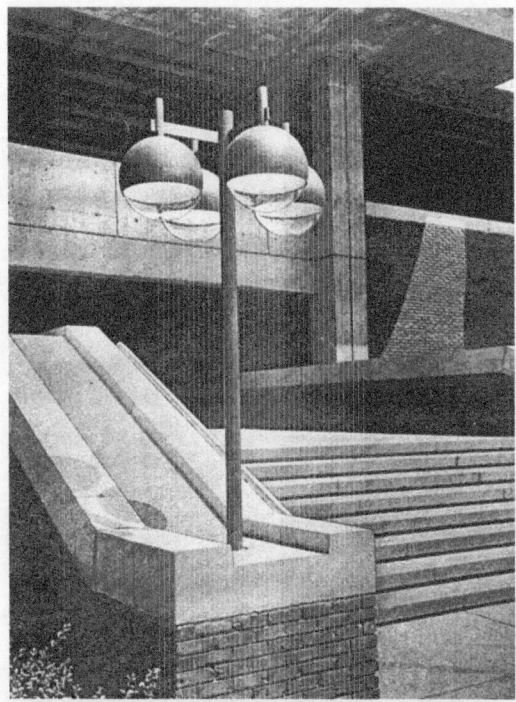

Fig. 4.140 Garden lighting: Lighting used in conjunction with the entrance to the Allegheny Centre in Pittsburgh

Lighting Outdoor lighting is used to illuminate pedestrian walk-ways, roads, and entry areas. It may also provide a dramatic effect when it lights up benches or fountains. The level of illumination varies with the extent of usage of the various external spaces. High intensity light is required for highways and other heavily used areas. Warm and coloured illumination is best suited for quiet areas along minor residential streets and parking areas. Figures 4.140–4.143 illustrate various outdoor lighting options.

Fig. 4.141 Outdoor lighting used in conjunction with the chapel at the Institute of Technology, Cambridge

Fig. 4.142 Light used to illuminate a roadway and a walkway on either side of a dividing island, Pittsburgh

Fig. 4.143 Light used near building entries at Illinois University

Planting Part of a landscape could be designed by studying the existing varieties of trees, shrubs, and flowers. Plant material is an important design element. It can articulate space, provide privacy, or act as a focal point. It can also provide shade or act as a wind break, surfacing material, or filter. It may enframe a view. It gives rise to shadow patterns, which add interest during daylight hours. Figures 4.144–4.147 illustrate tree planting detail and the various uses of plant material.

Fig. 4.144 Plant material—these rows of trees shade the seating area in between and provide shadow patterns (San Francisco)

Fig. 4.145 Trees may be used for their sculptural qualities

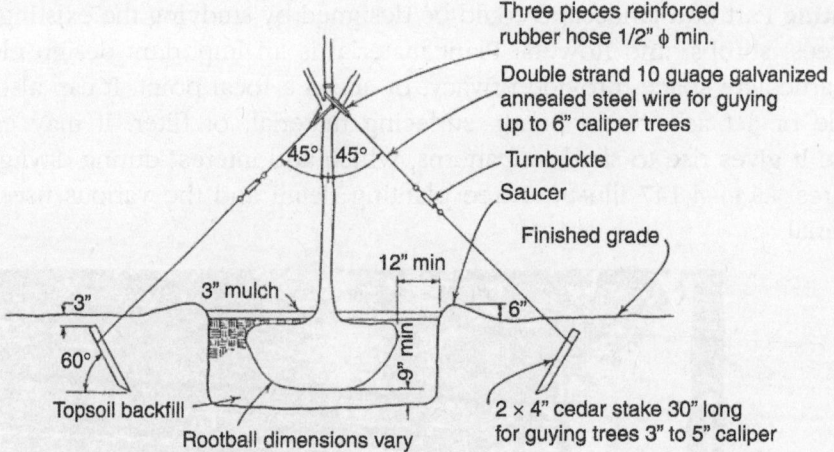

Three pieces reinforced rubber hose 1/2" φ min.

Double strand 10 guage galvanized annealed steel wire for guying up to 6" caliper trees

Turnbuckle

Saucer

Finished grade

45° 45°

12" min

3" mulch

3"

6"

60°

9" min

Topsoil backfill

Rootball dimensions vary

2 × 4" cedar stake 30" long for guying trees 3" to 5" caliper

Fig. 4.146 Tree planting detail

Fig. 4.147 Parking lots may be softened by the use of plant material. Here shrubs are used as ground cover.

Ground cover This is the landscape element that flows through the entire design and ties everything together. Grass is the best example of ground cover. It can easily be cut and maintained. Ground cover adds to the total appearance of the landscape. See Figs 4.148 and 4.149.

Fig. 4.148 Grass may be used as ground cover for ease of maintenance

Fig. 4.149 Ground cover—a moat filled with ground cover separates the building from the pedestrian walk system, California

Design of park and playfield

The main stages of the design process are collection of data, analysis, and solution.

Collection of data In the design of a park, the existing elements and the raw materials have to be understood first.

Topography The existing landforms, slopes, depressions, hillocks, rock outcrops, etc. should be identified (Fig. 4.150). This study helps in locating various activities. For example, a depressed region identified on the site could be used as an artificial lake, and an outcrop of rock could be used as a rock garden.

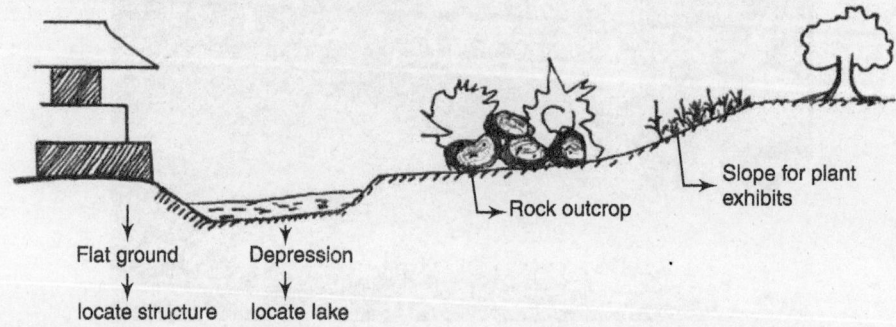

Fig. 4.150 Section across the park showing the existing elements

Geology The soil characteristics must be studied in detail. Knowledge of the soil's bearing capacity will help to identify whether the soil is sandy, clayey, or hard.

Hydrology Water availability must be examined. A study of the water table helps in identifying the details.

Existing forms The existing landscape elements—natural forms—such as trees, which could be indigenous species, can be identified and preserved. Distinct entities such as monuments and other distinctive structures should be identified (Fig. 4.151). Minor entities such as fences, retaining walls, paths, etc. could also be identified. In addition, other creative elements such as the following can be identified.

Fig. 4.151 Comparison of distinct structure versus general use areas

(a) Rock features that could be used for functional purposes.
(b) Distinct (visually prominent) structure versus generally useful areas.
(c) A quiet area or zone, which may be available at some higher point. The focus of research-oriented activities will automatically shift to that point and further useful areas could be located there.

Analysis In this stage, the data collected about the existing landforms (as explained in the preceding paragraph) are analysed. The analysis focuses on the following aspects of the various types of data collected.

- Functional aspirations
- Environmental orientation
- Scale of project
- Time to be allotted

By analysing these goals, the following can be satisfied.

- Functional needs
- Social needs
- Cultural background
- Commercial use (e.g., the park being visited by people and providing play facilities for children)
- Conservation

Zoning Zoning concerns the use of existing facilities and the plans for the future based on these. Such plans can be represented by layout designs (see Fig. 4.152).

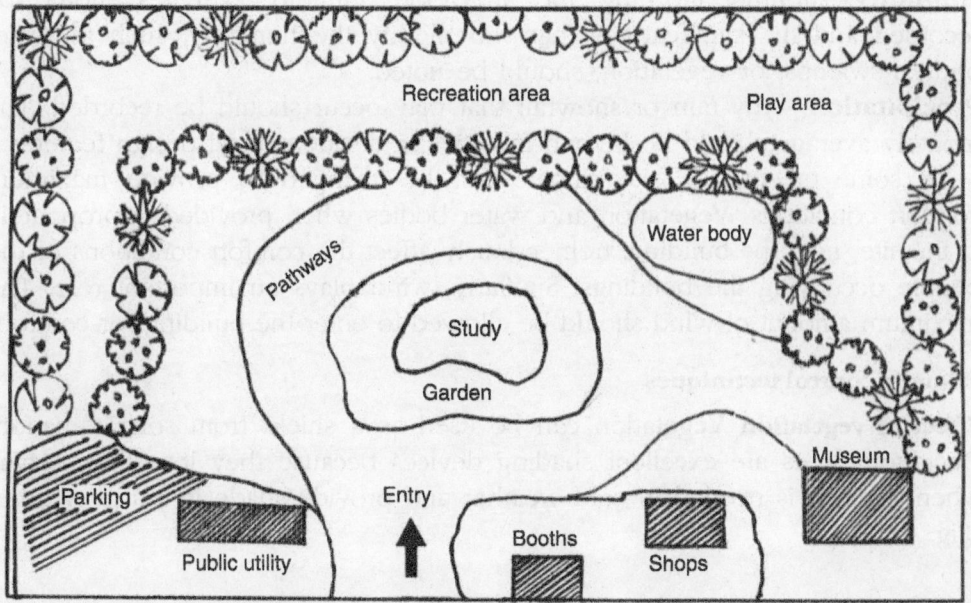

Fig. 4.152 Proposed park layout design

4.3.3 Climate

An architect is interested specifically in those aspects of the climate which affect human comfort and the use of buildings. These include averages, changes and extremes of temperature, temperature differences between the day and night (diurnal range), humidity, sky conditions, incoming and outgoing radiation, rainfall and its distribution, air movements, and special features such as trade winds, thunder storms, dust storms, and hurricanes.

An architect is concerned with the climate prevailing at a particular building site. Site climate is thus even more important than the local micro-climate. This will help in deciding the site planning, landscaping, and architectural form and construction.

To judge the site climate, the climatic data for that particular city should be collected and plotted. It is the designer's task to analyse climatic information and organize it in a form that allows him to identify features that are beneficial or harmful to the future occupants of a building on a particular site. The climatic data to be collected can be listed as follows.

Temperature—The normal, maximum, minimum, average, and extreme temperatures for the month should be determined to help in deciding the heating and cooling requirements.

Solar radiation—The hours of sunshine on clear and cloudy days, solar angle details to determine the solar gains, and the extent of exposure of the site to the sun will help in deciding the need for shading.

Wind—The direction and velocity of the wind will decide the ventilation possibilities and protection requirements.

Humidity—Monthly morning and afternoon humidity averages should be recorded and the site features that will modify the humidity, such as water bodies, swamps, or vegetation, should be noted.

Precipitation—Any rain or snowfall that may occur should be recorded. The monthly average should be known in order to decide special design features.

To some extent, the site climate can be changed to provide maximum comfort conditions. Vegetation and water bodies when provided appropriately at the site, near the building, tremendously affect the comfort conditions of the people occupying the buildings. Similarly, wind plays an important role. The maximum amount of wind should be allowed to enter the building for comfort.

Climatic control techniques

Effect of vegetation Vegetation can be used as a shield from solar radiation. Deciduous trees are excellent shading devices because they lose their leaves when sunlight is needed in cold weather and provide shade in warm weather (Fig. 4.153).

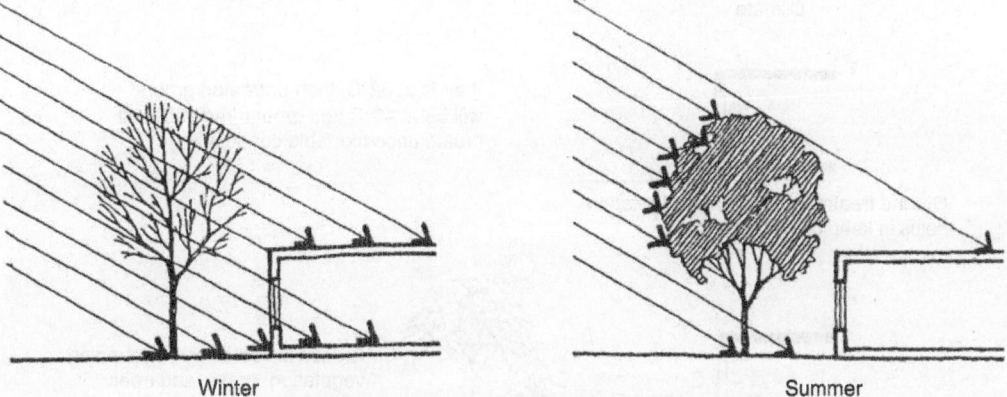

Winter   Summer

Fig. 4.153 Incident solar radiation and vegetation

Vegetation will cut either direct or reflected solar radiation before it strikes the building surfaces. Low shrubs or lawns can control reflections from water or paving surfaces. Vegetation also helps in air purification and wind protection. The landscaping shrubs not only improve the dramatic effects of the building but also serve as dust reducers, noise controllers, and provide thermal effects.

In a dry, tropical area, protection from sun rays has to be provided. Overheated dust, sand, and storms also have to be avoided. Plants can be used to absorb dust (Fig. 4.154) and sun-reflecting material such as asbestos and galvanized iron sheets should be used for roofs.

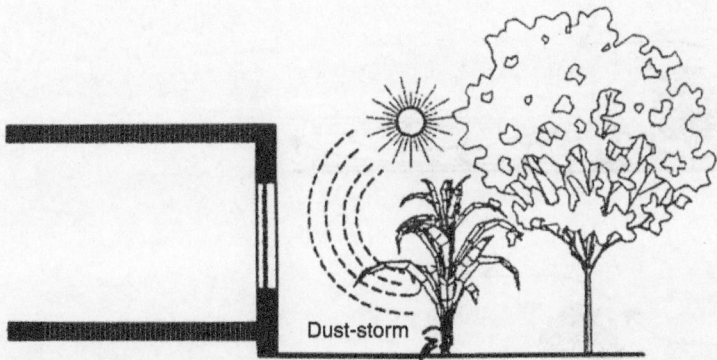

Dust-storm

Fig. 4.154 Vegetation prevents overheated dust, sand, and storm

If the air temperature is 32°C, then the untreated ground temperature will be 45°C and ground radiation will create uncomfortable conditions. Ground treatment with grass vegetation helps in keeping interiors cool due to breeze, shade, less hot surroundings, less light glaring surfaces, and visually cooler surfaces. The breeze is also cooled by the surrounding vegetation, grass, and trees. Figure 4.155 illustrates the concept of ground treatment with vegetation.

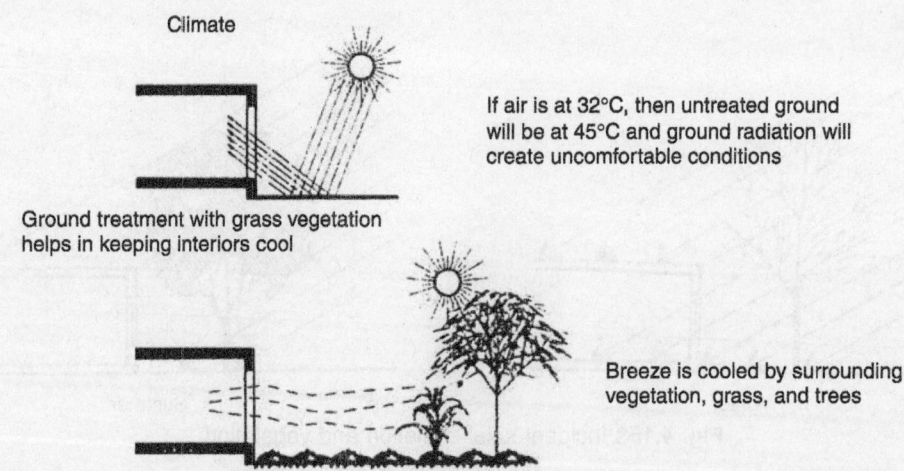

Climate

If air is at 32°C, then untreated ground will be at 45°C and ground radiation will create uncomfortable conditions

Ground treatment with grass vegetation helps in keeping interiors cool

Breeze is cooled by surrounding vegetation, grass, and trees

Fig. 4.155 Ground treatment with vegetation

The thermal comfort depends on the site conditions and the proposed building, its location, its orientation, its structure, its finish, and the activities that take place inside the building.

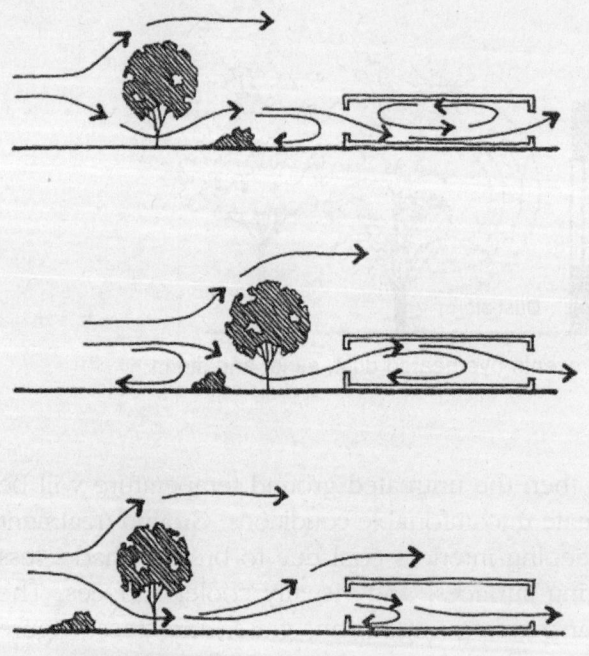

Fig. 4.156 Effect of vegetation on wind flow

Effect of vegetation on wind flow Thick vegetation can obstruct wind flow (Fig. 4.156). Careful placement of trees and hedges can direct and increase wind speeds (Fig. 4.157).

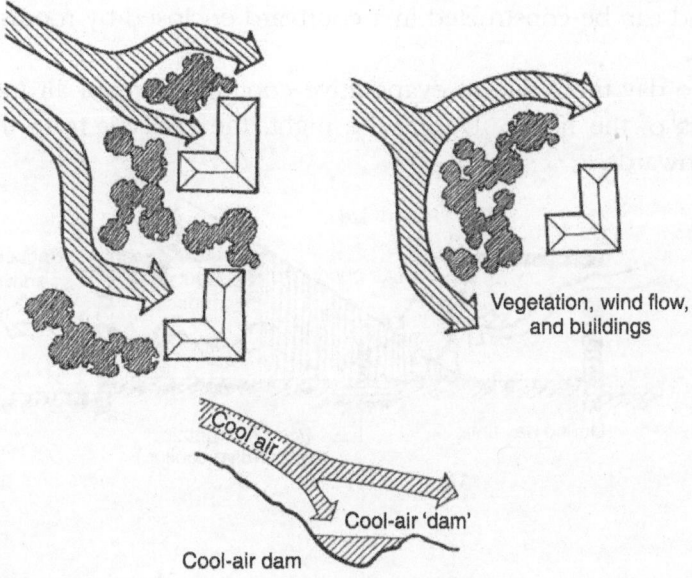

Cool air

Cool-air 'dam'

Cool-air dam

Vegetation, wind flow,
and buildings

Fig. 4.157 Vegetation, wind flow, and buildings: trees and hedges can direct and increase wind speeds

Effect of water ponds In arid regions increased use of water ponds in courtyards (Fig. 4.158) or atriums gives a cooling effect because of natural ventilation. This reduces exposure to dry winds and sun.

Fig. 4.158 Effect of water pond: Masjid-e-shah, Iran

A water pond can be constructed in a courtyard enclosed by rooms as shown in Fig. 4.159.

During the day time, due to evaporative cooling, the cool air moves through the openings of the house. During the night, the cool air from the sky (roof) moves downwards.

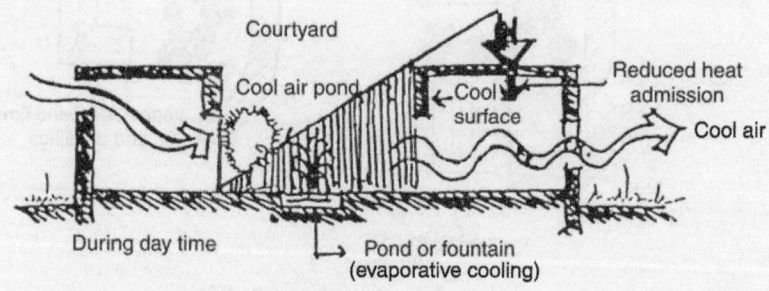

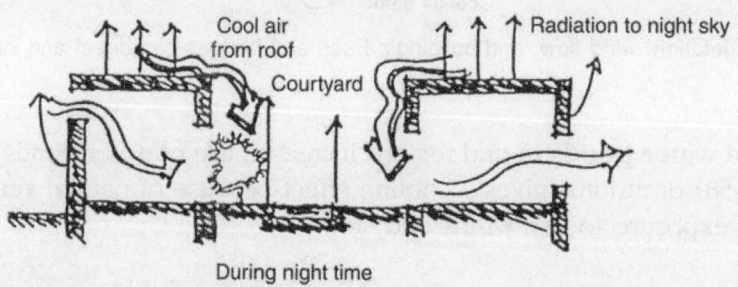

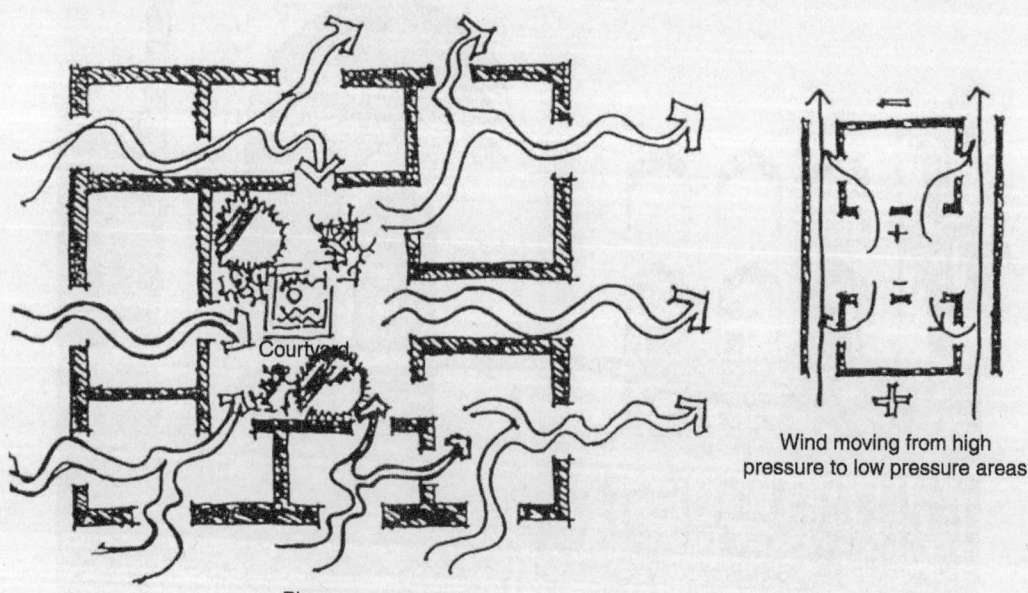

Fig. 4.159 Water pond created in a courtyard enclosed by rooms

Effect of wall openings The body loses heat by radiation and conduction if the air and the surroundings are at a temperature less than body temperature. It loses heat by evaporation if the air is dry, in order to absorb further moisture, which in turn depends upon the humidity of the air and the rate at which it passes over the body (Fig. 4.160). A fan or wind will help in cooling, provided that the air temperature is less than the skin temperature and not saturated with moisture.

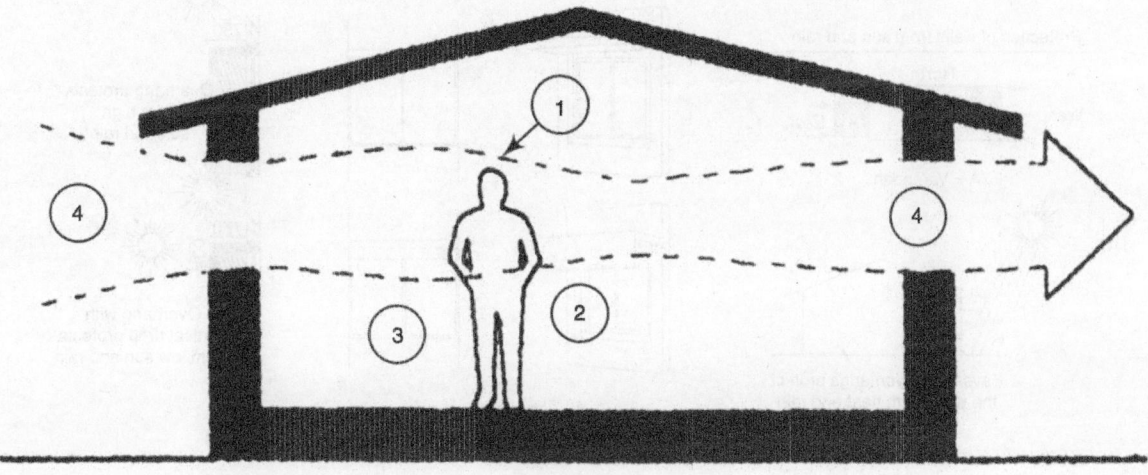

1. Humidity
2. Temperature
3. Radiation
4. Air movement

Fig. 4.160 Body heat loss

In a humid tropical area, comfortable conditions are created by moving air. So the wall openings should be made in such a way that the breeze is allowed to pass through freely (Fig. 4.161). The build-ing should be oriented in such a way that the maximum length is exposed to the breeze.

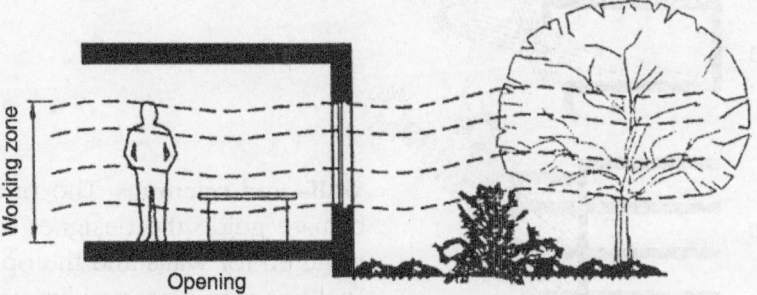

Fig. 4.161 Comfortable conditions created by moving air

Protection of walls from sun and rain Eaves and overhangs (sunshade or horizontal projection) protect walls from heat and rain. An overhang with a vertical drop protects from low sun and rain. An overhang without a drop protects from the high heat of sun and rain. Buffer rooms can also be used to protect habitable rooms from the sun. A verandah can be used as a buffer room. See Fig. 4.162.

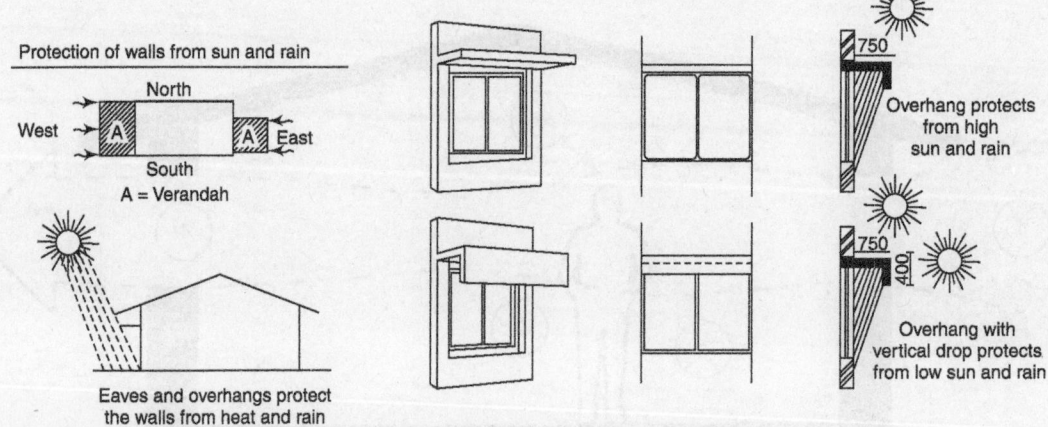

Fig. 4.162 Protection of walls from sun and rain (all dimensions in mm)

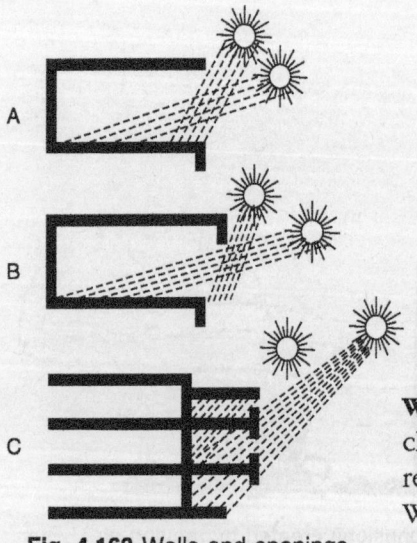

Fig. 4.163 Walls and openings

Walls and openings The orientation of the building and the climate guide the designer in deciding the type of protection required for walls and the openings in these walls (Fig. 4.163). Walls can be protected by overhangs, verandahs, sun breakers, projecting slabs over the balcony, and vertical drops.

Sunshades or overhangs are provided for window openings for protection against the sun. The projection of these depends on the angle of incidence of the sun's rays. This angle depends upon the latitude of the location. Vertical projections called louvers are also provided for protection against the sun. See Fig. 4.164.

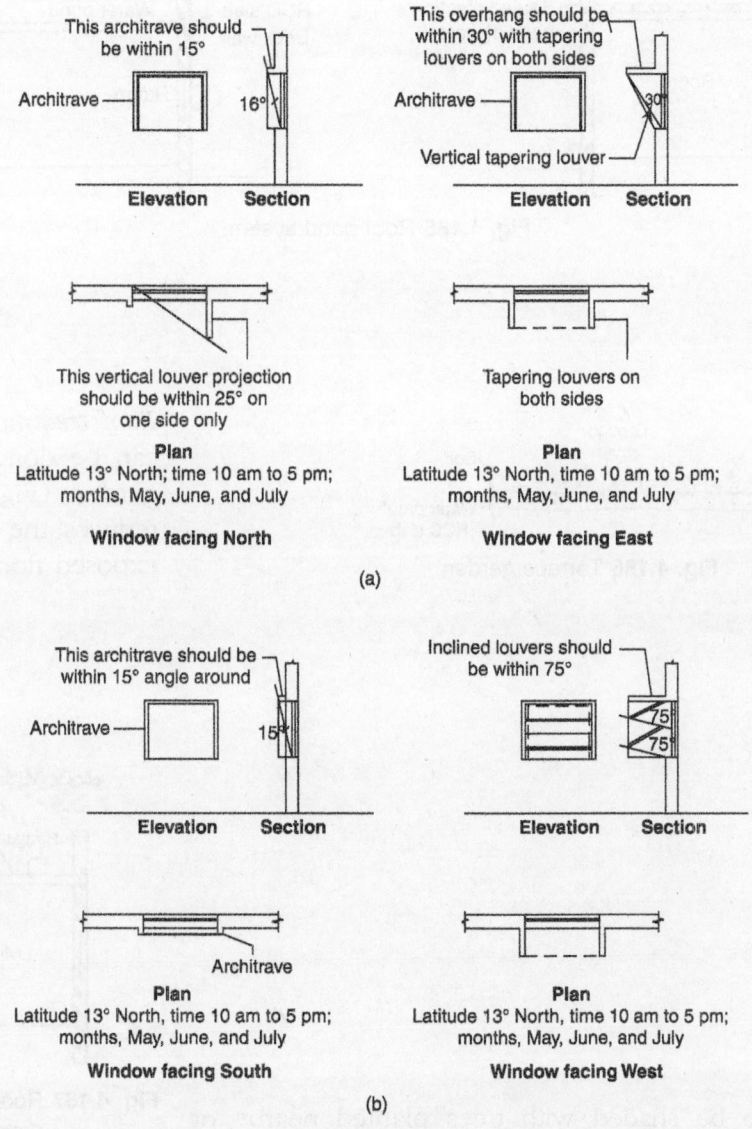

Fig. 4.164 Effect of shading devices

Effect of roof treatment In hot places, the roof pond system is very effective in reducing heat radiation from the roof into the building. In this system, a pond of water is made stagnant on top of the roof so that it cools the interior. Sealed plastic bags of water can also be used on top of the roof slab, over which a transparent cover and movable insulation can be laid. This system has been illustrated in Fig. 4.165.

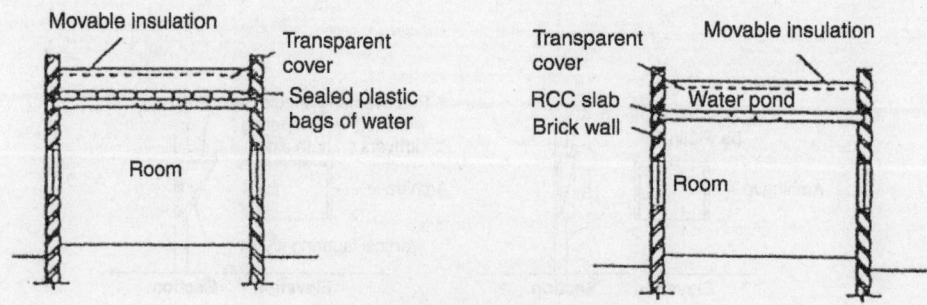

Fig. 4.165 Roof pond system

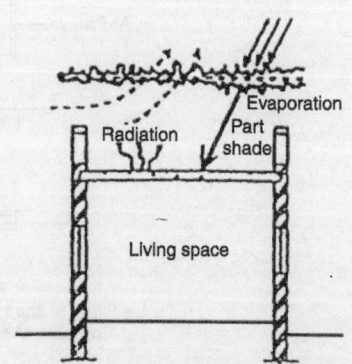

Ground cover, Soil, Water proof RCC slab

Fig. 4.166 Terrace garden

Roof treatment Roof shading can be done using a terrace garden (Fig. 4.166), which reduces the heat gain of an exposed floor.

Fig. 4.167 Roof shading can be done by trees or creepers planted nearby

Roofs can also be shaded with trees planted nearby or creepers, controlling solar radiation (Fig. 4.167).

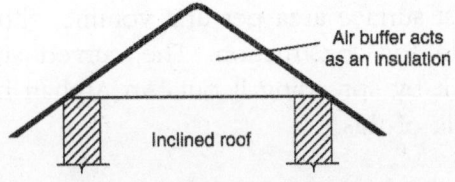

Fig. 4.168 Effect of sloping roof

Effect of sloping roof A sloping roof absorbs less heat than a flat roof due to its shape and less exposure to solar radiation (Fig. 4.168).

Roof temperature and treatment In a tropical climate the roof is exposed to the sun. Heat penetrates through the slab, which increases the heat inside the room. Increasing the thickness of the slab will only increase the time lag and is not economical. Time lag is the amount of time taken for the lower surface to attain the temperature of the upper surface of the slab. By providing a layer of hollow blocks on the top surface of the roof, which is exposed to sun its temperature is reduced. The best results would be obtained by providing an air gap below the roof. It will create comfortable conditions (Fig. 4.169).

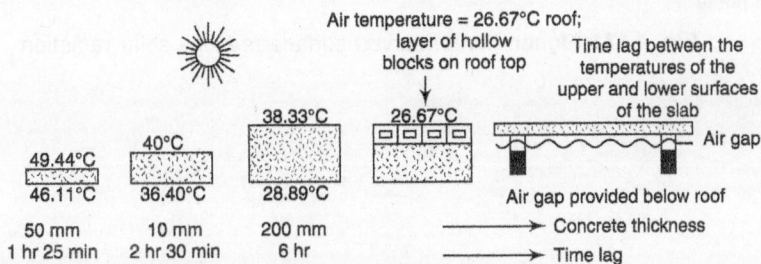

Fig. 4.169 Roof temperature and treatment

Form responses to natural environment Here the basic concept is the surface to volume ratio. If there are two equal volumes, the one with the least exposed surface area will gain or lose the least amount of heat to the surrounding (Fig. 4.170).

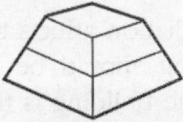  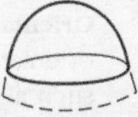 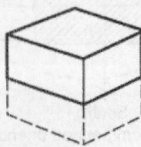

Fig. 4.170 Thermal retention: house forms based on climate conditions

Geometrically, a sphere has the least surface area per unit volume, although a curved object may not be suitable for construction. The curved surface reduces the solar radiation falling on it by spreading it out. An Afghan home, shown in Fig. 4.171, is a good example of this.

Afghan home

Fig. 4.171 Afghan home: curved surface reduces solar radiation

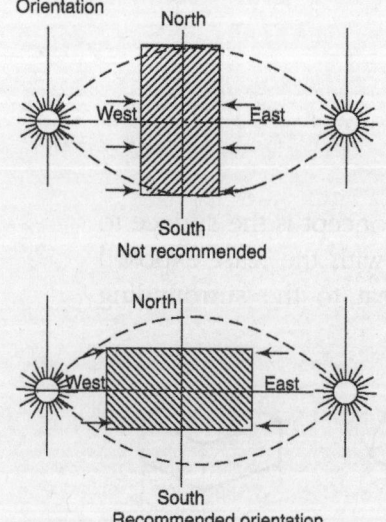

Orientation

North

West   East

South
Not recommended

North

West   East

South
Recommended orientation

Fig. 4.172 Building orientation

Orientation of building The sun's radiation affects the building orientation (Fig. 4.172). The building should be oriented in such a way that the longer side of the building is not exposed to direct solar radiation. It is preferable, instead, to expose the shorter side to direct solar radiation.

Fig. 4.173 Shading of glazed area, Secretariat Complex, Chandigarh, India

When the solar gain is more and when the glazed area in a building is large, shading the glazed area is very necessary, as in the Secretariat Complex in Chandigarh, India (Fig. 4.173). This shading device adds to the elevation of the building. In front of the window façade, certain horizontal and vertical fins are provided, which cover the total wall and provide a good sun shading device.

4.3.4 Integration of basic building services with architectural design

In the design of buildings, a provision should be made for essential services. The services which are considered to make the building design complete are
- Water supply and sanitation
- Drainage
- Electricity distribution
- Ventilation and air conditioning
- Lighting and illumination
- Fire safety

Water supply and sanitation

The main objectives of water services design are
- Protection of public health
- Maintenance of adequate availability of potable water
- Water availability for washing at required places
- Water conservation

Water requirement per person varies depending on the purpose of the building, such as factories, hospitals, hostels, cinemas, schools, and residences.

Water is also required for the following general uses:
- Gardening and landscaping
- Air-conditioning
- Swimming pool

Sanitation involves the provision of water closets, baths, and sinks. It also concerns the removal of waste water with contaminants. The required standards for sanitary fittings must be used in the design of buildings. The following table lists the standard number of sanitary fittings required in office buldings.

	Male	**Female**
1. Water closets	1 for every 25 persons	1 for every 15 persons
2. Urinals	4 for 70–100 persons; for > 200 persons add at the rate of 2.5%	
3. Wash basins	1 for every 25 persons	
4. Drinking water fountain	1 for every 100 persons or one on each floor	

Drainage

The pipe within one's premises (house compound) that carries sewage is called the drain. Beyond the boundary of the premises, the line connecting the house drain to the main sewer line is called the sewer. In laying out the drainage system, the aim should be to collect all the connections into one main pipe. Such connections are grouped together in manholes. Manhole covers can be lifted to find faults or stoppages in the flow.

Electricity distribution

Electricity is the lifeblood of any building. It is distributed by the electricity board to residential or commercial units using a single-phase or three-phase system.

Electric power supply systems installed in buildings should consist of the following.
- Main intake, metering and distribution equipment.
- High-, medium-, and low-distribution systems including, cabling, wiring, etc.
- Power outlets and switching system.
- Horizontal and vertical distribution points for distribution among rooms of the same floor and rooms of various floors, respectively.

Within a building, the electrical points in each space should be noted.

The position of the switchboard in each room must be taken into account and the electrical layout planned. Telephone points also have to be provided for. Heavy electric loads between 1000 W and 3000 W should be provided for

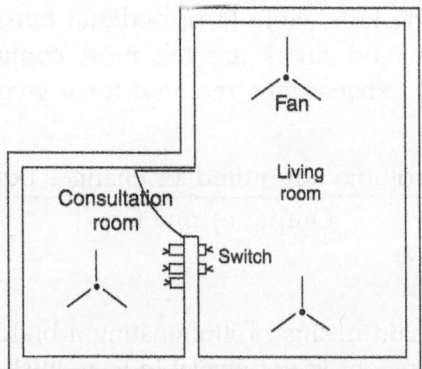

Fig. 4.174 Electrical layout of a consultation room attached to a residence: Fan, tubelight, bulb, two-way switch, 5 amp and 15 amp plug to be provided

appliances such as geysers and air conditioners. For safety and protection, alarms also have to be provided. Earthing terminals must be provided for appliances such as computers, grinders, refrigerators, and washing machines. Telecommunication points also must be provided. See Fig. 4.174.

Ventilation and air conditioning

Ventilation means effecting fresh air supply in adequate quantities and velocities as well as removal of objectionable odours and gases. These are achieved by air movement, which causes convection and a physiological cooling effect.

The desired ventilation can be obtained by natural or artificial means. About 12 to 28 cubic metres of fresh air per hour per person is needed depending on the physical activity in the room. The purpose of providing an air conditioning system is to control the temperature, humidity, and air quality and ventilation indoors simultaneously.

The capacity of air conditioning is decided by a general thumb rule [for human comfort, for every 80 to 120 sq. m of room area, one TR (Tonnage rating) of air conditioning must be provided].

To provide proper ventilation, fresh air must be brought indoors in adequate quantities and velocity. All rooms should have one or more openings, other than the door, and the sum of the areas should be at least 1/6 times the total area of the floor. The minimum area of an opening for any habitable space should be 1 sq. m. Staircases should have a minimum of 1 sq. m opening in an external wall or landing. No part of any habitable room should be more than 7.5 m away from the nearest source of ventilation. The minimum door opening for the bath should be 65 cm.

The following factors must be considered in mechanical ventilation.

- size of room
- number of occupants
- activities inside the room
- heat gains from equipment and solar radiation
- relative humidity and outside air temperature

Circulating fans, ceiling fans, table fans, pedestal fans, and air coolers along with sufficient windows and doors are the most commonly used ventilating devices. The number of exhaust fans required for a given area is given by the following ratio:

$$\frac{\text{Room volume} \times \text{required air changes per hour}}{\text{Output of one fan}}$$

Lighting and illumination

Natural daylight is the main means of illuminating a building. Artificial (electric) lighting is used when daylight is not available in requisite quantities to light up the interiors and exteriors. Lighting quality is a term used to describe the overall lighting scene in terms of

- luminance
- diffusion
- uniformity
- chromocity (colouring of light)

Lighting design is a combination of art and engineering. The daylight factor varies according to the kind of space being considered. The level of illumination, glare, colour, radiation, and property of the lamp are the main considerations for lighting design. Natural daylight must always be considered before planning for artificial light. Daylight penetrates up to 2.5 times the window height.

Glare Glare is excessive brightness in the field of vision. It can be controlled by using light sources of lower intensities and recessed, louvered, or diffused fixtures. By increasing the brightness of the surrounding areas, the reflected glare can be reduced.

Colour The colour rendering properties of lighting sources are of prime importance in hospitals, shop windows, sales floor area, etc.

Fire safety

The three objectives of building fire safety are

- protection of life
- protection of property
- continuity of operation

Building regulations specify two kinds of controls for fire protection in buildings.

(i) Passive control, resulting from the nature of the materials of construction, discourages fire spread.

(ii) Active control provides for fire-fighting equipment.

Fire protection requires the installation of an automatic detection device, a manual or automatic alarm, and a public address system to handle fire or panic situations. If the building is above four floors or 15 m in height, then a fire escape stasircase must be provided.

4.4 Litigation Factors Considered in Architectural Design

Rules and regulations in building design are laid down to avoid chaos. In creating a building design, the following rules and regulations must be taken care of. These are safety regulations, building rules and regulations, and layout regulations.

Safety in buildings is very important for their occupants. Building rules dictate the way the design should come up, otherwise each builder would construct a building in his own way, which may lead to confusion and chaos. The layout regulations set up rules for the control of the building line, density and height of buildings, use of special building materials, etc. They also decide the elevations in certain areas.

4.4.1 Safety regulations in buildings

Public buildings, by the nature of their occupancy and use, require higher standards of safety than other types of buildings. Provision of life safety takes top priority and affects the entire design/plan, the construction, and the selection of material.

When health and safety factors are taken into account in the design of a building, its complexity and cost might go up. The standards of the National Building Code, in which the fire safety code is included, should be followed strictly. Architects and engineers should obey the authorities and follow the applicable codes.

Safety considerations can be studied under the following heads.

Structural safety

- Material strengths and factors of safety.
- Provision of fire-proof and fire-resistant structures.
- Windstorm resistance: A high-rise building should be designed to withstand wind pressure and force.
- Earthquake resistance: For a high-rise building or a building in an earthquake prone area, the code for earthquake resistant buildings must be taken into consideration.

Fire safety

- Provision and protection of exits, corridors, and stairs.
- Provision of fire detector and alarm systems at short intervals of space.
- Provision of fire-fighting equipment such as sprinkler systems, fire extinguishers, sand-filled buckets, and water-filled buckets.
- Provision of materials and finishes with low flame spread rating and non-combustion characteristics.

Health safety

- Maintaining proper standards of ventilation systems; providing for proper natural ventilation, which is essential for good health.

- Lighting standards and electrical codes.
- Plumbing fixture requirements and plumbing code.
- Swimming pool and locker room requirements.

Emergencies

Adequate emergency lighting systems should be provided, in case of power and generator failure.

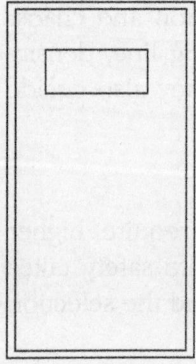

Fig. 4.175 Vision panel provided in door to avoid collisions

Accident protection

- Non-slip surfaces (especially in stairs, ramps, pool decks, and toilets).
- Vision panels in doors to avert accidents, such as people getting knocked back (Fig. 4.175).
- Properly designed door swings.
- Safety glass in doors and side-lights to help in easy opening.
- Hand rails in staircases.

Exits

- Exits should be properly planned and located to handle traffic flow without congestion.
- The EXIT sign should be clear and prominent and there should not be any doubt about its purpose. In cinema theatres the exits are marked with red light to be visible in the dark.
- A sign indicating the nearest exit should be visible from every point in the corridor.
- Two or more exits should be provided at any place of large gathering.
- If the space is a big classroom, more than one exit has to be provided for easy circulation.
- It should be possible to open every door from inside even after the building is closed, especially in a school.
- A well-defined exit will include a lighted red exit sign and a white emergency security light connected in the event of power failure.

Stairways

The overall circulation pattern has to be studied and the staircase should be critically located keeping in mind the following factors:

- load distribution
- safety
- destination of people

- elimination of cross traffic
- easy, fast, and safe movement

To avoid congestion and provide for easy movement up and down at the same time, the minimum length of the steps should be 1 m to 1.5 m. The staircase core should be fire-proof leading outdoors directly. Staircases should be provided with smoke control facilities separating the stairwells from the corridors.

Corridors

The function of corridors is to accommodate traffic flow without congestion. Codes should be checked to provide for proper corridor widths, corridor lengths, and smoke barriers at suitable intervals. The walls of the corridors should be free of all projections. AC units, drinking water fountains, fire extinguishers, doors, and any display cases should be recessed. The corridor should be clear of obstructions (Fig. 4.176).

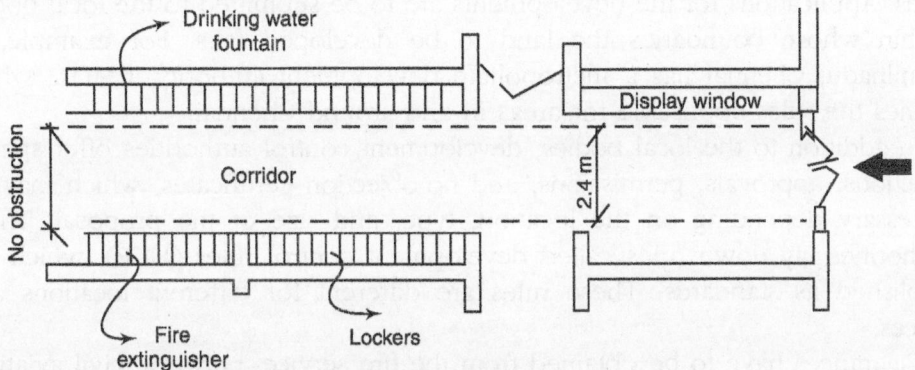

Fig. 4.176 Corridors should be clear of obstruction

Stair treads

The standard dimensions of staircase treads and risers must be used to avoid accidents. The tread should measure a minimum of 25 cm and riser 15 cm. The stair treads should incorporate anti-slip measures (Fig. 4.177).

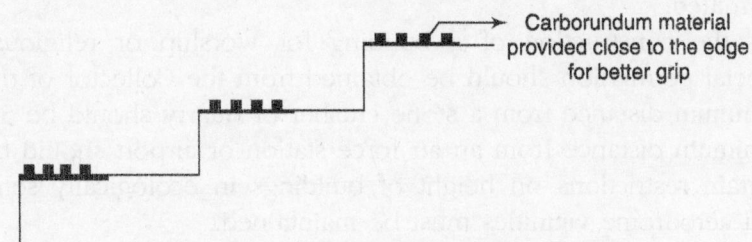

Fig. 4.177 Anti-slip measure provided in stair tread

Hand rails

Hand rails of minimum height 75 cm are necessary on both sides of stairways in accordance with the National Building Code (NBC).

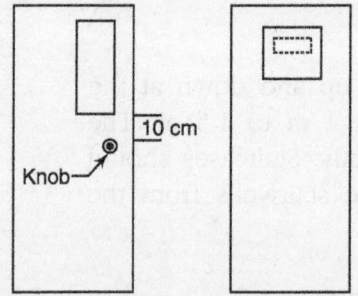

Fig. 4.178 Safety measures provided on a door

Doors

Doors should be opened and closed with caution. Recessing the door is important to keep the corridors clear. Vision panels should be provided (Fig. 4.178). The use of wired glass provides safety.

4.4.2 Building rules and regulations

The rules to be followed during a construction project are decided by the location of the project and the type of development to be carried out. Specific areas are largely covered by the Directorate of Town and Country planning rules. Applications for the developments are to be submitted to the local bodies within whose boundaries the land to be developed falls. For example, in Tamilnadu, Chennai has a metropolitan development authority, CMDA, which frames the rules applicable for areas in and around Chennai.

In addition to the local bodies, development control authorities offer special sanctions, approvals, permissions, and no-objection certificates, which may be necessary depending on the location, type, and size of the proposal. These authorities lay down rules called development control rules (DCR), which are published as standards. These rules are different for different locations and places.

Clearances have to be obtained from the fire service, railways, civil aviation, and pollution control authorities in the case of projects that come up in the vicinity of a railway station, airport, or industry.

The precautions to be observed for the selection of a site for development are as follows.

- The site should be at least 30 m away from the railway line.
- The set back from the tidal line should be 500 m. Deviations are not permitted.
- For the construction of a building for worship or religious purposes, special permission should be obtained from the Collector of the district.
- Minimum distance from a stone crusher or quarry should be 500 m.
- Minimum distance from an air force station or airport should be 900 m.
- Certain restrictions on height of buildings in ecologically sensitive areas and aerodrome vicinities must be maintained.

General building rules

1. Height of the building:
   - The height of the building should be 1.5 times abutting street width + 1 m for every 30 cm of the front set back of the building.

- The height of the building should not exceed 4 m, within the 500-m-high water mark of the sea.
2. The minimum size for a residential site should be 95 sq. m and the width of it should be at least 6 m.
3. The minimum distance from a water course or tank should be 15 m.
4. The set-off space surrounding the buildings should be at least 1.3 m from the boundary of the street.
5. If the adjoining buildings do not have any open space in between, a minimum area of 1/4 times the site should be left as vacant space open to the sky.
6. The minimum dimensions and requirements of rooms are as follows.
   - Rooms other than the kitchen, bath, and store should have an average height of 2.75 m (not less than 2.1 m) and a width of 2.5 m.
   - The bathroom area should be a minimum of 1.8 sq. m with dimensions 1.5 m × 1.2 m.
   - The water closet (toilet) area alone should be 1.08 sq. m with dimensions 0.9 m × 1.2 m
   - The bathroom cum water closet should have a minimum area of 2.7 sq. m. and minimum dimensions of 1.5 m × 1.8 m.
   - The ventilation area by means of windows or ventilators should be 1/8 times the room floor area.
7. Every domestic building should be so constructed that every room intended for human habitation should have at least one side abutting an open space (either external or internal) for a length of not less than 2.5 m.
8. Stairs
   - All storied buildings should be provided with sufficient number of staircases, depending on the number of occupants within a distance of not more than 18 m between the staircases.
   - Width of stairs: The clear width of all stairways should not be less than 60 cm excluding the hand rails (Fig. 4.179).

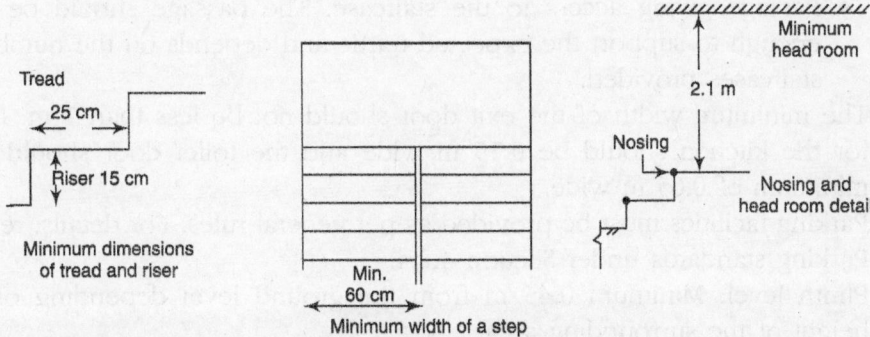

Fig. 4.179 Staircase details

- Head room: All stairways should have at least 2.1 m of clear head room, measured perpendicularly from the nosing (Fig. 4.179).
- Treads and risers: The tread should be a minimum of 25 cm and the riser should be 15 cm (Fig. 4.179).

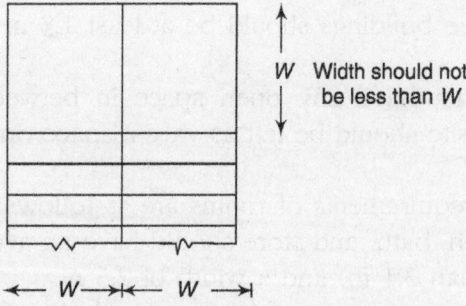

W Width should not be less than W

← W →|← W →|

Fig. 4.180 Minimum landing width

- Landings: No stairway should have a height of more than 3.75 m between landings, and the width of the landing should not be less than the width of the steps (Fig. 4.180).

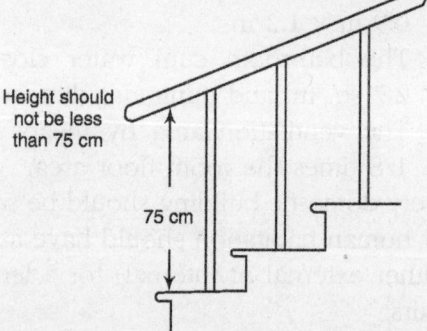

Height should not be less than 75 cm

75 cm

Fig. 4.181 Minimum height of hand rail

- Hand rails: Hand rails should be placed not less than 75 cm high (Fig. 4.181).

- Ventilation for staircase: The staircase should be properly lighted and ventilated. Windows can be provided along the external wall abutting the landing of the staircase to provide for good ventilation.
- Design load for stairs and landings: 390–394 kg/sq. m is taken as the standard load criterion for a staircase.
- Passage giving access to the staircase: The passage should be wide enough to support the expected traffic and depends on the number of staircases provided.

9. The minimum width of the exit door should not be less than 1 m. Doors for the kitchen should be 0.75 m wide and the toilet door should be a minimum of 0.65 m wide.

10. Parking facilities must be provided as per general rules. For details, refer to Parking standards under Section 4.2.4.

11. Plinth level: Minimum 0.45 m from the ground level depending on the height of the surrounding area.

12. Any building within 500 m from a sea or river should not let out sewage water into the sea or river without proper treatment.

4.4.3   Layout regulations

Role played by local authority

The local authority can regulate individual plot developments by regulating the building line, specifying the land use, controlling the density and building height, controlling the use of special building materials, and insisting on specific designs of elevations in certain areas.

Aim of layout regulation

The aim of layout regulation is the preparation of layouts which will satisfy standards with respect to open spaces outside and adequate area inside the building, and the provision of good light and ventilation.

Layout regulations

A good layout can be prepared by planning a high density of the order of 300 to 350 persons per acre by adopting a mixture of high- and medium-rise buildings and leaving a wide open space between them for the use of the entire community. The layout should be economical in the use of land and should provide parks near residences. The buildings should be located taking into account the prevailing breeze and sunlight.

To evolve a good layout design, the following aspects have to be considered:
- traffic engineering
- housing design
- recreational facilities
- social aspects
- architectural treatment

Large-scale layouts can use all the various shapes, forms, and types of buildings. In such a development it is not difficult to ensure proper access to the various buildings. Also, easy parking facilities can be provided. Such a unified grouping of the buildings allows for safety and easy pedestrian access. As large-scale developments take time for completion, any adjustments to be made in the course of development should be provided for.

In small-scale layouts, providing a solution for the planning problem is difficult, as the land is held by a number of owners, and buildings of different ages might still be existing. The individual owners may not be interested in the improvement of the vehicular or pedestrian traffic system or provision of open space, as no incentive is provided to them in such cases.

Large-scale development takes care of both social and individual needs. Safe roads, amenities for social life, parks and open spaces, educational and health services, etc. should be properly located in such development. When large-scale development is done, new sites are developed as old sites are improved. Sometimes these will relate to the realignment of existing roads, provision of shopping facilities, and creation of open spaces where needed.

Isolated plot-by-plot development should not be allowed, as this does not provide all the facilities for good living.

The main objective of preparing a layout has not changed over the years. The planning concepts alter with changes in living conditions. Traffic safety is the major issue to be solved by a planner. By adopting new planning trends, it is possible to evolve model layouts providing traffic and health safety. It is possible to design a good layout that is also economically viable. Figure 4.182 shows a planned residential design.

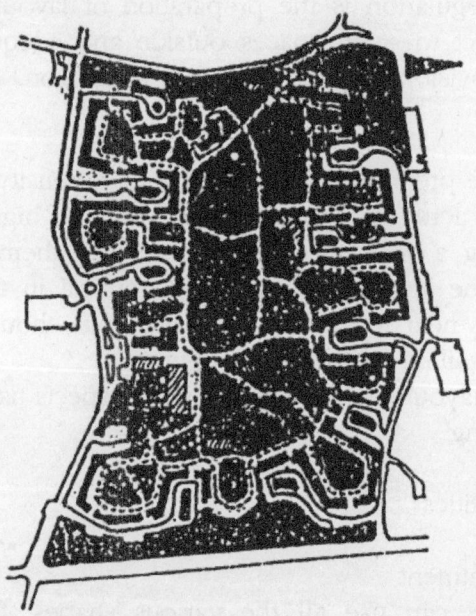

Fig. 4.182 Layout of Baronbackama housing scheme at Orebro, Sweden. This scheme has 1,200 flats in meander-shaped apartment houses at a density of 127.5 rooms per hectare (51 rooms per acre). The built-up area opens to courtyards, which are connected to a large central, green area. These inward courts are used for play space, while the outward courts are used for parking. This layout scheme has become internationally known as Sweden's outstanding contribution to planned residential design.

Summary

This chapter explains in detail how an architectural design evolves, step by step, from an idea into a physical construction.

Before going into the many factors that are involved in architectural design, it is essential to understand the relationship and balance between aesthetics and function—the two primary aspects of a building that have to be integrated into a design. Depending upon the purpose of the building, either of these will dominate the design. For instance, while constructing homes, hotels, shopping complexes, etc., an architect will attach great importance to aesthetics, whereas if he is designing hospitals, factories, railway stations, and airports, he will concentrate primarily on the functional aspects of the building.

The chapter then moves on to examine the many factors to be considered and analysed before the final design is produced. These are requirements, circulation, anthropometrics, space standards, site and landscape, climate, safety regulations, layout regulations, building laws, and basic services. An understanding of these factors, which are the foundation of all architectural designs, will help students understand and appreciate the work of leading international architects. The achievements of world-renowned architects, both western and Indian, will be discussed in Chapter 5.

REVIEW QUESTIONS

Part A (2 marks each)

1. What is architectural design?
2. What do you understand by design concept?
3. What is a building?
4. In an architectural design, how do you integrate function and aesthetics?
5. What are the factors to be considered in an architectural design?
6. What do you mean by requirements in an architectural design?
7. Write the requirements for a residence.
8. Write the requirements for a hospital.
9. Define circulation.
10. Keeping in mind the requirements of a residence, draw the circulation diagram for a residential building.
11. Define anthropometrics.
12. Draw sketches showing the furniture units of a living room in a residence. (Hint: Draw a living room plan showing sofas, tables, showcase, book shelf, etc.)
13. What are space standards?
14. Draw sketches showing the car parking standards in a public building.
15. Draw sketches to show the space standards required of a toilet. (A similar question could be asked for a kitchen, a bed room, etc.)
16. Draw the sketch of an office workstation, showing the space dimensions.
17. What is meant by site planning or site analysis?
18. What are the various factors to be considered in site analysis?
19. Define landscape design.
20. What are landscape elements?
21. What are the various climatic factors to be considered in an architectural design?
22. Mention the various climate control techniques that could improve the comfort conditions in a habitable building.
23. Name the basic services that have to be integrated in an architectural design.
24. Why is it necessary to provide safety in a public building?

25. Mention the various safety considerations that need to be taken care of in the design of a building.
26. What are the safety standards for a staircase? Describe with sketches.
27. What are the various fire-fighting devices?
28. Why are rules and regulations necessary to be followed in an architectural design?
29. Mention any five building rules to be considered in building design.
30. What are the various layout regulations that need to be taken care of in an architectural design?

Part B (16 marks each)

1. Explain how anthropometrics and space standards are used in the design of a residential building (or an office building or an institutional building).
2. What is anthropometrics? Explain with examples how it can be used to determine the size and shape of rooms for human activities.
3. Explain circulation in detail.
4. How will you design an institutional building taking into consideration the following requirements: circulation, anthropometrics, site, building rules, etc. (This question can be asked for all the various building types such as a residence, hospital, office, shopping complex, etc.)
5. Write about site analysis or site planning in detail.
6. Explain how landscape elements are integrated in a building design in detail with sketches.
7. What are the required safety considerations for buildings? Explain.
8. Explain layout regulations in detail.
9. Explain the building rules and regulations for your town and write about the safety of industrial buildings.
10. How will you design a park and playfield? Explain in detail.
11. Explain how services are integrated with a building design.
12. Explain the integration of aesthetics and function in an architectural design.
13. Explain in detail how various techniques can be adopted to improve the comfort conditions in a building.

Contemporary Architecture

Theme, Philosophy, and Works of Famous Architects

Every being carries from birth certain memories and with them associations and preferences. Every individual is thus unique in terms of behavioural patterns. However, certain needs are common to all beings, such as the need for shelter. Yet we see that the manifestations of need for shelter are varied owing to differing regional, economical, and technological conditions. There are also certain functional needs that are to be satisfied. These differences expressed through design represent the desire for identity. Each building has its own identity and style, which speaks of the architect. Hence, each architect has his own identity and style of design.

Having discussed the various principles of architecture in the preceding chapters, our study would be incomplete if we do not mention great architects who have imbibed these principles, concepts, and theories in their works and contributed and set a mark in the field of architecture. In this chapter, we will gain some insight into how great architects use architecture as a design process that results in functionally efficient, economically viable, and aesthetically pleasing buildings.

Contemporary Architecture

We have seen in Chapter 3 that contemporary or modern architecture is about style and character in traditional and modern architecture. Modern architecture is an advancement over traditional style.

The multi-storey buildings of modern architecture are repositories of structural marvel. In this chapter we will study the works of four renowned masters of contemporary Western architecture.

1. Frank Lloyd Wright
2. Le Corbusier
3. Mies Van der Rohe
4. Louis I. Kahn

Modern architecture in India

The history and criticism of modern architecture in India continue to reflect a Western bias. When a more balanced version is eventually written, India will deserve a special place in it. A special feature of Indian architecture is the mingling of traditional and modern design. There are many Indian architects who have been working towards this style. In this chapter, we will also study the pioneering works of two Indian architects who have deeply influenced modern Indian architecture.

1. Charles Correa
2. Balakrishna Doshi

Studying the achievements of these individuals helps us gain an insight into the spirit of architecture.

Frank Lloyd Wright (1867–1959), USA

Fig. 5.1 Frank Lloyd Wright
(1867–1959)

'A great architect is not made by way of a brain nearly so much as he is made by way of a cultivated, enriched heart'

—Frank Lloyd Wright

Frank Lloyd Wright was born on June 8, 1867 in Richland Centre, Wisconsin. Wright contributed the 'Prairie' and 'Usonian' styles to American residential architecture. Elements of his designs can be found in a large proportion of homes built today.

Wright studied civil engineering briefly at the University of Wisconsin. At 20 years of age he joined a Chicago architectural firm as a draftsman. Wright eventually became chief draftsman and supervised the firm's residential designs. Wright started his own firm in 1893, and began developing ideas for his 'Prairie House' concept.

In 1932, Wright published his autobiography and along with his wife, Olgivanna Lasovich, co-founded 'The Taliesin Fellowship'. The autobiography worked like an advertisement, inspiring many who read it to seek Wright out . Thirty apprentices came to live and learn under his tutelage. In 1936 Wright's most famous work, 'Falling Water', was designed and 'Taliesin West' was built in Arizona as a winter location for the school.

The years after World War II to the end of Wright's life were his most productive years. He received 270 residential commissions. Among his commercial designs were the Guggenheim Museum, the Marin Country Civic Centre, and the Price Tower. Wright died in Arizona at the age of ninety-two. He had never retired from his profession.

Theme and philosophy

The first project he executed was Winslow House in Illinois in 1894. This was the first example of a 'Prairie House'.

Prairie style

His work with space and construction possibilities in these houses had an effect on the design of his subsequent buildings. The typical characteristics of these free-standing houses were a broad, overhanging roof and horizontal window bands. A free plan was developed around a central chimney. The building had low proportions (not very high), which was related to the ground. It had a gently sloping roof. All these formed the characteristic features of domestic architecture.

The introduction of all these exterior features helped to form a new language in architecture.

I. The first step was the development of the *plan of the house* which had plenty of open spaces separated from one another by simple architectural devices instead of partitions, walls, or doors. This was called the *open plan.*

II. The integration of the building with nature was another innovation.

III. This helped Frank Lloyd Wright to develop the concept of *organic architecture.*

Organic archirecture

This school of thought holds that architecture should reflect nature and exhibit the same amount of unity as prevails in nature. F.L. Wright and Louis Sullivan were the pioneers of organic architecture. Wright defined organic architecture as that in which all the parts are related to the whole and the whole is related to the parts.

To explain the concept of unity in nature, the architect used a living organism as an example:

- Harmony of the part in relation to the whole
- The parts are made according to the function of the organism.
- The form (plan) of the organism decides the character of the organism.

Applying these concepts, his building designs emphasize the following principles:

- Integration of parts to the whole.
- Design of the parts controls the design of the whole.

Wright had a deep knowledge of and a lot of respect for natural materials such as stone and wood. These materials had hitherto been used in different ways—covered, painted, plastered, and altered to suit any particular fashion or taste. But in his works, these materials were always used in the natural form, by allowing, for instance, the use of masses of stone as the natural feature of the building.

Important works

Wright designed the Robie House in Chicago (1906–10). In 1909 he went to Europe and worked in European architecture for two years. Between 1915–22 he introduced a new construction technique to safeguard buildings against earthquakes and used this to build the Imperial Hotel in Tokyo. He developed certain pre-fabricated concrete parts which were used in Millard House.

Wright contributed new ideas on town planning in the title *Broadacre City*. This was a model for a community of car-driving families. One of his most intelligent works is the house 'Falling Water' in Pennsylvania (1935–1939). He designed the S.C. Johnson Wax Building, Wisconsin. Wright also used unusual designs like a spiral form for the Solomon R. Guggenheim Museum in New York.

Wright has 800 building designs to his credit, and has written many books and articles. He has brought out the concept of 'organic architecture' in his books *The Future of Architecture* (1953) and *The Living City* (1958).

Of all Wright's buildings, two are universally known and uniquely linked with his name—'Falling Water' and the Guggenheim Museum. Falling Water is recognized as a landmark of modern architecture and the work of a genius.

Kaufmann House, 'Falling Water', Pennsylvania, USA

Set in a very unusual location, the ideas implicit in the house are a highly dramatic and original combination of modern technology within a natural setting (see Figs 5.2 and 5.3). The notion of a house sitting over a waterfall evokes the imagination of the English romantic poet, Wordsworth. At the same time, scientific technology has been integrated with a modern concept. The cantilevered house which sits on a waterfall that is audible rather than visible was Wright's unique achievement.

Kaufmann House is an outstanding example of domestic architecture. In the 1930s, Wright developed the main theme of organic architecture, that is, combining modern techniques and natural landscape in a new way. 'Falling Water' is among Wright's most famous buildings and has become a symbol of the international movement. He used natural and organic elements in this building. This is a good example of organic architecture.

The house is located on a cliff with a waterfall. It is a weekend house. It consists of two levels of living areas. Both the living areas extend up to the waterfall and give a good view of the surrounding countryside (Fig. 5.4).

The entrance drive leads to the main living room, which extends in different directions in the ground floor. A staircase leads directly to the waterfall (Fig. 5.5). Terraces, balconies, kitchen, and dining area all extend in different directions. The bedroom on the second floor opens on to the terrace, which is cantilevered more than the terrace of the first floor. The second floor is much smaller than the first and has only one bedroom with an adjoining roof terrace.

Fig. 5.2 Kaufmann House, 'Falling Water', Pennsylvania, USA

All the three floor plans form a pattern in such a way that they are arranged round a single vertical element, which is a natural stone tower—*the staircase*. At the foot of the staircase is the supported ground floor and the slope of the hill. The base of the building is made of natural stone, the individual storeys are made of reinforced concrete, and the walls of glass (Fig. 5.6). The building literally combines

- nature and architecture
- the organic and the geometric
- natural stone and concrete
- exterior and interior
- nature and space

What Wright achieved in this building was to place its occupants in a close relationship to the surrounding beauty—the trees, the foliage, and wild flowers.

Fig. 5.3 Kaufmann House, combining architecture and landscape in a new way

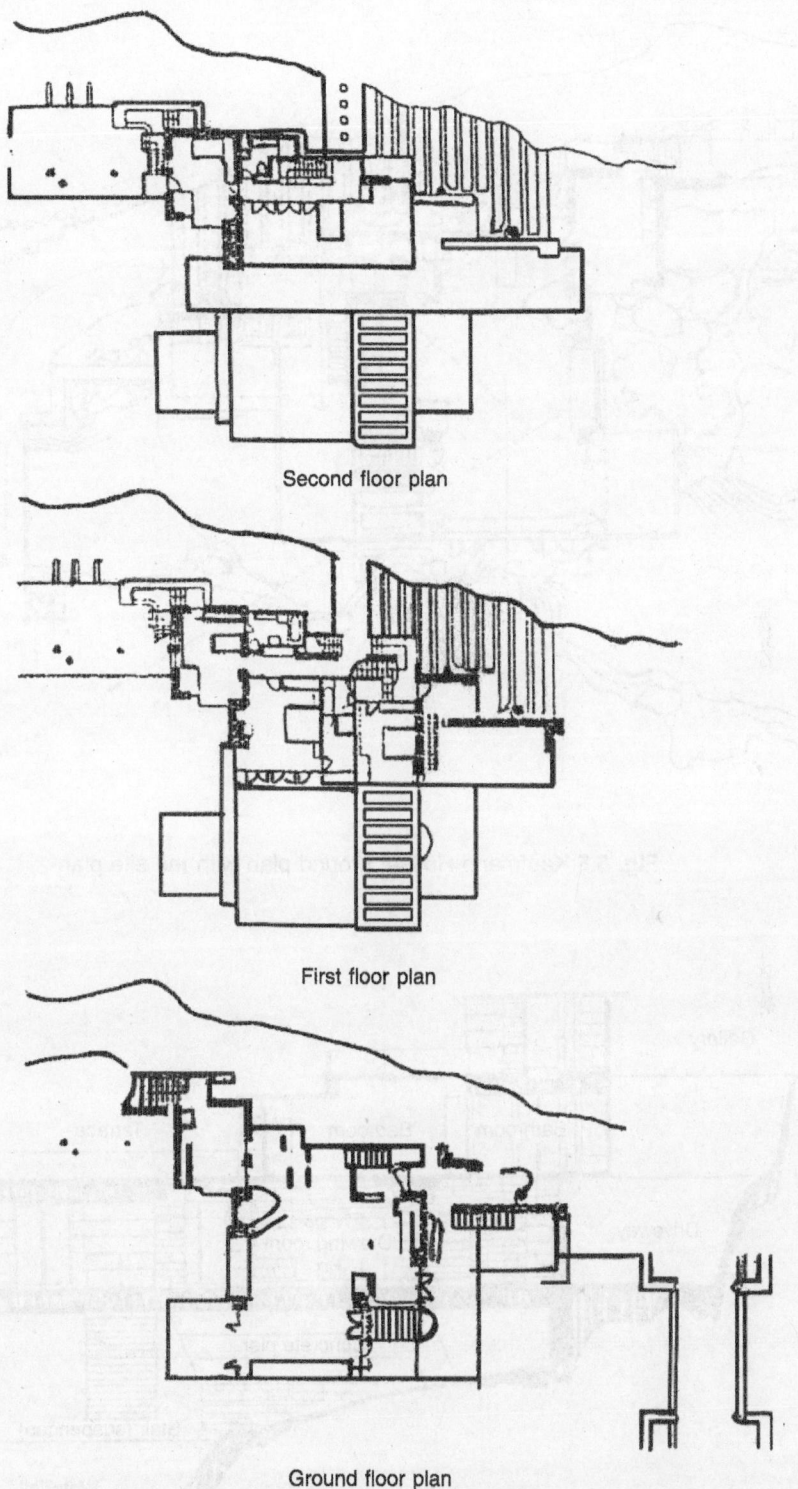

Second floor plan

First floor plan

Ground floor plan

Fig. 5.4 Falling Water floor plans

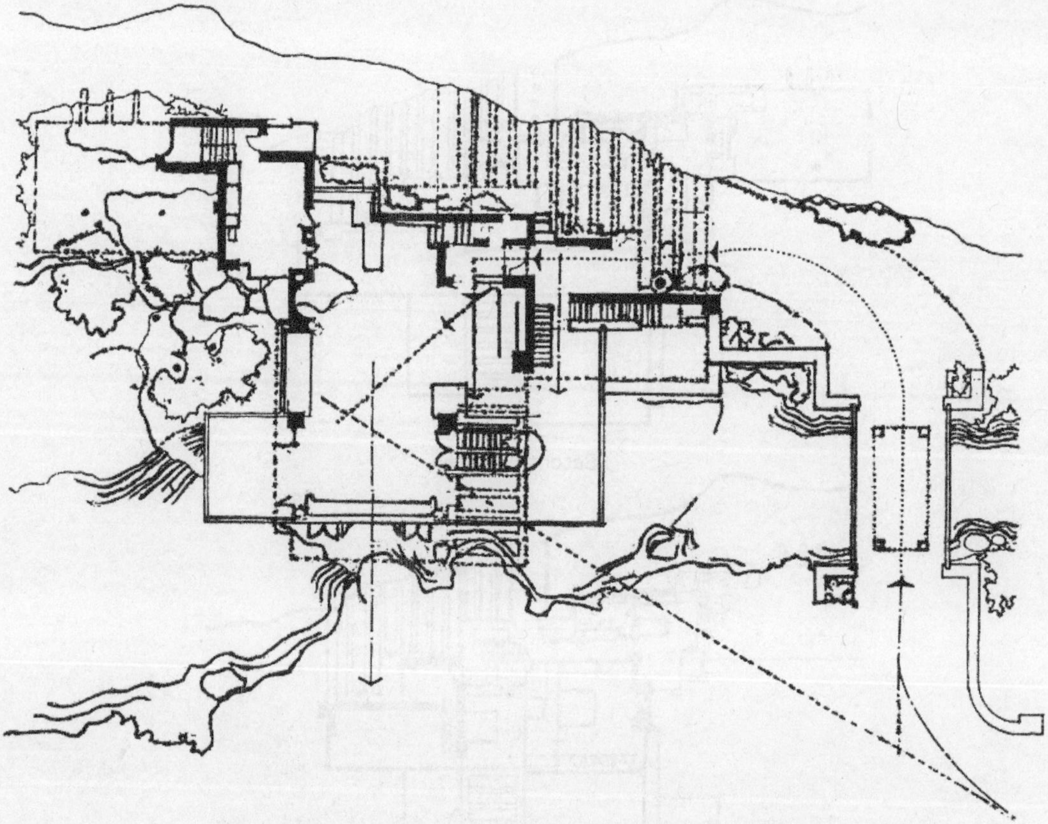

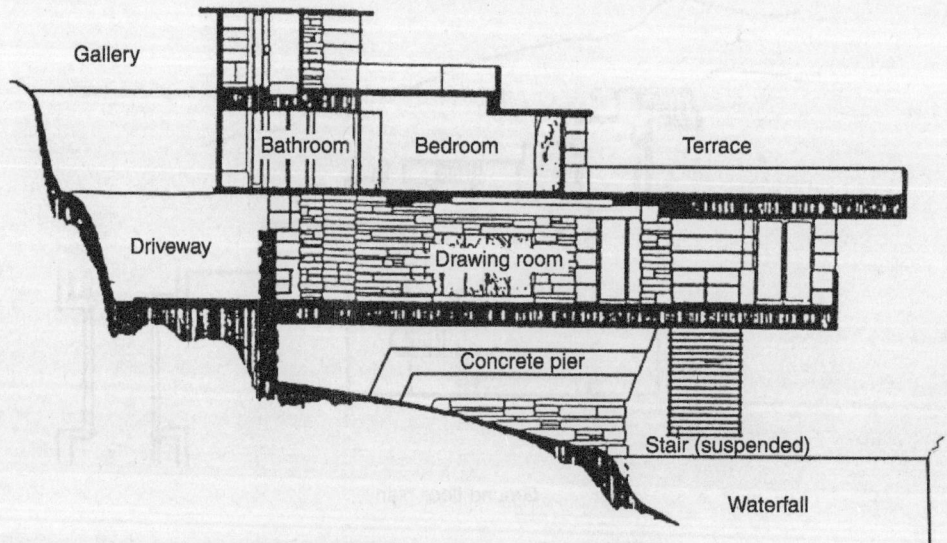

Fig. 5.5 Kaufmann House: ground plan with the site plan

Fig. 5.6 Falling Water: sectional elevation

Solomon Guggenheim Museum, New York (1946–59)

Fig. 5.7 Solomon R. Guggenheim Museum, New York

The Solomon Guggenheim museum was constructed to display the collected works of Guggenheim—his sculpture, paintings, and pictures (Fig. 5.7).

In order to give a completely original form for this building, Wright invented a radical, new shape for it. The seven-storey building that he developed has a spiral form around an open well, with a huge skylight as a roof (Fig. 5.8). The individual storeys project outwards at each level. The paintings are planned so that they hang on the external skin which follows the spiral.

The circulation pattern is very simple. Visitors are taken up to the top floor by a lift directly upon entering the museum, then they walk down the gentle slope of the spiral ramp to the ground floor, viewing the exhibits during their descent, with changing colours and forms (Fig. 5.9). There is a library and a book shop at the ground floor. The circular form of the spiral adds to the sense of movement (Fig. 5.10). The museum's administrative office is in a separate building to the north, which is also spiral shaped (Fig. 5.11).

Wright's purpose was to develop a new kind of museum in which each work of art did not merely form a part of the wall but stood out on a slightly curving surface. The unique feature of this design is that while looking at any one painting, all other paintings are always visible, since the whole exhibition can be seen from any point on the ramp. Wright gave special importance to lighting. By using both daylight and artificial light, a three-dimensional effect was created.

The main construction material is concrete in a variety of forms such as reinforced and sprayed concrete, as the spiral ramp design depends on the

![Close-up view of museum]

Fig. 5.8 Close-up view of the museum showing truncated cone form

Fig. 5.9 The view of the open well as one descends down the ramp

character of the material. The circular, inverted, and truncated conical main space is lit by a dome and the ground floor is left as an open circular court. A single, self-supporting, reinforced concrete spiral beam forms the structural system. The flat beam acts like a continuous ramp. This ramp is the architectural space.

Wright's concept is three-dimensional and links the visitor, the picture, and the environment into a single unit as structure, space and circulation come together to form a tremendous unity. The continuous seven-storey ramp is based on the principle of the unbroken wave (Fig. 5.12).

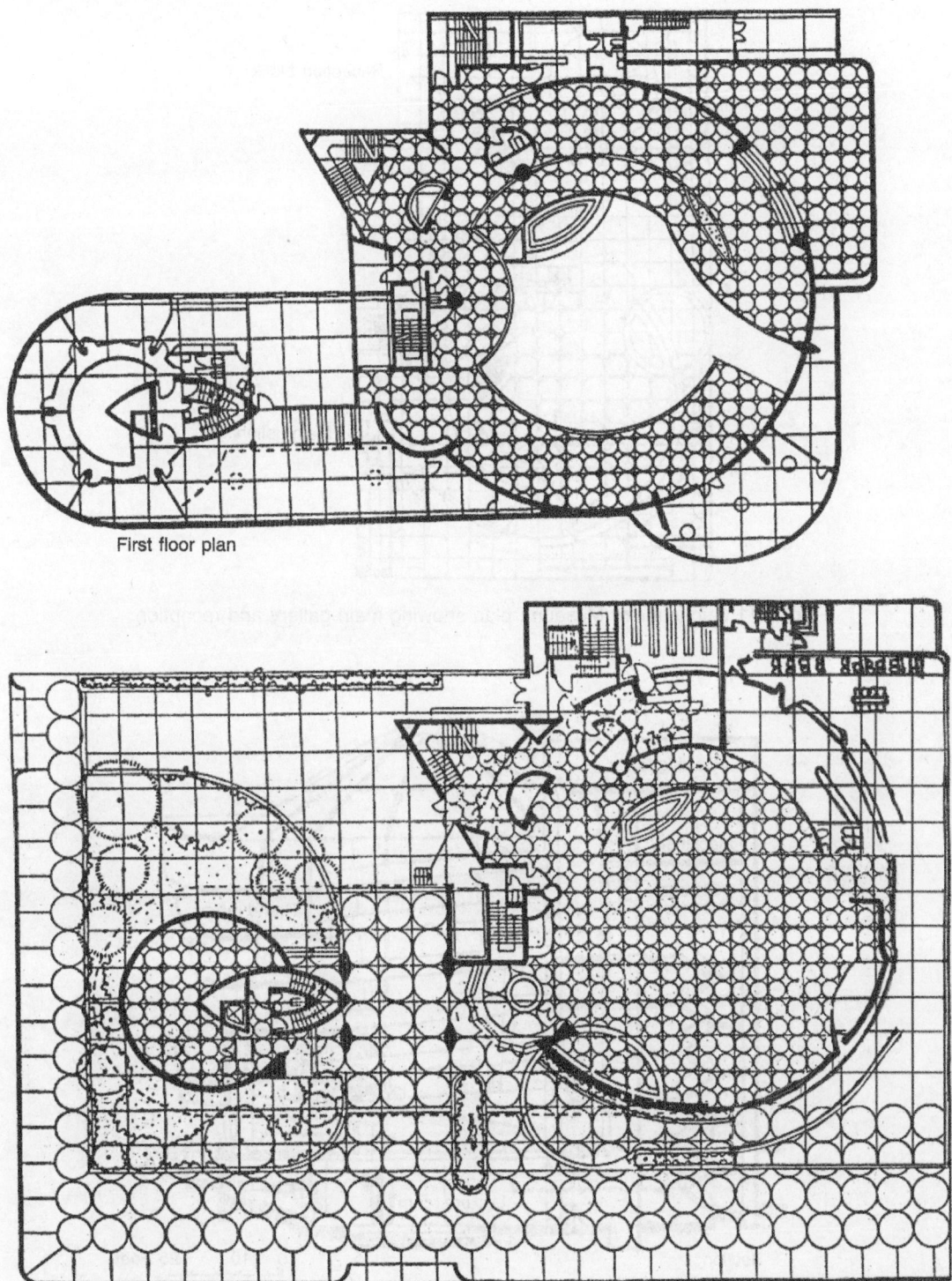

First floor plan

Ground floor plan

Fig. 5.10 Guggenheim Museum

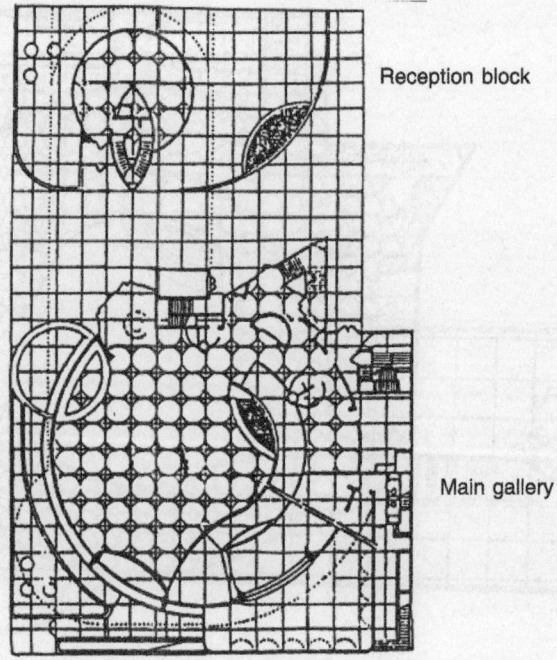

Reception block

Main gallery

Fig. 5.11 Guggenheim Museum: plan showing main gallery and reception

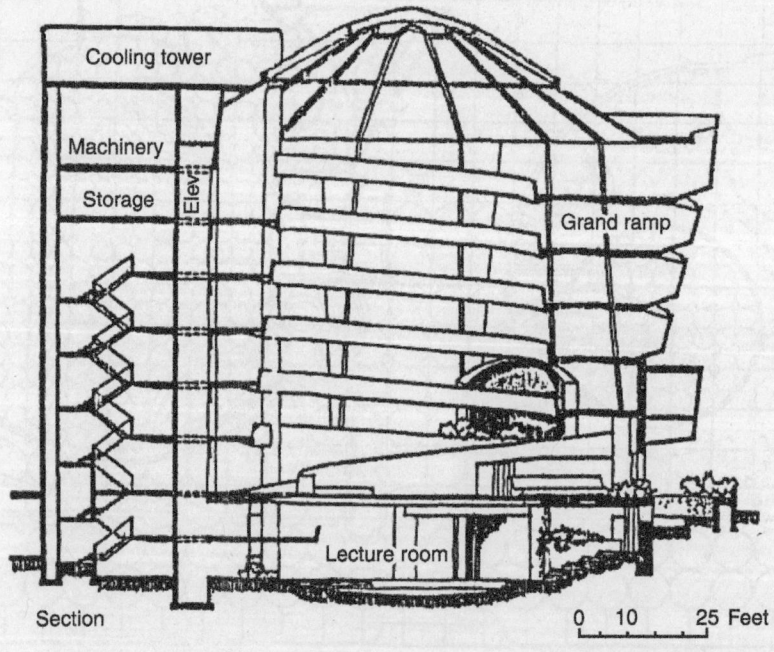

Cooling tower

Machinery

Storage

Elev.

Grand ramp

Lecture room

Section

0   10   25 Feet

Section of Solomon R. Guggenheim Museum, New York (1956–59)

Fig. 5.12 Section of Guggenheim Museum showing clearly the truncated cone, ramp that descends down

Le Corbusier (1887–1965), France

Fig. 5.13 Le Corbusier (1887–1965)

> **'Architecture is the play of forms under light'**
> —Le Corbusier

Charles-Edouard Jeanneret was born at La Chaux-de-Fonds on October 6, 1887 and later adopted the pseudonym Le Corbusier. He came from a family of watch engravers in Switzerland, and his mother was a musician. He grew to maturity in the intellectually stimulating city of Paris and adopted French nationality. He travelled extensively and learnt many lessons from the classical architecture of Greece and the European cities.

He became a powerful thinker of new urban theories and propounded a bold, modern architecture. In 1951 he was appointed Architectural Adviser to the Punjab Government for designing the new capital city, Chandigarh. This city represents the expression of his revolutionary ideas and is where his greatest monuments have been erected.

He lived an austere life and admired the simple and the useful. He was not only an architect and a planner but a painter, a sculptor, in secret a poet, a visionary whose view encompassed all that concerns man. Above all, Corbusier was a great humanist whose primary preoccupation was the welfare of man. He worked in India for a pittance and did not die a rich man. Le Corbusier died suddenly on August 27, 1965, while swimming at Cap Martin. Besides his legacy of architecture, he left behind a huge compendium of 'think sketches', futuristic ideas and books containing his theories on architecture and town planning.

Theme and philosophy

Le Corbusier was the most influential architect of the twentieth century. His works, ideas, and writings laid the foundation for modern architecture and planning. In 1942, he developed his modular system of measurements based on the human scale and the golden section.

Five-point programme

Le Corbusier developed a five-point programme for a new and contemporary architecture:
- The Pilotis (posts)
- The roof garden
- The free plan
- The horizontal strip windows
- The free façade.

Le Corbusier famously said that 'architecture is the play of forms under light'.

Important works

Corbusier built several important houses. The Villa Savoye at Poissy is one such house, built during 1929–31. The house is supported on slender posts with ramps. Among his important buildings was Unite d' Habitation in Marseilles, which is based on his concept of Modular. The Chapel of Notre Dame at Ronchamp, Paris, was a religious structure designed by him. Le Corbusier put many of his planning ideas into practice when he was commissioned to design the city of Chandigarh in India, where he also designed the government buildings, the Palace of Assembly, and the High Court. In India, he also designed private houses. The best known ones are the Sarabhai residence and the Shodan residence at Ahmedabad. Both were designed during 1951–56. His only project in the USA was Harvard University's Carpenter Centre at Cambridge (Visual Arts Centre).

Chapel of Notre Dame, Ronchamp, France (1950–5)

Le Corbusier's pilgrimage chapel of Notre Dame at Ronchamp was developed through a series of sketches through which he expressed his ideas about volume, image, light, and plan. This chapel, which is on top of a hill, consists of a rolling, dark-coloured, pointed roof on smooth, whitewashed, concrete walls, with small punched-in openings (Figs 5.14–5.16). The composition has three towers of different sizes.

Light enters the chapel through small, carefully placed windows and through the roof–wall junction and illuminates the interior with the dramatic effect of light and shadow (Fig. 5.17). This sculptural effect captured the imagination of the world at large. This was not the usual frame type of building but had a dynamic and imaginative spatial form (Fig. 5.18).

The interior space is 25 m long and 13 m wide and can seat 200 people. Three towers, one 22 m and two 15 m, perform the dual functions of crowning the architecture externally and bringing light into the interior. The light which comes directly from the three towers falls on to the altars of the three chapels inside the church (Fig. 5.19). The dominant feature is the roof, which is made

Fig. 5.14 Ronchamp chapel, Notre Dame, France: The sculptural effect captured the imagination of world at large.

Fig. 5.15 Ronchamp chapel: The rolling, dark coloured, pointed roof and the smooth, white-washed, concrete walls with punched small openings are seen.

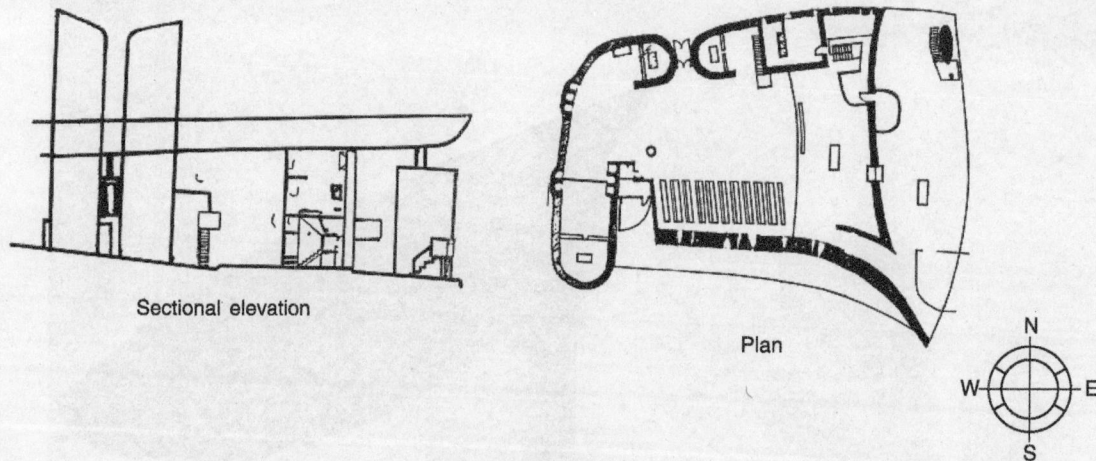

Sectional elevation

Plan

N
W · E
S

Fig. 5.16 Ronchamp chapel

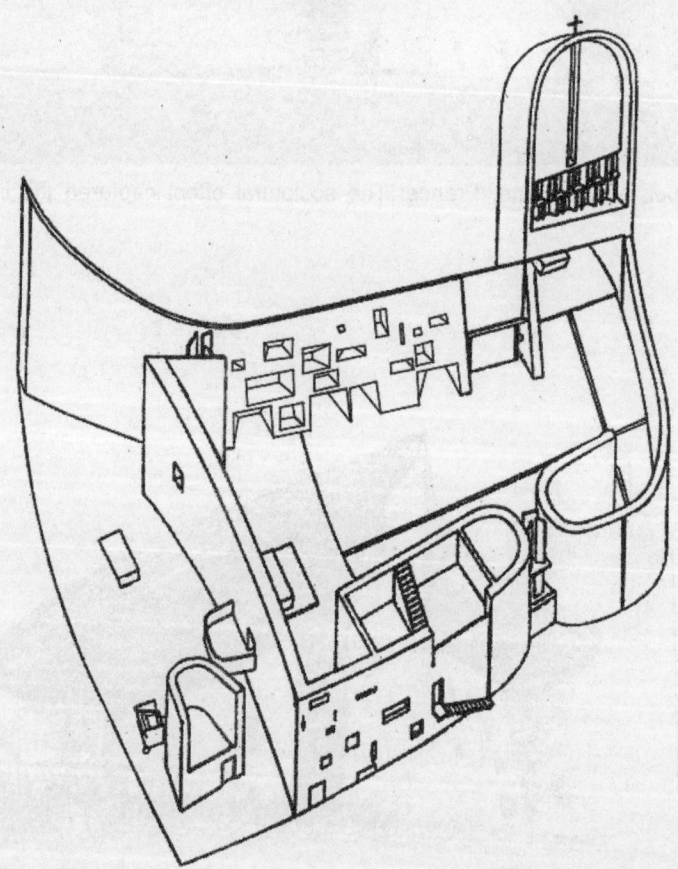

Fig. 5.17 Axonometric view of Ronchamp chapel showing altar and the punched openings in the wall that bring light in the interior

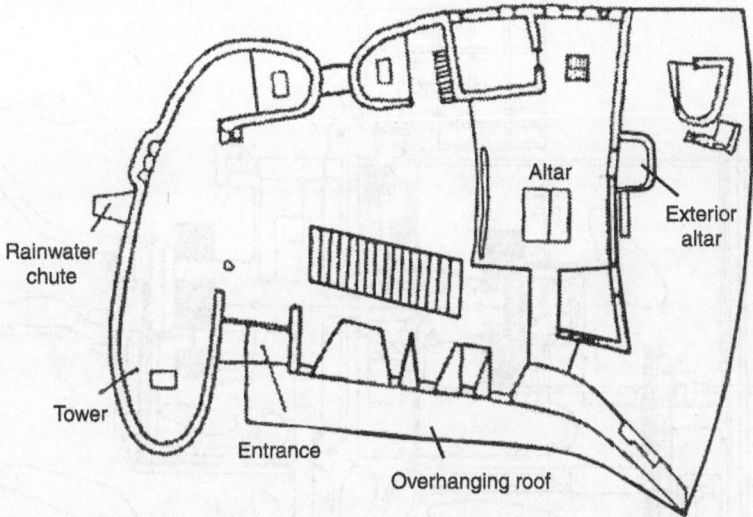

Fig. 5.18 Concept plan of Le Corbusier's pilgrimage chapel

Fig. 5.19 View of Ronchamp showing the three towers

of two roof shells. The height of the ceiling varies from 4.5 m to 10 m. Towards the outside, the roof projects over the east wall and makes a covered space which has an open altar.

The sculptural hand of Le Corbusier is seen in the bold, startlingly new design of Ronchamp. The church is visible for miles around. It is comparatively small but its effect is one of monumentality.

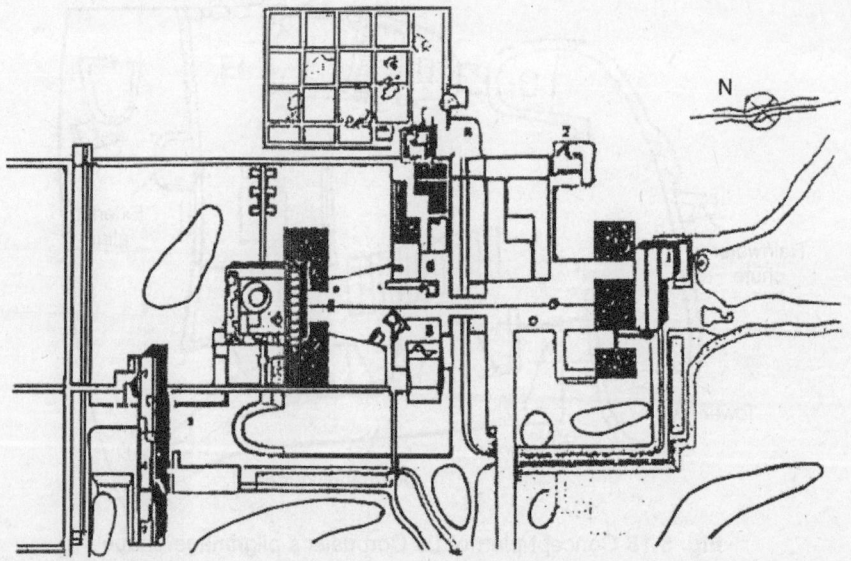

Fig. 5.20 Plan of the Capital Complex, Chandigarh, India, 1956: The capital is laid out as a series of buildings and monumental plazas, each of which is a defined entity or object in the landscape. This drawing, made after the buildings had commenced construction, shows the layout as finalized. Although grand, the plaza concept takes no note of (and does not function well in) the hot Indian climate

Le Corbusier's modern image of Chandigrah

One of Le Corbusier's important works was the planned development of Chandigarh, provincial capital of Punjab. It became symbolic of the newly independent Indian nation (Figs 5.20 and 5.21). This capital city was developed on a site selected by Jawaharlal Nehru. Commencing work from 1951 until Le Corbusier's death in 1965, he shaped the city and gave it its image. The city had a grid plan based on the hierarchy of movement from highways to pedestrian walkways. The metaphor of a human being was employed in the plan—the 'head' contained the capital complex, the 'heart' the commercial centre, and the 'arms', which were perpendicular to the main axis, had the academic and leisure facilities. The plan incorporated Le Corbusier's principles of light, space, and greenery.

Fig. 5.21 View of the Capital Complex, Chandigarh

The Parliament or Palace of Assembly (1951–62)

Fig. 5.22 The Parliament or Palace Assembly: the centrepiece of the capital, 1951–62

The Parliament or Assembly was designed as a large box with the entrance portico on one side, concrete piers on the other, and a repetitive pattern on the façade. Sculptural forms on the roof, a dramatic 'funnel' top light over the Assembly, and a tilted pyramid over the Senate chambers completed the composition (Fig. 5.22).

The Assembly Hall has a square plan (Fig. 5.23). The Assembly chamber, in the form of a hyperbolic shell, is surrounded by ceremonial space. This circulation space is planned as a dimly lit, triple height, columned hall for informal meetings and discussions. The side of the hall facing the high court has a great portico and has eight thin piers. These piers frame a view of glimpses of the Shivalik hills.

A ceremonial pivoting door is placed in an off-centre bay of the portico. Le Corbusier was inspired by the form of the cooling towers of a power station near Ahmedabad. The architect designed the hyperbolic shell of the Assembly chamber with a base diameter of 39.6 m. This shell is 38 m high and terminates in an oblique section with a metallic framework at the top. This framework directs the interplay of natural and artificial lighting, ventilation, and acoustics.

The hyperbolic shell is only 15 cm thick, which helped in reducing the cost and the weight of structure. The Assembly chamber has a seating capacity for 252 persons. Additional galleries are provided for ladies, journalists, and officials. An attempt has been made to modulate or control the acoustics resulting from such a form, by providing sound absorbing panels in bright colours and random curvilinear shapes.

The Council chamber, with a capacity of 70 seats, is crowned by a pyramid, which admits light from the north into its interior. A ladies' gallery with 90 seats, a men's gallery with 104 seats, and a press gallery with 24 seats are also provided in this chamber. Staircases, lifts, and ramps provide various means of circulation and access to different levels of the building. The construction of the entire structure is in exposed reinforced concrete.

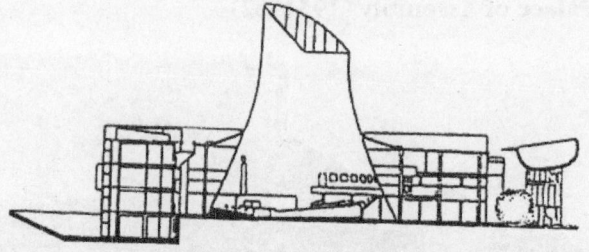

Section showing the hyperbolic shell

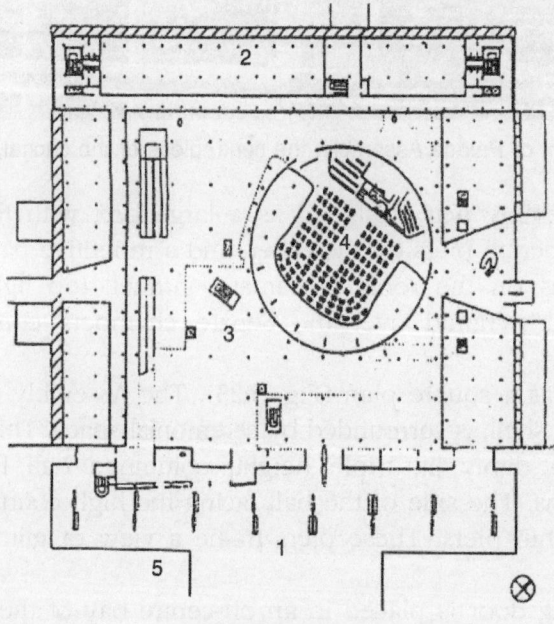

Ground floor plan

1. Entrance
2. Office
3. Assembly hall
4. Assembly chamber
5. Pool

Capital complex view from Secretariat roof

Fig. 5.23 The Assembly Hall, Chandigarh

Ludwig Mies van der Rohe (1886–1969), Germany

Fig. 5.24 Mies van der Rohe (1886–1969)

Mies explicitly challenged Sullivan's dictum that 'form follows function'.

'We do the opposite. We reverse this, and make a practical and satisfying shape, and fit the functions to it. Today this is the only practical way to build, because the functions of most buildings are continuously changing, but economically the structure cannot change.'

—Mies van der Rohe

Mies was born in Aachen, Germany, on March 27, 1886, son of Michael Mies and Amalie Rohe. After having trained with his father, a master stonemason, he moved to Berlin at the age of 19, where he worked for Bruno Paul, the Art Nouveau architect and furniture designer. At 20 he received his first independent commission, to plan a house for the philosopher Alois Riehl. In 1908 he began working for the architect Peter Behrens. He studied the architectural works of the Prussian Karl Friedrich Schinkel and Frank Lloyd Wright. He opened his own office in Berlin in 1912, and married in 1913.

After World War I, he began studying the concept of skyscrapers and designed two innovative steel-framed towers encased in glass. One of them was the Friedrichstrasse Skyscraper, designed in 1921 for a competition. It was never built, although it drew critical praise and foreshadowed his skyscraper designs of the late 1940s and 1950s.

In 1921, when his marriage ended, he changed his name, adding the Dutch 'van der' and his mother's maiden name 'Rohe'; Ludwig Mies became Ludwig Mies van der Rohe.

In the 1920s, he was active in a number of the Berlin's avant-garde circles (the magazine 'G' and organizations such as the 'Novembergruppe', 'Zehner Ring', and 'Arbeitsrat für Kunst') that supported modern art and architecture along with artists like Hans Richter, El Lissitzky, and Theo van Doesburg, among others. He made major contributions to the architectural philosophies of the late 1920s and 1930s as Artistic Director of the Werkbund-sponsored Weissenhof project, a model housing colony in Stuttgart, Germany. The modern apartments and houses were designed by leading European architects, including a block by Mies.

In 1927 he designed one of his most famous buildings, the German Pavilion at the international exposition in Barcelona. This small hall, known as the Barcelona Pavilion (for which he also designed the famous chrome and leather Barcelona chair), had a flat roof supported by columns. The pavilion's internal walls, made of glass and marble, could be moved around as they did not support the structure. The concept of fluid space with a seamless flow between the indoors and outdoors was further explored in other projects he designed for decades to come. Mies began working with Lilly Reich, who remained his collaborator and companion for more than ten years.

In 1930, Mies met New York architect Philip Johnson, who included several of his projects in the first International Exhibition of Modern Architecture held in 1932, thanks to which Mies's work began to be known in the United States.

He was the director of the Bauhaus school from 1930 until its disbandment in 1933, shut down under pressure from the new Nazi government. He moved to the United States in 1937. From 1938 to 1958 he was head of the Architecture Department at the Amour Institute of Technology in Chicago, later renamed the Illinois Institute of Technology. In the 1940s, he was asked to design a new campus for the school, a project in which he continued to refine his steel-and-glass style. He had also formed a new relationship with Chicago artist Lora Marx, which lasted for the rest of his life.

By 1944, he had become an American citizen and was well established professionally. Around this time, he designed one of his most famous buildings, a small week-end retreat outside Chicago, a transparent box framed by eight exterior steel columns. The 'Farnsworth house' is one of the most radically minimalist houses ever designed. Its interior, a single room, is subdivided by partitions and completely enclosed in glass.

In the 1950s he continued to develop this concept of open, flexible space on a much larger scale.

In 1953, he developed the convention hall. The structural system that spanned large distances was highly innovative. During this period he also realized his dream of building a glass skyscraper.

The twin towers in Chicago were completed in 1951, followed by other high-rises in Chicago, New York, Detroit, Toronto, culminating in 1954 with the 'Seagram Building' in New York, hailed as a masterpiece of skyscraper design.

For his achievements, he earned the 'Order Pour le Mérite' (Germany) in 1959 and the 'Presidential Medal of Freedom' (USA) in 1963.

In 1962, his career came full circle when he was invited to design the 'New National Gallery' in Berlin. His design for this building achieved his long-held vision of an exposed steel structure that directly connected interior space to the landscape. He returned to Berlin several times while the gallery was under construction, but was unable to attend the opening in 1968. He died in Chicago on August 17, 1969.

Theme and philosophy

Mies always maintained an intellectual affinity with medieval and Greek architecture by integrating the concepts of modern art into his work. Mies believed in designing buildings according to the dimensional constraints characteristic of a plot of land and fitting functions accordingly. He created radical and conservative structures: radical because his constructions reflected scientific and technological character and conservative because they were based on the external laws of architecture—'Order, space and proportions'. He was a follower of the New Bauhaus style. Mies believed in flowing of spaces. He introduced steel and glass as design elements in his buildings. His dream was to build a glass skyscraper. According to Mies, for a building, form is primary to function. The only practical way to build (in present times) is to construct a satisfying structure and fit the functions to it.

Important works

Among Mies' works, some of the most important are the Alois Riehl house in Berlin; Crown Hall, Illinois Institute of Technology (1950–6); Lake Shore Drive apartments, Chicago (1948–51); Seagram building, New York (1954–8); and the IBM Regional Management Building, Chicago (1966–9).

Alois Riehl house, Berlin

At the age of 21, Mies received his first commission, a house for philosopher Alois Riehl in Berlin, which effectively showcased his talents. The house is rectangular in plan. The kitchen, the book cases, and the radiator grill reflect Mies's interest in built-in furniture (Fig. 5.25). The planning of the house is dominated by the furniture. Another dominant feature is the right angle between the axis of entry and the downhill view (Fig. 5.26).

The entrance hall

Upper floor alcove

Fig. 5.25 The interior of the house shows Mies's interest in built-in furniture

The entrance

The gable roof

Overall view

Fig. 5.26 Alois Riehl House in Berlin

In 1937, the famous architect Frank Lloyd Wright received him in USA. He was made director of Armour Institute's School of Architecture. The Illinois Institute of Technology (IIT) was created in 1940 from the amalgamation of Armour and Lewis Institute. Mies van der Rohe worked for the new site plan of IIT. He designed 22 buildings in the Illinois Institute Campus, all of them bearing his distinctive, slick, and lucid style. Mies built the dimension of time into the whole enterprise. He created radical yet conservative structures—radical because he accepted scientific and technological character and conservative because the construction was based on the eternal laws of architecture, i.e., order, space, and proportion. Only nine laboratory and dormitory buildings resort to a concrete skeleton, with brick infill. In IIT at Chicago, Mies developed steel frame construction. The first post-war building at IIT was Alumni Memorial Hall. From 1949-50 Mies designed many buildings in reinforced concrete as well.

The Crown Hall, Illinois Institute of Technology, Chicago (1950–56)

General view showing the projection of steel member

Front view

View of the south-west corner

Fig. 5.27 Crown Hall, Illinois Institute of Technology

At the end of his tenure at IIT, Mies built Crown Hall for the School of Architecture and Urban Planning, and the Institute of Design, a successor to the New Bauhaus, which was taken over by IIT in December 1949. The Crown Hall is a single uninterrupted space 120′ × 220′ (36.6 m × 67 m) with a height of 505 m, roofed by a spectacular structure, consisting of four large girders of welded steel, under which the ceiling suspends. The main floor is completely open and suitable for a combination of tables (Fig. 5.27).

The main entrance is approached by a steel stairway. For Mies, this building was to remain the clearest structure, the best to express his philosophy. Mies aspired to vast spaces as if they represented an ideal habitation. Though the steel skeleton was exposed internally in some previous buildings, it is now invisible. Crown Hall is the first complete realization of Mies's idea of a big space where anything is possible (Fig. 5.28).

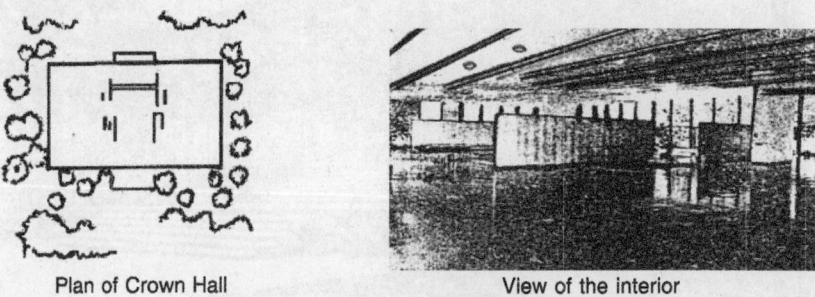

Plan of Crown Hall   View of the interior

Fig. 5.28 Crown Hall

The skyscraper apartment block—Lake Shore Drive apartments, Chicago (1948–51)

Alongside the theme of the great covered space, Mies concerned himself with the elaboration and improvement of another architectural type that is characteristic of his American work—the skyscraper apartment block (Fig. 5.29).

In total, Mies constructed six high-rise buildings for Greenwald, in addition to the fourteen others that he built in Chicago between 1948 and 1969. The two 26-storey blocks of 860–880 houses in the Lake Shore Drive develop the original steel frame version of previous towers constructed by Mies (Fig. 5.30). These two blocks were set at right angles, aligned with the checker board urban pattern of Chicago. Through their partially open ground floors, the main steel skeleton forms three bays on the end wall and five on the main façade, each bay filled by four aluminium window panels. This 5 × 3 proportion became a recurrent feature in the later projects of Mies (Fig. 5.31). The steel skeleton supports a secondary structure into which the panels of the façade are inserted. Inside the apartments, the open plan initially proposed was replaced by partitioned rooms. In the finished blocks, silvery curtains keep out the sun and discreetly hide the interiors. The grey, natural aluminum window frames, the glossy, dark, linoleum floor, and the white walls come together in a

Fig. 5.29 Lake Shore Drive, Chicago: general view

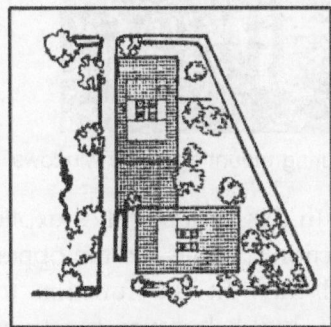

Plan of the ground floor

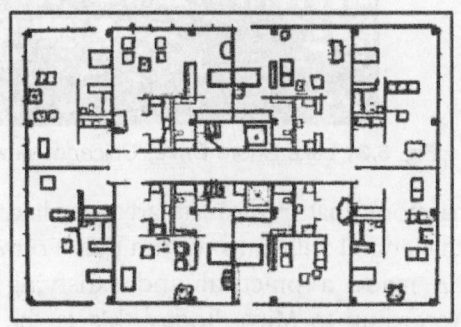

Typical plan of a standard floor

Fig. 5.30 Lake Shore Drive apartments

Fig. 5.31 Lake Shore Drive, Chicago: view of the site during mounting of the windows

harmony that frames the adjacent lake (Fig. 5.32). In this context, Mies explicitly challenged Sullivan's dictum that 'Form follows function...'. Mies did the opposite. He made a practical and satisfying shape, and then fit the function to it. According to Mies, 'today this is the only way to build, because functions of most buildings are continuously changing, but economically the structure cannot change.'

An entrance   Part of the facade

Entrance hall photos

Fig. 5.32 Lake Shore Drive

Seagram Building, New York (1954–58)

At the age of 68, Mies designed an office building (see Fig. 5.33). As a piece of urban design, Mies's design was radically new. Set on a granite platform, the Seagram Building, completed in 1958, dominates the surroundings. Flanked by a symmetrical pair of pools, this 38-storey building seems to hover on the ground. The mass of the building gives an impression of geometric unity, consisting of two T-shaped configurations.

A view

Ground floor plan, Seagram building

Fig. 5.33 Seagram Building, New York

Louis Isadore Kahn (1901–74), USA

Fig. 5.34 Louis Isadore Kahn (1901–74)

'Ruins of historical buildings always
talk to us about the life they lived,
I thought of beauty of ruins …
the absence of frames … of things
which nothing lives behind.'

—Louis Kahn

Louis Kahn was born in Sqarama, Estoniq in 1901. His family emigrated to USA in 1905. He graduated from the University of Pennsylvania with a thorough grounding in the Beaux-Arts School of Art and Architecture. During the 1920s and 1930s he worked as a draughtsman and later as a head designer for several Philadelphia-based firms.

In 1925–6 Kahn acted as the chief of design for the Sequin Centennial Exhibition. During the Depression he was active in the design of public-assisted housing. Beginning in 1935, Kahn worked with a series of partners; but from 1948 until his death in 1974, Kahn worked alone. From 1947 to 1957 he was design critic and Professor of Architecture at Yale University, after which he was the Dean of the University of Pennsylvania. Kahn's architecture was notable for simple platonic forms and compositions. Through the use of brick and poured-in-place concrete masonry, he developed a contemporary and monumental architecture that maintained a sympathy for the site, while being rooted in the International style. Kahn's architecture was an amalgam of his Beaux-Arts education and a personal aesthetic impulse to develop his own architectural forms.

Considered one of the foremost architects of the late twentieth century, Kahn received the AIA gold medal in 1971 and the RIBA gold medal in 1972. He was elected a member of the American Academy of Arts and Letters in 1971.

Theme and philosophy

Kahn developed what is called a *classically romantic* style, in which functional areas such as stair wells and air ducts feature prominently, often as tower-like structures surrounding the main living and working areas. Louis Kahn talked much of light and of the need to express the consciousness of man. He spoke of the beauty of the *arch* and stated that he did not want to use concrete beams, but instead make huge arches of brick circles. In his opinion, light was to be present everywhere through the very structure of the buildings. He always seemed to think of rooms flowing into each other as opposed to participation design that flowed from one end to the other and spaces where the mind would speak. Though all buildings are made of bricks, the way the bricks serve the cause to make space is remarkable.

Important works

Louis Kahn's famous projects are Dacca Assembly House, Richards Medical Research Building, University of Pennsylvania (1958–61) and the Centre of British Art and Studies at Yale University (1969–74). Some of his designs outside USA are Indian Institute of Management, Ahmedabad, India (1962–74), and the United States Consulate in Luanda, Angola (1952).

Assembly House in Dacca

Fig. 5.35 Dacca Assembly House

Constructed on 200 acres of land, the 'Sansad Bhaban' (Parliament Building) sits brilliantly in an enormous pool of water, which gives the whole complex the look of a post-modern castle, complete with a moat (Fig. 5.35). The buildings do not just include the Parliament itself but also the members' residences and spaces for housing their functionaries and dignitaries. The complex sits peacefully amidst a beautiful expanse of green fields. Lining the ground is a forest of trees. The main structure is designed with 9–storey atriums. These are bordered by offices hidden behind huge walls of raw, grey concrete lined with white marble and cut with enormous geometric openings.

Kahn has explained in a famous passage 'I thought of the beauty of ruins.... the absence of frames... of things which nothing lives behind.' It was in Dacca that Louis Kahn was able to realize this conception. The 'ruins' are huge, free standing, concrete screens pierced by circles, squares, and triangles whose function is to stand between the windows of the Assembly building and the glare of the sun. He was able to invest the Assembly with megalithic grandeur and an extraordinary simplicity.

Richards Memorial Buildings, University of Pennsylvania, Philadelphia (1958–61)

This building in Philadelphia, designed in 1958 and completed in 1961, is Kahn's most important building (Fig. 5.36). It is designed according to his theories and is a brilliant expression of his approach to modern architecture.

Fig. 5.36 Richard Memorial Building, University of Pennsylvania, Philadelphia

The vertical, form-defining towers contain rooms essential for supplying and ventilating the rest of the building, where the actual research is carried out. The plan is based on the square. The towers are built on a smaller square, and are attached to the main research buildings which are built on a large square (Fig. 5.37). The verticals of the towers frame the horizontals, relieved by glass windows. The buildings repetitive design means that it can easily be extended. In this building, there were some remarkable achievements in prefabrication and onsite assembly of parts.

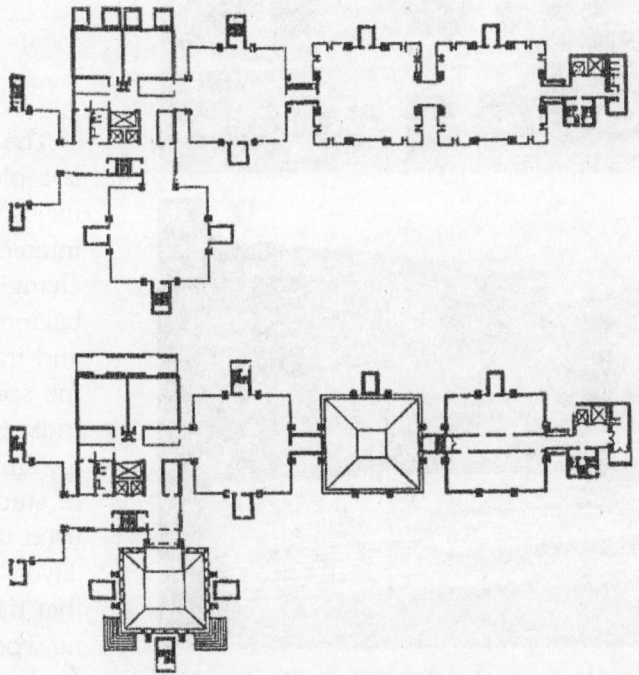

Fig. 5.37 Richards Memorial Building
Floor plans

Indian Institute of Management Ahmedabad

The site for the Indian Institute of Management Ahmedabad (IIMA) at Vastrapur, to the west of the city, was a flat piece of land. Kahn grouped the buildings close together and angled them to catch the prevailing wind, which comes from the south-west for most of the year. Circular openings were cut into bold brick forms, and concrete-relieving beams allowed a façade language of shallow arches over loggias.

Kahn distinguished dormitories from teaching zones by contrasts of diagonal and rectangular geometry, and linked the functions together with a network of streets allowing vistas through sequences of light and shade. He seemed to consider this institution a 'citadel of learning' replete with bastions, ramps, and huge cylindrical stair towers. It is quite possible that he was influenced by Mughal prototypes, such as the Fort in Agra. The emblematic squares, diamonds, and circles and the distinctions between fringe and central spaces predisposed Kahn towards Islamic prototypes.

The 60-acre campus of IIM is a blend of austerity, majesty, spaces for casual interaction, frequently changing perspectives and a balance between modernity and tradition that captures the spirit of contemporary India (Fig. 5.38). It stimulates the imagination and creativity of students. In this, Kahn used the idea of overlapping 'layers'. This plan suggests that the institution is a network of streets, open spaces, and lecture theatres (Fig. 5.39).

Fig. 5.38 Indian Institute of Management Ahmedabad: View showing the circles pierced on the concrete screens

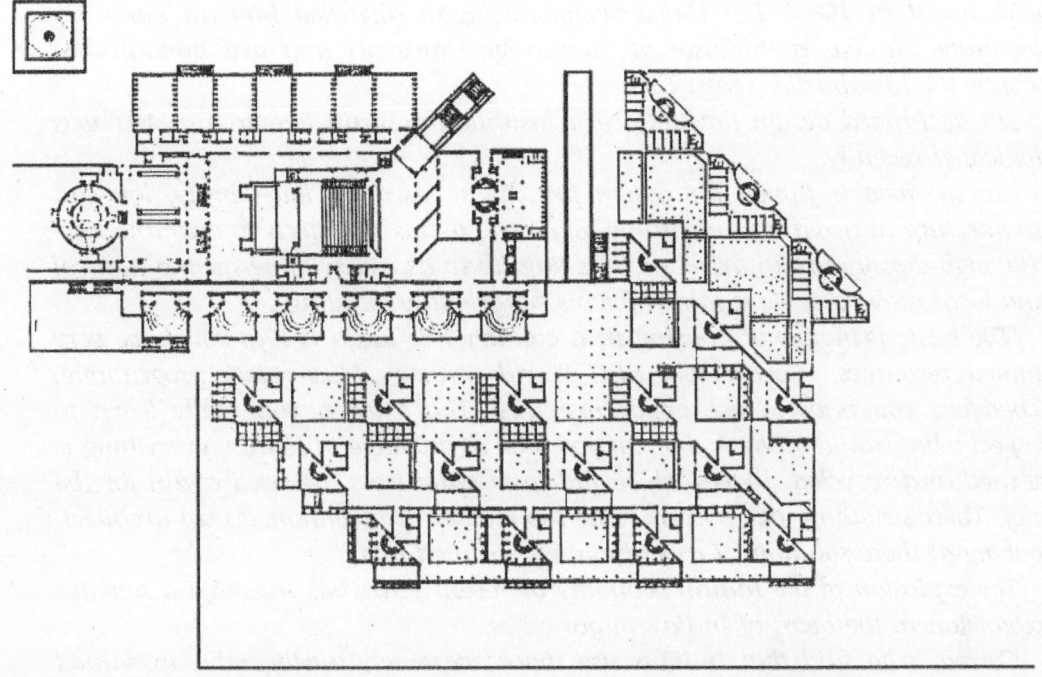

Fig. 5.39 Indian Institute of Management Ahmedabad (IIMA)

Charles Correa (1930), Hyderabad, India

Fig. 5.40 Charles Correa

> **'Housing designs from my home country offer the key to eco-friendly buildings of the future'**
>
> —Charles Correa

Charles Correa (born in Hyderabad, India, on 1 September 1930) is an Indian architect, planner, activist, theoretician, and a fundamental figure in the worldwide panorama of contemporary architecture. He studied architecture at the University of Michigan and at Massachusetts Institute of Technology, after which he established a private practice in Bombay in 1958.

All of his work—from the planning of New Bombay to the carefully detailed memorial to Mahatma Gandhi at the Sabarmati Ashram in Ahmedabad—has placed special emphasis on prevailing resources, energy, and climate as major determinants in the ordering of space.

Over the last four decades, Correa has done pioneering work on urban issues and low-cost shelter in the Third World. In 1972, he was awarded the Padma Shri. In 1985, Prime Minister Rajiv Gandhi appointed him Chairman of the National Commission on Urbanization. He was awarded a Royal Gold Medal in 1984 by RIBA, London. The Indian Institute of Architects (IIA) awarded him a gold medal in 1987. The University of Michigan awarded him an honorary doctorate. He was a Professor at Cambridge University and was honoured to occupy the Jawaharlal Nehru Chair.

His acclaimed design for McGovern Institute for Brain Research at MIT was dedicated recently.

Correa, who is famed for design principles based on low-density, low-cost architecture at a reduced environmental cost, wants architects to examine low-rise, high-density urban areas such as Rajasthan as a way of best using natural and local resources. He is famed for his eco-friendly buildings.

'The basic principle of housing in a country like India is that you have very limited resources,' Correa told BBC World Service's Masterpiece programme. 'Therefore you have to use great ingenuity. That's when you really learn to respect what traditionally is done. If you look at a village in Kerala, everything is re-used and recycled. Leaves which fall from palm trees are used again for the roof. There's nothing like poverty to be the mother of invention. As an architect, looking at those solutions, I was absolutely stunned by it.'

The explosion of the Indian economy in recent years has triggered a massive expansion in the heart of India's major cities.

Correa, who said that Indians use space 'extremely intelligently', explained that in India, tower blocks—'going high'—do not attract many people, and therefore better use of space in low-rise buildings has to be achieved. He has played a part in designing some of the large number of developments which have begun springing up. He said that this had been a chance to put his principles into practice—not only environmentally sound buildings, but ones that fit with their surroundings too.

Correa conceded that in the West, sustainable architecture is not cheap. He said that one environmentally friendly element on one building could pay for electricity for a Kerala village for a year.

'What I've learned, living here in India, is that the most wonderful traditional solutions exist which exemplify all the concerns of the environmentalist today. 'We don't have to invent these things again' he said.

Theme and philosophy

Charles Correa's work in India shows a careful development, understanding, and adaptation of modernism to a non-western culture. Correa's early works attempt to explore a local vernacular architecture within a modern environment. His land use planning and community projects continually try to go beyond typical solutions of Third World problems.

Charles Correa designed energy-efficient buildings. He wrote, 'In a poor country like India, we cannot simply afford to squander the kind of resources required to air-condition a glass tower under a tropical sun.' What he meant was that the building must itself, through its very form, create the controls for the user's needs. It should create its own micro-climate without air conditioning. The energy-conscious administrative complex of Electronics Corporation India Limited is a practical example, which has been given an innovative design by Correa. The building consists of a number of three-storied office units grouped around a courtyard and sheltered by a single roof on giant columns, which is partly scattered and partly covered with a sheet of water reflecting the sun back into the sky. The amount of solid surface which absorbs heat and transmits it to the space below is therefore reduced to a minimum, the spaces themselves consisting of garden courts and balconies as well as indoor rooms.

Correa views a house as a part of a larger problem of housing and regards housing as an integral part of town planning. He was appointed Chief Architect of New Bombay from 1971 to 1974 and Consulting Architect to the Government of Karnataka from 1975 to 1978. He quickly gets down to the fundamentals of city planning. In the case of Bombay, he tried to increase Bombay's holding capacity by opening up new growth centres across the harbour, public ownership of land, low-rise, high-density housing on a linear pattern, and a *mass transport system* to get people to their place of work in the shortest time possible.

The aim of creating new growth centres across the harbour was to restructure the whole city of Bombay, so that the existing north–south linear structure of the island would become a circular, polycentred one.

The idea of New Bombay arose in 1964, when the population was 4 million. It was predicted to increase to 15 million in 2000. He suggested new centres of growth, neither rural nor urban, but kind of quasi-rural, in which population densities are high enough to have a bus service and a school system, but low enough for people to keep buffaloes and therefore have another source of income. In particular, Charles Correa envisaged a role for Third World architects in formulating the programme and design for such centres. The National Commission of Urbanization, of which he is Chairman, is promoting this idea.

Important works

Of the many buildings Correa has designed, such as ECIL office complex (Hyderabad), Jawahar Kala Kendra (Jaipur), Vidhan Bhavan (Bhopal), Gandhi Smarak Sangrahalaya, Kovalam Beach Resort, Kanchanjunga Apartments (Mumbai), we will discuss *Kovalam Beach Resort* and *Kanchanjunga Apartments* in this section.

Kovalam Beach Resort, Thiruvananthapuram (1969–74)

This resort at Kovalam was designed by Charles Correa without disturbing the beauty of the surroundings (Fig. 5.41). The resort provides 300 beds as well as specialized facilities such as centres for Yoga and ayurvedic massage, and water sports. The master plan scatters the facilities over the site, rather than concentrating them in one area, thus creating a number of potential growth centres and allowing for a flexible response to future demands (Fig. 5.42). In order to

Fig. 5.41 Kovalam Beach Resort

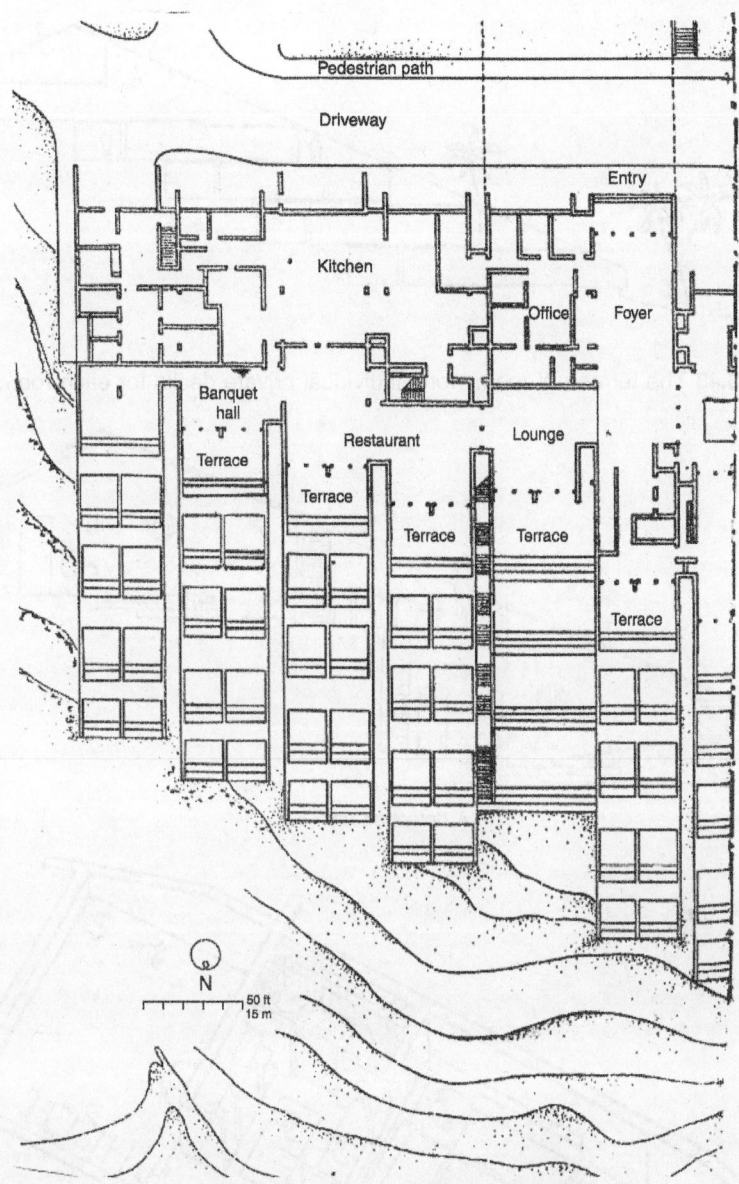

Fig. 5.42 Plan of the terraced floors

preserve the site's natural beauty, the buildings follow the natural hill slope. This also means that each room gets its own private terrace for sunbathing and relaxing (Fig. 5.43). There are also a number of detached units and independent units or 'kudils' (Fig. 5.44).

Although the design is contemporary, not directly derived from local forms, [except for the beach centre pavilions, which are light-weight bamboo chatris (Fig. 5.45)], the buildings refer to the vernacular with the plastered white walls, red tiled roofs and sundecks. The interiors utilize light furnishings, floor mats, and simple, Indian, crafted finishes.

Fig. 5.43 The terrace of rooms form individual private decks for each room

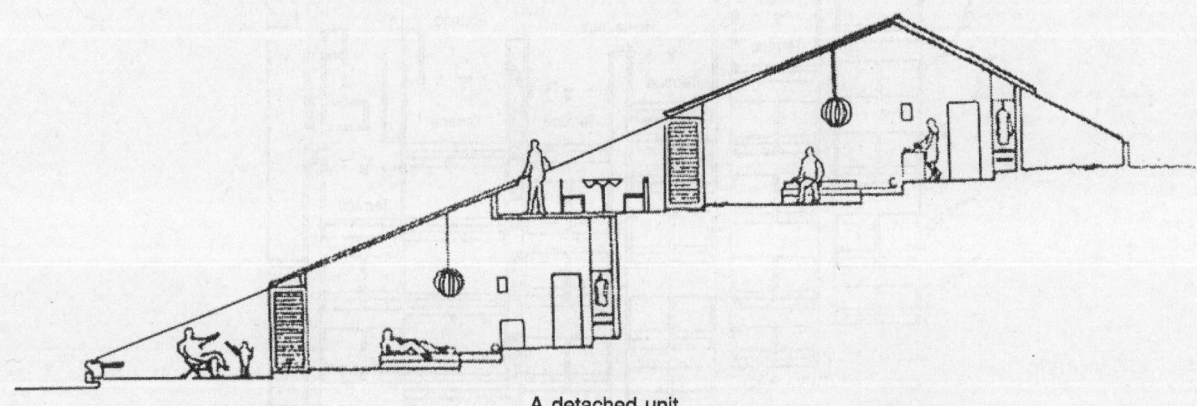

A detached unit

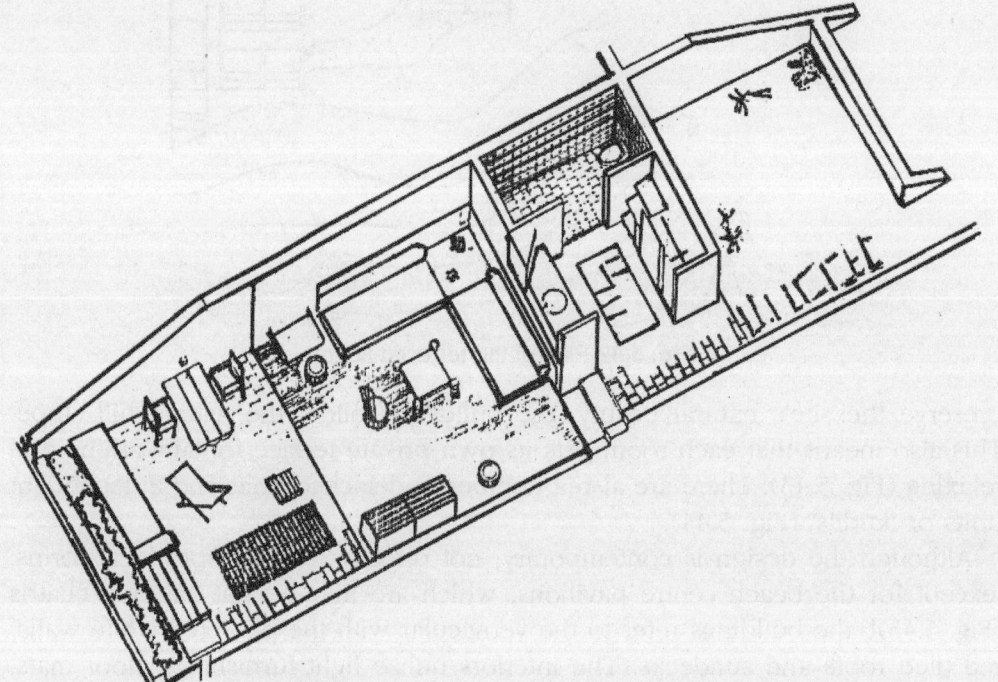

Fig. 5.44 A typical kudil at Kovalam Beach Resort

Bamboo chatri in the Beach Centre, Kovalam
Fig. 5.45

Kanchanjunga Apartments, Cumbala Hills, Bombay (1970–83)

Named after the second highest mountain of the Himalayan range, Kanchanjunga, is a condominium of 32 luxury apartments, each having three to six bedrooms. The building is 28 storeys (85 m) high and square (21 m × 21 m) in plan. The basic interlock is that of three- and four-bedroom apartments; the larger flats are formed by the addition of another half level as shown in Fig. 5.46. The structures are built around a central service core, which was constructed first. Each of the flats have large, usable, garden terraces offering dramatic city views.

Fig. 5.46 Kanchanjunga Apartments, Cumbala Hills, Bombay—28-storey apartment block

The building stands out in Bombay's urban landscape. The apartments are well ventilated and appear to suit the contemporary lifestyle of the city's elite. From within the flats themselves, there are views out from the living rooms and bedrooms. The terraces overlook the city, presenting the inhabitants with an ever-changing panorama (Fig. 5.47).

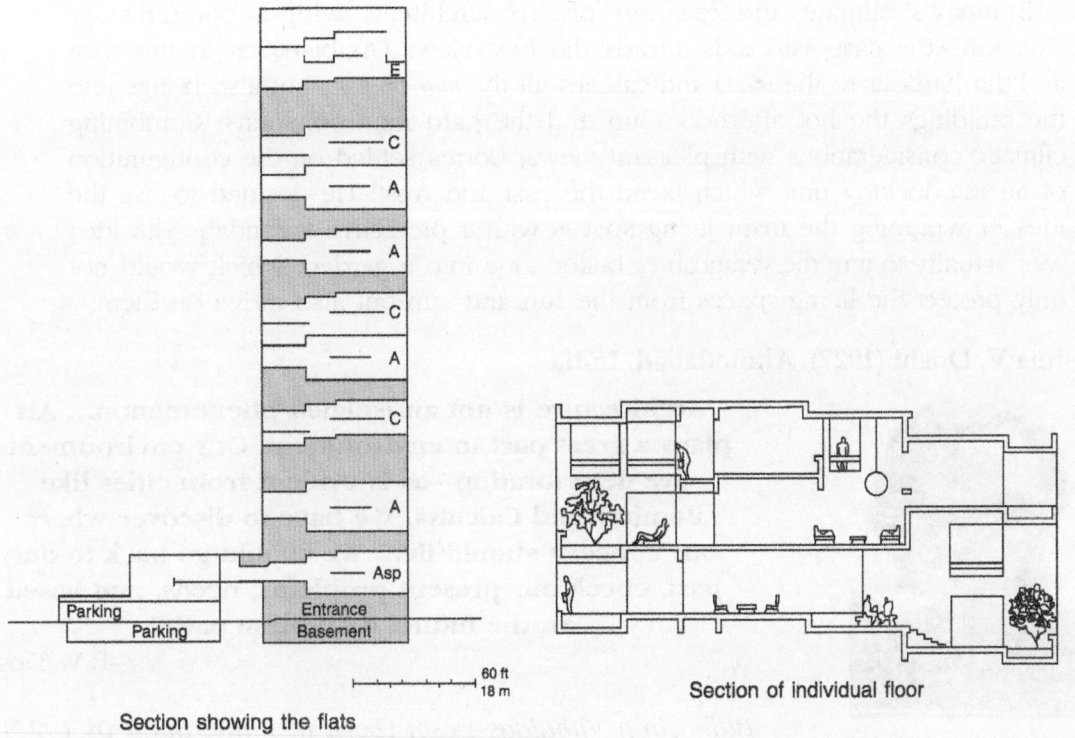

Section showing the flats

Section of individual floor

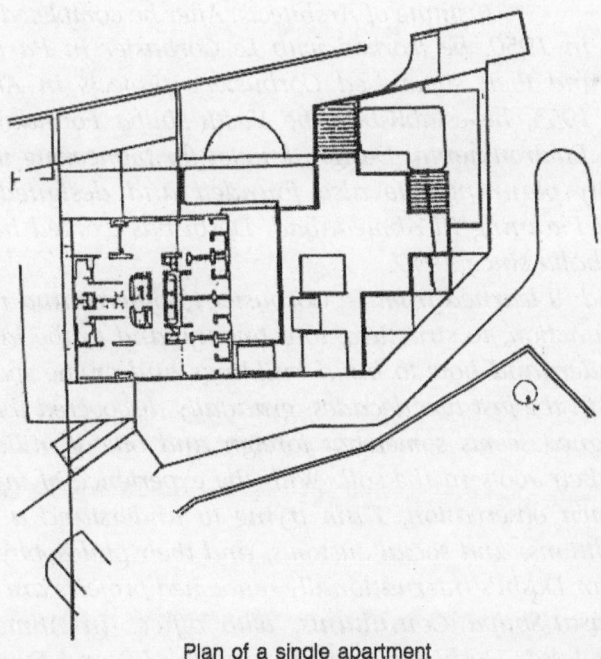

Plan of a single apartment

Fig. 5.47 Kanchanjunga Apartments

Bombay's climate and location present architects with a contradictory situation—the east–west axis affords the best views (Arabian Sea to the west and the harbour to the east) and catches all the sea breeze, but also brings into the buildings the hot afternoon sun and the hard monsoon rains. Combining climatic considerations with pleasant views, Correa settled on the configuration of an interlocking unit which faced the east and west. He decided to use the idea of wrapping the main living spaces with a protective verandah. The idea was actually to turn the verandah or buffer zone into a garden, which would not only protect the living spaces from the sun and rain but also thrive on them.

Balkrishna V. Doshi (1927), Ahmedabad, India

'Architecture is not an isolated phenomenon... Art plays a great part in environment. Our environments are deteriorating—as is evident from cities like Bombay and Calcutta. We have to discover where our concern should lie... We should go back to our past, check our present problems, needs, and based on the future, determine path.'

—B.V. Doshi

Fig. 5.48 Balkrishna Vithaldas Doshi

Balkrishna Vithaldas Doshi (born in Pune, India in 1927) is an Indian architect, educator, and academician. He is a fellow of the Royal Institute of British Architects as well as the Indian Institute of Architects. After he completed his studies at J.J. School of Art, Bombay, in 1950, he worked with Le Corbusier in Paris (1951–54) as senior designer and then supervised Corbusier's projects in Ahmedabad and Chandigarh. In 1955, he established the Vastu-Shilpa Foundation for Studies and Research in Environmental Design, known for pioneering work in low-cost housing and city planning. He also founded and designed the School of Architecture and Planning in Ahmedabad. Doshi has worked in partnership as Stein, Doshi & Bhalla since 1977.

B.V. Doshi said: 'I learned from Le Corbusier to observe and react to climate, to tradition, to function, to structure, to economy, and to the landscape. To an extent, I also understand how to build buildings and create spaces and forms. However, I have in the last two decades, gradually discovered that the buildings that I have designed seems somewhat foreign and out of milieu; they do not appear to have their roots in the soil. With the experience of my work over the years and my own observation, I am trying to understand a little about my people, their traditions, and social customs, and their philosophy of life.'

Today, Professor Doshi's internationally renowned projects are designed under the name of Vastu-Shilpa Consultants, with offices in Ahmedabad. As an academician, Professor Doshi has been visiting the US and Europe since 1958, and has held important chairs in American universities. He has received numerous international awards and honours, including the Padma Shri from the

Government of India and an honorary doctorate from the University of Pennsylvania, USA. Professor Doshi served as a member of the 1992 Award Master Jury and was presented the 1995 Aga Khan Award for Architecture for the Aranya Community Housing in Indore, India.

Over the years Doshi has created designs that rely on a sensitive adoption and refinement of modern architecture within the Indian context. The relevance of his environmental and urban concerns make him unique as both a thinker and teacher. Architectural scale and massing as well as a clear sense of space and community mark most of his work. Doshi's architecture provides one of the most important models for modern Indian architecture.

Important works

The important buildings designed by B.V. Doshi are Sangath, his office building at Ahmedabad; his residence at Ahmedabad (1959–61); the Indian Institute of Management Bangalore; Institute of Indology, Ahmedabad (1957–62); Gandhi Labour Institute, Ahmedabad; School of Architecture and Planning, Ahmedabad; and the township for Gujarat State Fertilizers, Baroda (1964–9).

Design of Doshi's residence (1959–61)

Doshi designed his own house in Navrangpura, a suburb of Ahmedabad. This small but compact structure is built on a 1000 sq. m plot. The plan is simple, based on a square divided into four quadrants by central slots in the structure. These are the unchanging elements consisting of the kitchen, corridor, bathrooms, and stairs, which form the base of the plan. The changing spaces are living, eating, and resting, and were placed in the quadrants, and the unchanging elements were placed in the slots (Fig. 5.49).

Doshi's house was designed for the hot climate of North India. It was also a mini testing ground for trying out his various ideas and devices that might be used in low- to medium-cost building.

The walls were rough brick with cavity walls between piers. The attached shading panels were made of bare concrete. One enters the site through a gate and then is guided onto the plinth (Fig. 5.50). An overhanging balcony in the north signals the front door. A large wooden door (with subsidiary apertures containing insect screens) can be swung open to extend the public area still further over a verandah into a garden in the rear. Stairs lead up to rooms that extend into small areas which make ideal study corners.

Natural climate control and economic construction were a major consideration. The design also provided for least exposure to the sun and easy passage of the south-west breeze.

Design of Doshi's own office (studio)—Sangath (1979–81)

Constructed from 1979 to 1981 in the western part of Ahmedabad, Sangath—a Sanskrit word meaning 'moving together through participation'—is more than just an architectural office. At Sangath, Doshi has drawn together a number of themes of his own earlier work—vaults on walls, platforms, terraces, maze-like interiors, and ambiguous edges.

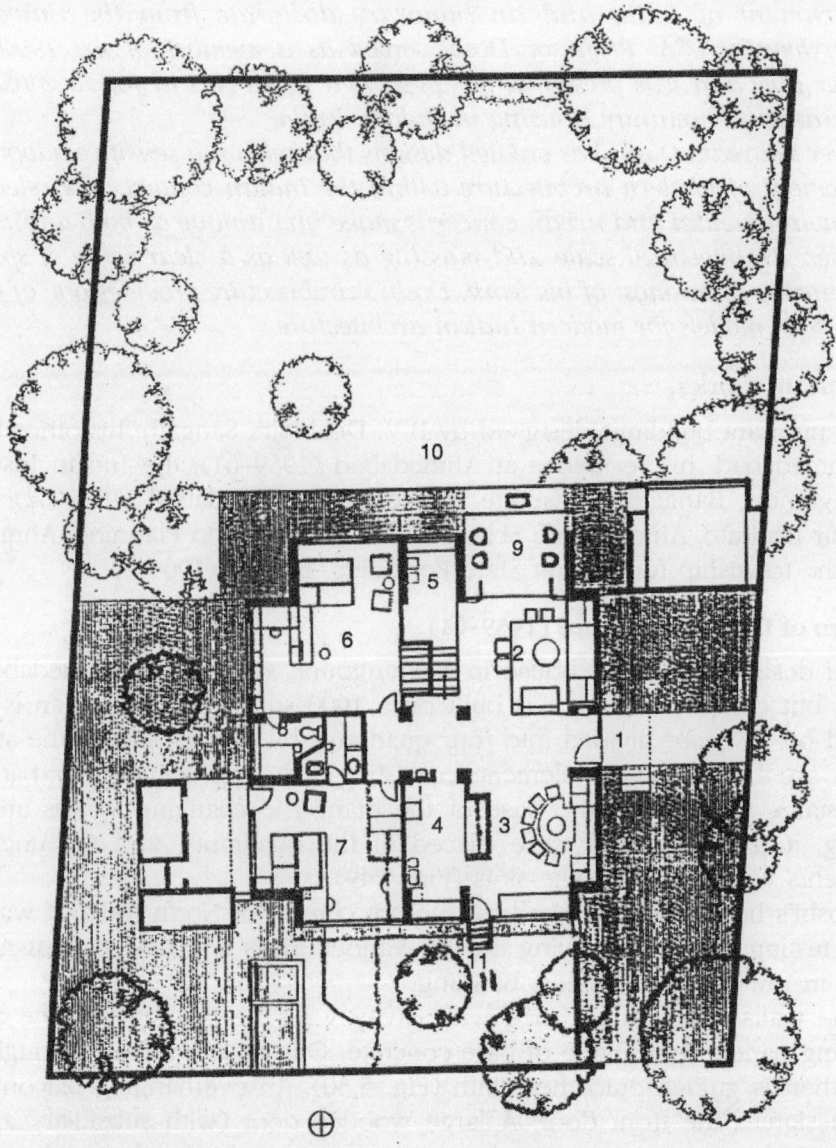

1. Entrance
2. Living room
3. Dining room
4. Kitchen
5. Music alcove
6. Master bedroom
7. Children's bedroom
8. Toilet and bath
9. Verandah
10. Garden
11. Kitchen yard

Fig. 5.49 Plan of Doshi's house

Fig. 5.50 Balkrishna Doshi's own house, Ahmedabad, India

There are dynamic sequences of structure to serve a rich blend of ideas (Fig. 5.51). The complex is a green enclave of grassy mounds, steps, terraces, water cascades, and earth-hugging vaults covered in chips of china mosaic to reflect glare and heat. The visitor is welcomed to Sangath by a shallow cascade of grassy steps that make an informal amphitheatre (Fig. 5.52). A longitudinal section would show how single, double, and triple height spaces flow into each other.

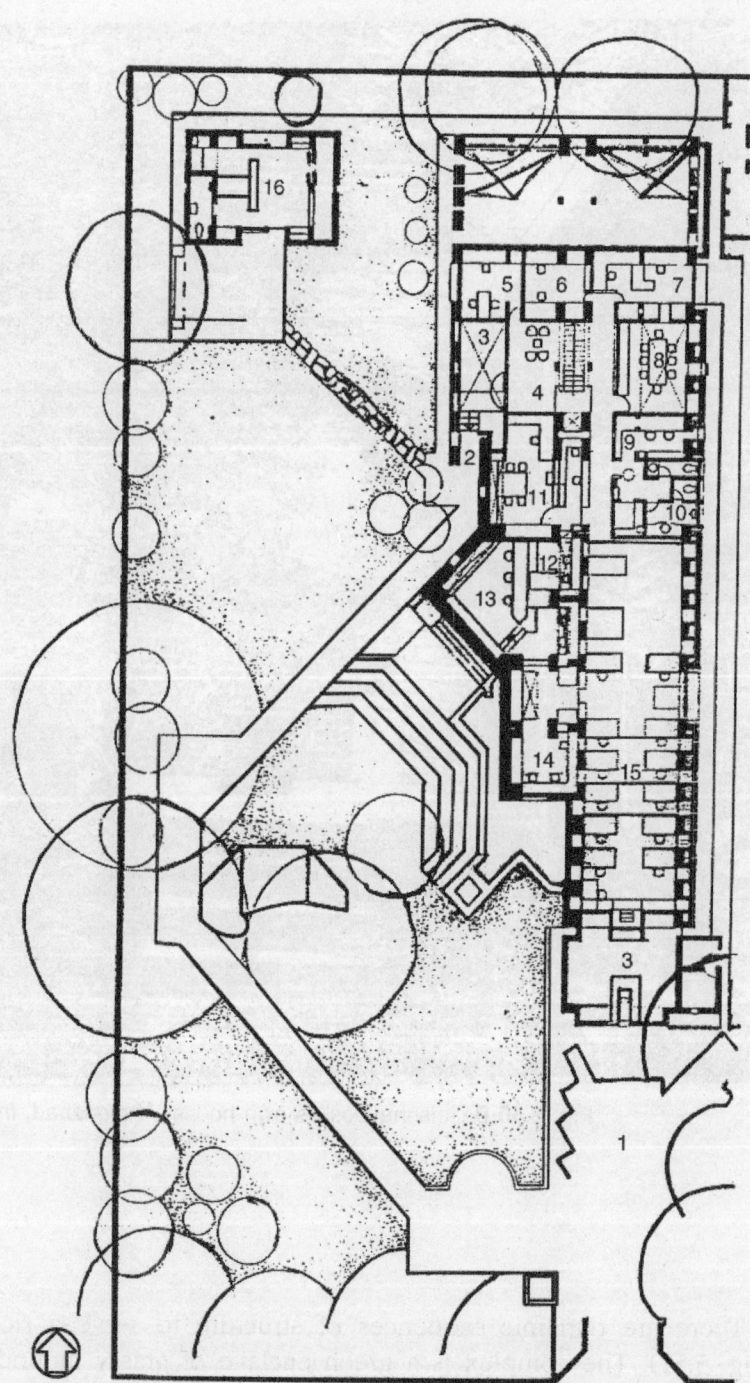

1. Forecourt
2. Pond
3. Amphitheatre
4. Entrance court
5. Reception
6. Workshops
7. Conference room
8. Subterranean meeting room
9. Architect's office
10. Toilets
11. Engineers' room
12. Design studio
13. Library

14. Administrative office
15. Design studio
16. Out house

Fig. 5.51 Plan of the studios in Doshi's office, Sangath

Fig. 5.52 Water channels and grass steps of the amphitheatre at Sangath, Doshi's office in Ahmedabad, India (1979–81)

Indian Institute of Management Bangalore (1977–1985)

A major turning point in Doshi's search for an Indian architectural expression was the Indian Institute of Management at Bangalore built between 1977 and 1985.

Bangalore has a comfortable climate. The city is full of lush greenery and trees. Therefore, in this project, the building includes external spaces and links between the various wings of the building.

The functional and physical attributes of the design relate to the local tradition of pavilion-like spaces and courtyards with ample provision for plantation.

The design also includes long, unusually high corridors (three-storeyed) with innumerable vistas or focal points for generating a mood for quiet introspection. Some of these corridors are open, some covered with pergolas, and some topped by glazed skylights (Fig. 5.53). The width of the corridors was modulated in many places to allow for casual eating and interaction to take place. Access to classrooms and administrative offices was provided through these links to generate constant activity.

External view

View through covered walkway

Fig. 5.53 A view of Indian Institute of Management Bangalore (IIMB)

Fig. 5.54 The corridor in IIM Bangalore covered with pergolas forming an excellent light and shadow effect

In the morning and evening, the sun's golden rays are reflected in the glazed windows and the long corridors surrounded by the classrooms (Fig. 5.54).

Doshi's inspiration for the design of IIM Bangalore was the capital of the Mughal Empire under Emperor Akbar, Fatehpur Sikri, near Agra, and southern temple cities such as Madurai. In this design, he explicitly brougfht brought out a contrast in the layouts of the teaching courts and the dormitories (Fig. 5.55).

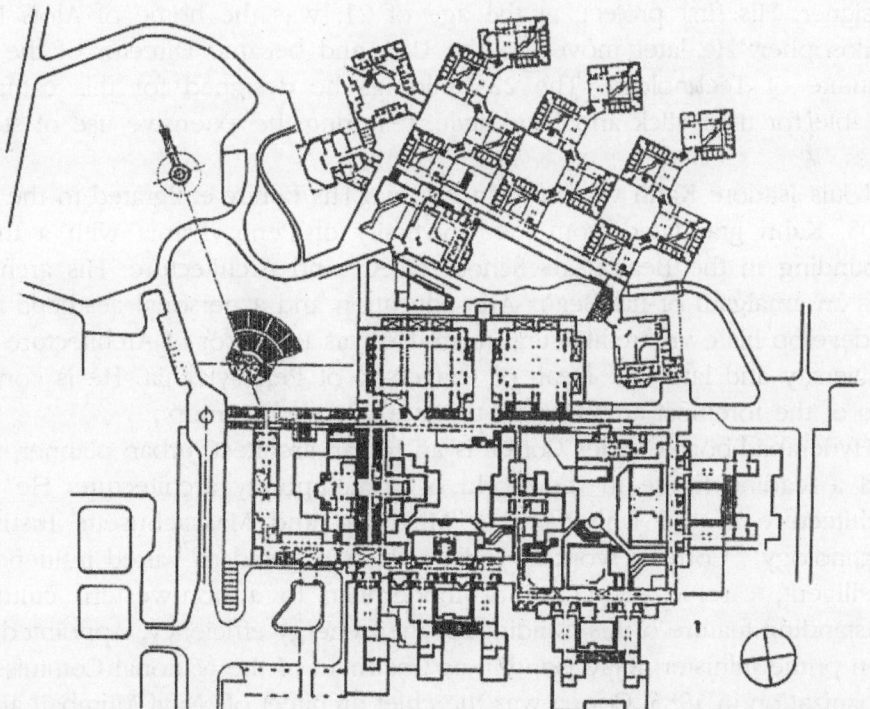

Fig. 5.55 Plan of IIM Bangalore with rectangular teaching courts and diagonal dormitories

Summary

This chapter briefly discusses the life history and works of six well-known architects—four pioneered radical changes in western architecture and two are leading Indian architects. Some of these individuals, even though they were not trained architects, had their early studies in creative arts and went on to create landmarks in modern architecture through their mastery of building design.

Frank Lloyd Wright, the leading American architect of the late nineteenth to mid-twentieth century, contributed the 'Prairie' home design to American residential architecture. The typical features of this design can be noticed in a large proportion of homes even today. Wright developed the concept of organic architecture—this style emphasises integration of building with nature. He had a great regard for natural materials such as wood and stone and used them extensively in his famous building 'Falling Water', Kaufmann House. Another landmark construction by Wright is the Guggenheim Museum in New York.

Le Corbusier, a French architect, was perhaps the most influential architect of the twentieth century. His works, ideas, and writings came to form the basis for modern architecture and city planning. He developed the modular system of measurements based on the human scale and the golden section. He is best known in India for his design of Chandigarh, the post-independence capital of Punjab.

Mies van der Rohe, a German architect by birth, trained with his father, a stonemason, and later with Bruno Paul, the art nouveau architect and furniture designer. His first project, at the age of 21, was the home of Alois Riehl, a philosopher. He later moved to the USA and became Director of the Illinois Institute of Technology. The 22 buildings he designed for this campus are notable for their slick and lucid style, featuring the extensive use of steel and glass.

Louis Isadore Kahn was born in Estonia. His family emigrated to the USA in 1905. Kahn graduated from the University of Pennsylvania with a thorough grounding in the Beaux-Arts School of Art and Architecture. His architecture was an amalgam of his Beaux-Arts education and a personal aesthetic impulse to develop his own architectural form. He was Professor of Architecture at Yale University and later the Dean of University of Pennsylvania. He is considered one of the foremost architects of the late twentieth century.

Hyderabad-born Charles Correa is an Indian architect, urban planner, activist, and a leading figure in the world of contemporary architecture. He studied architecture at the University of Michigan and Massachusetts Institute of Technology. Correa's work in India reflects a blend of varied influences—an intelligent, careful adaptation of modernism to a non-western culture. An outstanding feature of his buildings is their energy efficiency. Appointed by the then prime minister, Rajiv Gandhi, as Chairman of the National Commission on Urbanization in 1985, Correa was the chief architect of Navi Mumbai, an urban centre for 2 million people across the harbour, some distance from the existing city. Correa's land use planning and community projects continually try to go beyond conventional, western-oriented solutions in solving Third World problems. His designs are based on using natural, local resources and ingeniously incorporating elements of traditional architecture into modern design, in order to cut costs and create environment-friendly buildings.

Balakrishna V. Doshi hails from Pune, India. After completing his studies at J.J. School of Art, Bombay, he worked with Le Corbusier first in Paris and later in planning the capital complex of Chandigarh. Like Correa, Doshi has, over the years, created architecture that relies on a sensitive adoption and refinement of modern architecture to suit the Indian context. A major turning point in his search for an Indian architectural expression was the design of the Indian Institute of Management, Bangalore. This chapter takes a detailed look at the design of Doshi's own home in Navrangpura, Ahmedabad, and his office-cum-studio, Sangath. Doshi has also done pioneering work in low-cost housing and city planning.

REVIEW QUESTIONS

Part A (2 marks each)

1. Name five buildings designed by the following architects
   (a) F.L. Wright
   (b) Le Corbusier
   (c) Mies van der Rohe
   (d) Louis I. Kahn
   (e) Charles Correa
   (f) B.V. Doshi
2. Name five famous architects.
3. Name some famous Indian architects.
4. Which foreign architects worked in India?

Part B (16 marks each)

1. Describe the theme and philosophy of F.L. Wright. Write about his buildings in details. (The same question can be repeated for all the six architects.)
2. Explain the theme and philosophy of any Indian architect who has set an expression for Indian architecture.

Glossary

Aesthetics Set of principles concerned with beauty

Anthropometrics Measurements based on the dimensions or proportions of the human body

Architecture Art, science, or profession of building structures for human use

Balance Visual equilibrium in architecture

Building A structure with a roof and walls

Building services Essential requirements (water supply, electricity, plumbing, etc.) to fulfil the functions of a building

Character The ability of a building to express its particular function or use

Circulation Movement through space

Climax A visual phenomenon in which interest is slowly built up till the actual structure comes into view

Contemporary Belonging to the present time

Contrast The perceptual effect of the juxtaposition of very different traits

Creating Executing an architectural plan

Datum The basic standard or level for a comparison

Designing The process in which a structure to be constructed later is first visualized and the details worked out

Dominance One effect (colour, tone, or texture) being stronger than the remaining

Harmony Combination of many parts to make a good composition

Hierarchy The order of importance of a form or space based on its size, shape, and place relative to other forms or spaces

Modern Relating to a recently developed fashion or style

Planning Formulating a series of systematic steps to be carried out to achieve an objective or target

Punctuation Pause in visual continuity

Rhythm The use of recurring patterns to organize a series of similar forms and spaces

Style A quality that gives distinctiveness to artistic expression

Symmetry Balanced distribution of forms and spaces about a line or point

Traditional Long established and generally accepted

Transformation Principle using which an architectural concept can be retained or strengthened

Unity Achieving singleness in a visual composition

Bibliography

Ching, Francis D.K., *Architecture: Form, Space, and Order*, Van Nostrand Reinhold, New York, 1979.

Curtis, William, *Balakrishna Doshi: An Architect for India*, Rizzoli, New York, 1988.

Fletcher, Bannister, *A History of Architecture*, revised by CBS Publishers & Distributors, Delhi, 1986.

Iengar, Keshavram N., *Composing Architecture*, Academy of Art and Architecture, Mysore, 1996.

Koenigsberger, O.H., T.G. Ingersoll, A. Mayhew, and S.V. Szokolay, *Manual of Tropical Housing and Building, Part 1—Climatic Design*, Orient Longman Ltd, Madras, 1973.

Pramar, V.S, *Design Fundamentals in Architecture*, Somaiya Publications, Bombay, 1973.

Rubenstein, Harvey M., *Environmental Planning*, John Wiley & Sons, New York, 1969.

Smithies, K.W., *Principles of Design in Architecture*, Van Nostrand Reinhold, New York, 1981.

Snyder, James C. and Anthony J. Catanese, *Introduction to Architecture*, McGraw-Hill, New York, 1979.

Vaidyanatha, G., I. Kulasekaran, and G. Sathish Kumar, *Building Planning and Construction Companion*, Edifice Institute of Building Services, Madras.

Model Question Papers

MODEL QUESTION PAPER—I

BE/BTech Degree Examination
Second Semester
Civil Engineering

CE 1151: PRINCIPLES OF ARCHITECTURE

Time: 3 hours

Maximum marks: 100

Answer all questions
Draw neat sketches wherever necessary

Part A (10 × 2 = 20 marks)

1. What are the basic elements of aesthetics?
2. Explain the erecting of a building structure with a sketch.
3. What do you understand by unity as an aesthetic quality in building?
4. What is meant by character in buildings?
5. Draw sketches to show the safety standards that have to be considered in the design of a staircase.
6. State any four climatic zones in India.
7. Define anthropometrics.
8. What do you understand by integration of building services?
9. List five buildings designed by Mies van der Rohe.
10. Write the names of two foreign architects who have done pioneering works in India.

Part B (5 × 16 = 80 marks)

11. Explain the integration of aesthetics and function in architectural design.
12. (a) Explain in detail how the aesthetic qualities of scale, balance, symmetry, and rhythm influence a building design.

or

(b) Explain the site studies and analysis that you would conduct to understand a site and use it efficiently for an architectural project.

13. (a) What is the meaning of style and character in buildings and explain how it is realized in traditional and modern architecture?

or

(b) Explain in detail the integration of building services in architectural design.

14. (a) What is anthropometrics? Explain with examples how anthropometrics can be used to determine the size and shape of rooms for human activities. Substantiate with several sketches.

or

(b) Explain the basic principles (elements) of landscape design with sketches.

15. (a) Discuss the various building rules and regulations of your town. What are the safety standards followed in the design of industrial buildings?

or

(b) Explain the theme and philosophy of Charles Correa by describing his works in detail.

MODEL QUESTION PAPER—II

BE/BTech Degree Examination
Second Semester
Civil Engineering

CE 1151: PRINCIPLES OF ARCHITECTURE

Time: 3 hours   Maximum marks: 100

Answer all questions
Draw neat sketches wherever necessary

Part A (10 × 2 = 20 marks)

1. Define architecture.
2. What do you understand by scale in architecture?
3. Distinguish between architecture and civil engineering.
4. What is meant by symmetry?
5. Differentiate between plinth area and carpet area.
6. State the importance of anthropometry in building design.
7. State any four elements of climate.
8. Name the various landscape design elements.
9. Name the building services that are integrated with architectural design.
10. What is meant by site analysis?

Part B (5 × 16 = 80)

11. Explain the following:
    (i) Axis
    (ii) Symmetry
    (iii) Hierarchy
    (iv) Rhythm
12. (a) Explain harmony, contrast, dominance, punctuation, and climax in detail with examples from history.

    or

    (b) Explain the various kinds of proportion and how the form of a building is influenced by the proportioning system.
13. (a) Explain the factors to be considered in the design of public buildings.

    or

    (b) Explain the various building services and their integration in architectural design.
14. (a) Discuss the process of site analysis in detail

    or

    (b) Discuss the works of pioneering architect F.L. Wright, bringing out his theme and philosophy.

15. (a) Explain the various aspects of landscape design with neat sketches.

or

(b) Explain in detail the theme, philosophy, and works of any Indian architect who gave an expression to Indian architecture.

ANNA UNIVERSITY PREVIOUS YEAR'S QUESTION PAPER

BE/BTech Degree Examination
May/June 2005
Second Semester
Civil Engineering

CE 1151: PRINCIPLES OF ARCHITECTURE

Time: 3 hours   Maximum marks: 100

Answer all questions
Draw neat sketches wherever necessary

Part A (10 × 2 = 20 marks)

1. Define the term aesthetics.
2. Distinguish between balance and symmetry.
3. What do you understand by the term style?
4. Explain the golden section.
5. Explain how the idea of harmony is expressed in buildings.
6. List any two types of climate zones prevalent in India.
7. What does the term DCR refer to?
8. What is the height of a standard Indian door?
9. In an auditorium, in which direction should the doors open and why?
10. What is punctuation in architecture?

Part B (5 × 16 = 80 marks)

11. Character and style are not synonyms, one represents the surface decoration while the other is holistic. Do you agree with this distinction between character and style? Explain your viewpoint with illustrations and examples.

12. (a) Explain the difference between the proportioning system proposed by Vitruvius and the Modulor proposed by Le Corbusier.

or

(b) Explain with suitable drawings how symmetry and rhythm play a role in Hindu and Islamic religious buildings.

13. (a) Using the example of residences, explain how space requirements, circulation, and climate affect design. Illustrate your arguments.

(b) Explain how the idea of harmony, contrast, and climax are used in traditional and modern buildings. Compare the different kinds of usage.

14. (a) Discuss the ideas and works of F.L. Wright.

or

(b) Discuss the ideas and works of Le Corbusier.

15. (a) Explain how international modern architecture can be contextualized to suit Indian ethos and express 'Indian identity' in the works of Charles Correa.

(or)

(b) Discuss the architectural ideas, techniques, and processes involved in the works of B.V. Doshi.

Notes

Notes